T0293326

Monographs on Statistics and Applied Probability 155

Asymptotic Analysis of Mixed Effects Models
Theory, Applications, and Open Problems

Jiming Jiang
University of California, Davis, USA

CRC Press
Taylor & Francis Group
Boca Raton   London   New York

CRC Press is an imprint of the
Taylor & Francis Group, an **informa** business

A CHAPMAN & HALL BOOK

MONOGRAPHS ON STATISTICS AND APPLIED PROBABILITY

General Editors

F. Bunea, V. Isham, N. Keiding, T. Louis, R. L. Smith, and H. Tong

CRC Press
Taylor & Francis Group
6000 Broken Sound Parkway NW, Suite 300
Boca Raton, FL 33487-2742

© 2017 by Taylor & Francis Group, LLC
CRC Press is an imprint of Taylor & Francis Group, an Informa business

No claim to original U.S. Government works

Printed on acid-free paper
Version Date: 20170510

International Standard Book Number-13: 978-1-4987-0044-3 (Hardback)

Visit the Taylor & Francis Web site at
http://www.taylorandfrancis.com

and the CRC Press Web site at
http://www.crcpress.com

To Thuan

Contents

List of Figures

List of Tables

Preface

Large sample techniques play fundamental roles in all fields of statistics. As quoted from [1], "Large-sample techniques provide solutions to many practical problems; they simplify our solutions to difficult, sometimes intractable problems; they justify our solutions; and they guide us to directions of improvements." Mixed effects models, including linear mixed models, generalized linear mixed models (GLMMs), non-linear mixed effects models, and non-parametric mixed effects models, are complex models by nature, yet, these models are extensively used in practice. See, for example, [2], [3], [4]. For such reasons, there have been substantial growing interests in finding, simplifying, and justifying the solutions for this important class of models.

This monograph provides a comprehensive account on asymptotic analysis of mixed effects models. Historically, large sample theory has provided guidelines for methodology developments as well as applications of these models. For example, restricted maximum likelihood (REML) has established itself as the preferred method of linear mixed model (LMM) analysis largely due to an asymptotic theory that has been developed in the 1990s. As another example, penalized quasi-likelihood (PQL) has been gradually giving way to (full) maximum likelihood (ML) inference partially due to inconsistency of the PQL estimators that has been discovered; on the other hand, major computational developments, such as Monte-Carlo E-M algorithm and data cloning, have made the ML inference feasible.

A large portion of the monograph is related to the author's research publications over the past 20 years on linear and generalized linear mixed models. These include the author's work in the 1990s on REML asymptotics; the author's 15-year pursuit in proving consistency of the ML estimator (MLE) in GLMM involving crossed random effects; the author's long-standing interest in small area estimation (SAE; e.g., [5]); and other topics of mixed model analysis such as mixed model diagnostics and mixed model selection. Of course, the monograph also covers research work of many other authors in this area. In particular, there has been growing interest in semi-parametric and non-parametric mixed effects models and related asymptotics. Asymptotic theory has been used as guidelines, for example, in the analysis of longitudinal data, where semi/non-parametric mixed effects models have been used. Another area in which extensive asymptotic studies have been done is estimation of mean squared prediction errors (MSPE) and construction of prediction intervals in SAE.

The author intends to build a strong connection between the asymptotic theory and real-life applications. Interesting real-data examples are provided to motivate the theoretical development. For example, the salamander-mating data [6] has been a driving force for the computational developments for GLMM, and a primary motivation for proving the consistency of the MLE for GLMMs involving crossed random effects. Some recent advances in genome-wide association study have motivated asymptotic analysis of misspecified LMM. The estimation of MSPE offers another excellent example where asymptotic study is highly motivated by practical problems.

In a way, mixed effects model asymptotics are difficult problems due to the complexity of these models and the fact that observations are typically correlated under these models. As a result, classical limit theory regarding independent random variables (e.g., [1], Chapter 6) does not apply, or at least cannot be directly used. There have been some vast developments in the asymptotic theory for mixed effects models over the past decades. On the other hand, some important, and challenging, problems regarding the mixed effects model asymptotics remain unsolved. The author provides a number of such challenges as open problems. A main goal of this monograph is to encourage young researchers to devote their talents, and energy, to asymptotic studies that are motivated by practical problems. For this purpose, we have supplemented eah chapter (starting Chapter 2) with a section of additional bibliographical notes and open problems. The problems range from theoretical and methodological to computational and practical issues, and challenges that are important, as well as interesting.

The main topics of this monograph are asymptotics, which are often treated as theoretical properties. Thus, in a way, we are theoreticians. However, we are theoreticians driven by practical problems, and this is made clear throughout the entire monograph. It should also be noted that the monograph is not written in a theorem-proof style. Instead, we focus on interpretation of the results as well as assumptions, and applications to problems of practical interest. Of course, interested readers can always explore the detailed theorems and proofs in the references provided. In this regard, we have adopted a strategy similar to [1].

More importantly, asymptotic analyses often guide us to directions of improvements, as noted earlier. The monograph intends to provide a comprehensive account of such improvements in mixed model analysis.

Finally, it should be noted that the monograph is not a "collection of results." Instead of trying to cover all of the relevant results, we focus on important methods that have had impacts in the fields, have been used in other work, and are potentially useful in future studies. In spite of our efforts, there is a nonzero probability that some of the important methods in these categories are missing. We apologize for such omissions, and promise to catch up in the next edition of the monograph.

The author wishes to thank Professors Ronghui (Lily) Xu, Jimin Ding, and Jianhua Huang for communications and suggestions. The author also would like to express his appreciation to Professors Thuan Nguyen, Can Yang, and Dr. Cecilia Dao for graphical assistance. In addition, the author is grateful to a number of anonymous reviewers at various stages of the manuscript for their comments and suggestions.

Jiming Jiang
Davis, California

About the Author

Jiming Jiang is a professor of statistics at the University of California, Davis. He received his Ph. D. in statistics in 1995 from the University of California, Berkeley. His research interests include asymptotic theory, mixed effects models, model selection, small area estimation, analysis of longitudinal data, and big data intelligence. He is the author/co-author of two previously published books and one monograph: *Linear and Generalized Linear Mixed Models and Their Applications* (Springer 2007), *Large Sample Techniques for Statistics* (Springer 2010), and (with T. Nguyen) *The Fence Methods* (World Scientific, 2015). He is an editorial board member of several major statistical journals, including *The Annals of Statistics* and *Journal of the American Statistical Association*. He is a Fellow of the American Statistical Association, a Fellow of the Institute of Mathematical Statistics, and an Elected Member of the International Statistical Institutes. Jiming Jiang is a co-recipient of Outstanding Statistical Application Award (American Statistical Association, 1998), the first co-recipient of the NISS Former Postdoc Achievement Award (National Institute of Statistical Sciences, 2015), and a Yangtze River Scholar (Chaired Professor, 2016), which is widely considered the highest academic honor awarded (to a foreigner) by the Ministry of Education of the People's Republic of China.

Chapter 1

Introduction

Statistics is nowadays often referred to as *data science*. Today's data are much bigger, and more complex, than what a statistician had to handle decades ago; hence, not surprisingly, today's statistical models are also much more complex. From a classical i.i.d. (independent and identically distributed) model to a linear regression model; from a linear regression model to a linear mixed model; and from a linear mixed model to a nonlinear mixed effects model, we are witnessing an ever-changing story about our data, in good ways.

The complexity of statistical modeling allows practitioners to better understand the nature of their data, such as correlations among the observations, as more and more people have come to realize that the independent-data assumption is often unrealistic. On the other hand, the advances in statistical modeling also present new challenges. For example, given the importance of large sample techniques in statistics (e.g., [1]), how much do we (really) know about the large sample properties of an estimator derived under a complex statistical model?

In this monograph, we focus on one particular class of complex statistical models, known as *mixed effects models*. A common feature of these models is that the observations are correlated. The correlation may be modeled by the so-called random effects, in the form of a hierarchical model, or in terms of a variance-covariance structure. The detailed definitions of different types of mixed effects models will be given in the later chapters, but here we would like to first offer a few examples as a preview for what we are going to do.

1.1 Example: Generalized estimating equations

In medical studies, repeated measures are often collected from patients over time. Let y_{it} be the measure collected from patient i at time t. Here $1 \leq i \leq m$, m being the total number of patients that are involved, and $t \in T_i$, where T_i is a subset of times which may be patient-dependent. Let $y_i = (y_{it})_{t \in T_i}$ denote the vector of measures collected from patient i. Quite often, a vector of explanatory variables, or covariates, are available for each patient and at each observational time. Let x_{itk} denote the kth component of the covariates for patient i at time t, $1 \leq k \leq p$, and $X_i = (x_{itk})_{t \in T_i, 1 \leq k \leq p}$ the matrix of covariates for patient i. Data with such kind of layout is often referred to as *longitudinal data*, and are encountered in many fields well beyond medical

1

studies (e.g., [7]). In general, we shall use the term subject, which in the case of medical study would correspond to a patient.

A class of models, known as *marginal models*, for the longitudinal data are defined by assuming that $y_i, 1 \leq i \leq m$ are independent such that $\mathrm{E}(y_i) = \mu_i = \mu_i(\beta)$, where β is a vector of parameters, and $\mathrm{Var}(y_i) = V_i$, where V_i is an unknown covariance matrix. Usually, v_i is a known function of β that is associated with X_i, for example, $\mu_i = X_i\beta$ in case of a linear model. On the other hand, so far V_i is assumed completely unknown except, perhaps, that it is finite, so the assumption on the variance-covariance structure among the data within the subject is very weak. However, in many cases, the main interest is the mean function, $\mu_i = (\mu_{it})_{t \in T_i}$. For example, in a medical study, μ_{it} corresponds to the mean turnover of a marker of bone formation, *osteocalcin*, for subject i at time t [8]. Nevertheless, the covariance matrix is closely related to any reasonable measure of uncertainty of an estimator of the mean function. Thus, at some point, one will have to deal with V_i, even though it is not of direct interest.

Because $\mu_i(\cdot)$ is a known function, to assess the mean function, one needs to estimate the parameter vector, β. A well-known method of estimation is called generalized estimating equations, of GEE, defined by solving the equation

$$\sum_{i=1}^{m} \frac{\partial \mu_i'}{\partial \beta} V_i^{-1}(y_i - \mu_i) = 0, \tag{1.1}$$

where the (k,t) element of $\partial \mu_i'/\partial \beta$ is $\partial \mu_{it}/\partial \beta_k$. The motivation of (1.1) will be given in Section 3.2. In order to solve (1.1) for β, one needs to know the V_i's, which are often difficult to estimate. One reason is that there are many unknown parameters involved in $V_i, 1 \leq i \leq m$. In fact, if the V_i's are completely unspecified, there are $\sum_{i=1}^{m} n_i(n_i+1)/2$ unknown parameters involved in all of the V_i's, where $n_i = |T_i|$, the cardinality of T_i. Of course, one may consider modeling the V_i's by assuming some parametric forms for the V_i's, and thus reduce the number of unknown parameters involved. A challenging problem, in the latter regard, is to determine which parametric form is reasonable to use. Again, this topic will be discussed later. Here, let us first introduce an idea, first introduced by Liang and Zeger ([9]; also see [10]) that, in a way, may have surprised many. Let $\tilde{V}_i$ be a "working covariance matrix" (WCM), for each i. Here by WCM it means that $\tilde{V}_i$ is not necessarily the true V_i, or even an estimator of the latter. But, the WCM is known and ready to use for the computation. For example, the simplest WCM is the identity matrix, $\tilde{V}_i = I_{n_i}$, where I_n denotes the n-dimensional identity matrix. Liang and Zeger suggested to estimate β by solving (1.1) with V_i replaced by $\tilde{V}_i, 1 \leq i \leq m$. A reader's first response to the proposal might be something like "Wait a minute. The V_i's in (1.1) are supposed to be the true covariance matrices; replacing them by the $\tilde{V}_i$'s, which may not be even close to the V_i's?" Nevertheless, Liang and Zeger [9] provided a solid argument in support of their proposal by proving that the GEE estimator, obtained by solving (1.1) with V_i replaced by $\tilde{V}_i$ ($1 \leq i \leq m$), is consistent (e.g., [1], p. 33) for any $\tilde{V}_i$ satisfying some mild conditions.

Although the result of Liang and Zeger may sound surprising, it does not mean that knowing the V_i's (or not) does not matter. In fact, it matters, but only to the "second-order". Roughly speaking, a first-order asymptotic property states that the estimator is consistent, while a second-order (asymptotic) property states that the estimator is efficient in the sense that its asymptotic variance is the smallest (e.g., [11], Section 6.2). A related issue is that, because the GEE estimator with WCM is likely to be inefficient, its variance may be large. Therefore, it is important to have a good assessment of the variance of the GEE estimator. Liang and Zeger [9] proposed the following asymptotic method, widely known as the "sandwich" estimator. Denote the right side of (1.1), with V_i replaced by $\tilde{V}_i$, by $g(\beta)$, a vector-valued function with the same dimension as β. Then, the GEE estimator, $\hat{\beta}$, satisfies

$$0 = g(\hat{\beta}) \approx g(\beta) + \frac{\partial g}{\partial \beta'}(\hat{\beta} - \beta),$$

where β denotes the true parameter vector, and the partial derivatives are evaluated at the true β. Here the approximation is due to the Taylor expansion, and consistency of $\hat{\beta}$, as noted above. More precisely, the approximation is in the sense that the difference is of lower order in terms of convergence in probability. See, for example, Section 3.4 of [1]. One can make a further approximation (in the same sense) by replacing $\partial g/\partial \beta'$ by its expectation, and thus obtain the following approximation: $0 \approx g(\beta) + \mathrm{E}(\partial g/\partial \beta')(\hat{\beta} - \beta)$, or

$$\hat{\beta} - \beta \approx -\left\{\mathrm{E}\left(\frac{\partial g}{\partial \beta'}\right)\right\}^{-1} g(\beta), \qquad (1.2)$$

assuming non-singularity of $\mathrm{E}(\partial g/\partial \beta')$. By taking the covariance matrix on both sides of (1.2), one expects that, naturally,

$$\mathrm{Var}(\hat{\beta}) \approx \left\{\mathrm{E}\left(\frac{\partial g}{\partial \beta'}\right)\right\}^{-1} \mathrm{Var}\{g(\beta)\} \left\{\mathrm{E}\left(\frac{\partial g'}{\partial \beta}\right)\right\}^{-1}. \qquad (1.3)$$

The right side of (1.3) already looks like a "sandwich," but it is not yet an estimator. To convert it to an estimator, write $g_i = (\partial \mu_i'/\partial \beta)\tilde{V}_i^{-1}(y_i - \mu_i)$, and note that

$$\mathrm{E}\left(\frac{\partial g}{\partial \beta'}\right) = \mathrm{E}\left(\sum_{i=1}^{m} \frac{\partial g_i}{\partial \beta'}\right), \qquad (1.4)$$

$$\mathrm{Var}\{g(\beta)\} = \sum_{i=1}^{m} \mathrm{Var}(g_i)$$

$$= \sum_{i=1}^{m} \mathrm{E}(g_i^2)$$

$$= \mathrm{E}\left(\sum_{i=1}^{m} g_i^2\right). \qquad (1.5)$$

Here, in deriving (1.5), we have used the fact that the g_i's are independent with mean zero, provided that the WCM are constant matrices. We then approximate the left sides of (1.4) and (1.5) by their right sides, respectively, with the expectation signs removed. The latter is a standard technique used in asymptotics (e.g., [1], Section 3.1). Thus, combined with (1.3), we obtain the following estimator of $\mathrm{Var}(\hat{\beta})$, the sandwich estimator:

$$\widehat{\mathrm{Var}}(\hat{\beta}) \;=\; \left(\sum_{i=1}^{m} \frac{\partial \hat{g}_i}{\partial \beta'}\right)^{-1} \left(\sum_{i=1}^{m} \hat{g}_i^2\right) \left(\sum_{i=1}^{m} \frac{\partial \hat{g}_i'}{\partial \beta}\right)^{-1}, \tag{1.6}$$

where $\hat{g}_i$ and $\partial \hat{g}_i/\partial \beta'$ are g_i and $\partial g_i/\partial \beta'$, respectively, with β replaced by $\hat{\beta}$. The sandwich estimator can be used to obtain, for example, the standard errors of $\hat{\beta}_k$, $1 \leq k \leq p$ by taking the square roots of the diagonal elemenets of $\widehat{\mathrm{Var}}(\hat{\beta})$.

1.2   Example: REML estimation

REML—restricted or residual maximum likelihood, is a method of estimating dispersion parameters, or variance components, involved in a linear model. Here the linear model typically has a more complicated variance-covariance structure than the linear regression model, although the latter may serve as one of the simplest examples to illustrate REML (see below).

  Let us begin with the simplest of all. Suppose that $Y_1, \ldots, Y_n$ are i.i.d. observations with mean μ and variance σ^2. A standard estimator of the variance, σ^2, is the sample variance, defined as

$$\hat{\sigma}^2 \;=\; \frac{1}{n-1} \sum_{i=1}^{n} (Y_i - \bar{Y})^2, \tag{1.7}$$

where $\bar{Y} = n^{-1} \sum_{i=1}^{n} Y_i$ is the sample mean. If, furthermore, the Y_i's are normal, an alternative estimator of σ^2 is the maximum likelihood estimator (MLE), which is

$$\hat{\sigma}^2 \;=\; \frac{1}{n} \sum_{i=1}^{n} (Y_i - \bar{Y})^2. \tag{1.8}$$

Comparing (1.7) with (1.8), the only difference is the denominator: $n-1$ for the sample variance and n for the MLE. In fact, this is the difference between REML and maximum likelihood (ML) in this very simple case. In other words, the sample variance is, indeed, the REML estimator of σ^2 in this case. Of course, if n is large, there is not much difference relatively between n and $n-1$, so the REML estimator and MLE should be close in values. However, this may not be the case when the situation is more complicated.

  A case that is more general than the i.i.d. situation is when the observations are independent but not identically distributed. One of such cases is linear regression. Before we move on, we would like to make a note on the notation we use. Classical statistics and probability theory (e.g., [11]) were built on the i.i.d. case, where a standard notation for an observation is the same as for a random variable, which is in a capital letter, say, Y_i. Sometimes, to distinguish a random variable from a realized value of the random variable, a lower case letter is used for the latter, say, y_i. In some areas of modern statistics, however, such a distinction is no longer important. Instead, the lower case letters are used both for the observations considered as random variables, and for the actual observations which are realizations of the random variables. These areas include regression (e.g., [12]), generalized linear models (e.g., [6]), and mixed effects models (e.g., [2]). In this monograph, we follow the modern trends while respecting the classical convention, in a classical situation. So, for example, when dealing with the i.i.d. case, the capital letters are usually used; in a regression situation, the lower case letters are used. Specifically, in the latter case, the observations $y_1, \ldots, y_n$ are assumed to be independent, with mean $x_i'\beta$, where x_i is a vector of known covariates and β a vector of unknown regression coefficients, and variance σ^2. Suppose that the matrix $X = (x_i')_{1 \leq i \leq n}$ is of full rank p, and let $y = (y_i)_{1 \leq i \leq n}$. Then, a standard estimator of σ^2 is $\hat{\sigma}^2 = \text{MSE} = \text{SSE}/(n-p)$, where

$$\text{SSE} = \sum_{i=1}^{n}(y_i - x_i'\hat{\beta})^2, \tag{1.9}$$

also known as the residual sum of squares (RSS), $\hat{\beta} = (X'X)^{-1}X'y$ is the least squares (LS) estimator of β, and $n - p$ is the degrees of freedom (d.f.) of SSE. On the other hand, if normality is assumed, the MLE of σ^2 is $\tilde{\sigma}^2 = \text{SSE}/n$. Again, we see the difference between the MSE and MLE is only in the denominator. In other words, MSE has taken into account the loss of d.f. in estimating β, p, by subtracting this number from the total d.f., n, in the data, while the loss of d.f. is ignored in MLE. One may guess that, similar to the sample variance, MSE is the REML estimator of σ^2, and the guess is exactly right. In fact, the sample variance is a special case of MSE when $x_i = 1, 1 \leq i \leq n$, in which case the LS estimator of β is the sample mean.

Once again, if the sample size, n, is large, there is not much difference between the REML and ML estimator of σ^2, if p is fixed. So, when will a difference emerge, asymptotically? An answer is given by the following well-known example.

Example 1.1 (Neyman–Scott problem). Neyman and Scott [13] gave the following example which shows that, when the number of parameters increases with the sample size, the MLE may not be consistent. Suppose that two observations are collected from m individuals. Each individual has its own (unknown) mean, say, μ_i for the ith individual. Suppose that the observations are independent and normally distributed with variance σ^2. The problem of interest is to estimate σ^2. The model may be expressed as the following, $y_{ij} = \mu_i + \varepsilon_{ij}$, where ε_{ij}'s are independent and distributed as $N(0, \sigma^2)$. It can be shown that the MLE of σ^2 is inconsistent. It is not difficult to understand why this is happening. There are $m + 1$ unknown parameters, of which the means $\mu_1, \ldots, \mu_m$ are nuisance, while the sample size is $2m$. Apparently, there is not sufficient information for estimating σ^2 after having to estimate all of the nuisance parameters.

Now let us try a different strategy. Instead of trying to estimate the nuisance parameters, consider the difference $z_i = y_{i1} - y_{i2}$. It follows that $z_1, \ldots, z_m$ are independent and distributed as $N(0, 2\sigma^2)$. What makes a key difference is that the nuisance parameters are gone—they are not involved in the distribution of the zs. In fact, the

MLE of σ^2 based on the new data, $z_1, \ldots, z_m$, is consistent. Note that, by taking the difference, one is now in a position with a single parameter, σ^2, and m observations. The MLE based on z_i, $1 \le i \le m$ is, once again, the REML estimator of σ^2. This time, there is a huge difference, asymptotically, between the REML and ML estimators.

REML estimation and related asymptotic theory will be discussed, in detail, in the context of linear mixed models in Section 2.2.

1.3   Example: Salamander mating data

A well-known example, the first published in the context of generalized linear mixed models (GLMM), was given by McCullagh and Nelder [6]. The example involved data from mating experiments regarding two populations of salamanders, Rough Butt and Whiteside. These populations, which are geographically isolated from each other, are found in the southern Appalachian mountains of the eastern United States. The question whether the geographic isolation had created barriers to the animals' interbreeding was thus of great interest to biologists studying speciation. Three experiments were conducted during 1986, one in the summer and two in the autumn. In each experiment there were 10 males and 10 females from each population. They were paired according to the design given by Table 14.3 in [6]. The same 40 salamanders were used for the summer and first autumn experiments. A new set of 40 animals was used in the second autumn experiment. For each pair, it was recorded whether a mating occurred, 1, or not, 0.

A mixed logistic model, which is a special case of GLMM, was proposed by Breslow and Clayton [14] for the salamander data that involved crossed random effects for the female and male animals. The model has since been used (e.g., [15], [16], [17]). Some alternative models, but only in terms of reparametrizations, have been considered (e.g., [18]). Jiang and Zhang [19] noted that some of these models have ignored the fact that a group of salamanders were used in both the summer experiment and one of the fall experiments; in other words, there were replicates for some of the pairs of female and male animals. Nevertheless, all of these models are special cases of the following, more general setting. Suppose that, given the random effects $u_i, v_j, (i,j) \in S$, where S is a subset of $\mathscr{I} = \{(i,j) : 1 \le i \le m, 1 \le j \le n\}$, binary responses y_{ijk}, $(i,j) \in S$, $k = 1, \ldots, c_{ij}$ are conditionally independent such that, with $p_{ijk} = P(y_{ijk} = 1 | u, v)$, we have $\text{logit}(p_{ijk}) = x'_{ijk}\beta + u_i + v_j$, where $\text{logit}(p) = \log\{p/(1-p)\}$, $p \in (0,1)$, x_{ijk} is a known vector of covariates, β is an unknown vector of parameters, and u, v denote all the random effects u_i and v_j that are involved. Here c_{ij} is the number of replicates for the (i,j) cell. Without loss of generality, assume that S is an irreducible subset of $\mathscr{I}$ in that m, n are the smallest positive integers such that $S \subset \mathscr{I}$. Furthermore, suppose that the random effects u_i's and v_j's are independent with $u_i \sim N(0, \sigma^2)$ and $v_j \sim N(0, \tau^2)$, where σ^2, τ^2 are unknown variances. One may think of the random effects u_i and v_j as corresponding to the female and male animals, as in the salamander problem. In fact, for the salamander data, $c_{ij} = 2$ for half of the pairs (i,j), and $c_{ij} = 1$ for the rest of the pairs. It can be shown that the log-likelihood function for estimating β, σ^2, τ^2 can be expressed as

$$
\begin{aligned}
l_1 &= c - \frac{m}{2}\log(\sigma^2) - \frac{n}{2}\log(\tau^2) + \left\{ \sum_{(i,j)\in S}\sum_{k=1}^{c_{ij}} x_{ijk} y_{ijk} \right\}' \beta \\
&\quad + \log \int \cdots \int \left[\prod_{(i,j)\in S}\prod_{k=1}^{c_{ij}} \{1+\exp(x'_{ijk}\beta + u_i + v_j)\}^{-1} \right] \\
&\quad \times \exp\left\{ \sum_{(i,j)\in S} y_{ij\cdot}u_i + \sum_{(i,j)\in S} y_{ij\cdot}v_j - \frac{1}{2\sigma^2}\sum_{i=1}^m u_i^2 - \frac{1}{2\tau^2}\sum_{j=1}^n v_j^2 \right\} \\
&\quad du_1\cdots du_m dv_1 \cdots dv_n,
\end{aligned}
\tag{1.10}
$$

where c is a constant, and $y_{ij\cdot} = \sum_{k=1}^{c_{ij}} y_{ijk}$. The multidimensional integral involved has dimension $m+n$, which increases with the sample size, and the integral cannot be further simplified. In fact, not only does the likelihood not have an analytic expression, it is also difficult to evaluate numerically. See, for example, [2], p. 126, for some discussion on the computational difficulty. For years, the salamander data has been a driving force for computational developments in GLMM. Virtually, every numerical procedure that was proposed used this data as a "gold standard" to evaluate, or demonstrate, the procedure. See, for example, [14–24].

The fact that the random effects are crossed, as in the case of salamander data, presents not only a computational challenge but also a theoretical one, that is, to prove that the MLE is consistent in such a model. In contrast, the situation is very different if the GLMM has clustered, rather than crossed, random effects. For example, consider the following.

Example 1.2. Suppose that, given the random effects $u_1, \ldots, u_m$, binary responses $y_{ij}, i = 1, \ldots, m, j = 1, \ldots, n_i$ are conditionally independent such that, with $p_{ij} = P(y_{ij} = 1|u)$, we have $\mathrm{logit}(p_{ij}) = x'_{ij}\beta + u_i$, where x_{ij} is a vector of known covariates, β a vector of unknown coefficients, and $u = (u_i)_{1\le i\le m}$. Furthermore, suppose that the u_i's are independent with $u_i \sim N(0, \sigma^2)$, where σ^2 is unknown. The log-likelihood for estimating β, σ^2 can be expressed as

$$
l_2 = c - \frac{m}{2}\log(\sigma^2) + \sum_{i=1}^m \left(\sum_{j=1}^{n_i} x_{ij} y_{ij} \right)' \beta + \sum_{i=1}^m \int \frac{\exp(y_{i\cdot}z - z^2/2\sigma^2)}{\prod_{j=1}^{n_i}\{1+\exp(x'_{ij}\beta + z)\}} dz,
$$

where $y_{i\cdot} = \sum_{j=1}^{n_i} y_{ij}$. It is clear that the log-likelihood function for estimating β, σ^2 only involves one-dimensional integrals. Not only that, a major theoretical advantage of this case is that the log-likelihood can be expressed as a sum of independent random variables. In fact, this is a main characteristic of GLMMs with clustered random effects. Therefore, limit theorems for sums of independent random variables (e.g., [1], Chapter 6) can be utilized to obtain asymptotic properties of the MLE.

Generally speaking, the classical approach to proving consistency of the MLE (e.g., [11], Chapter 6; [1]) relies on asymptotic theory for the sum of random variables, independent or not. However, one cannot express the log-likelihood in (1.10) as a sum of random variables with manageable properties. For this reason, it is very

8   INTRODUCTION

difficult to tackle asymptotic behavior of the MLE in the salamander problem, or any GLMM with crossed random effects, assuming that the numbers of random effects in all of the crossed factors increase. In fact, the problem is difficult to solve even for the simplest case, as stated in the open problem below.

> _Open problem_ (e.g., [1], page 541): _Suppose that $x'_{ijk}\beta = \mu$, an unknown parameter, $c_{ij} = 1$ for all i, j, $S = \mathscr{I}$, and σ^2, τ^2 are known, say, $\sigma^2 = \tau^2 = 1$. Thus, μ is the only unknown parameter. Suppose that $m, n \to \infty$. Is the MLE of μ consistent?_

It was claimed ([1], pp. 541, 550) that even for this seemingly trivial case, the answer was not known but expected to be anything but trivial.

The problem regarding consistency of the MLE in GLMMs with crossed random effects began to draw attention in early 1997. It remained unsolved over the next 15 years, and was twice cited as an open problem in the literature, first in [2] (p. 173) and later in [1] (p. 541). The problem was eventually solved in 2012 (see [20]). See Section 3.3 for details.

1.4 A roadmap

We shall begin with the simplest type of mixed effects models, the linear mixed models (LMM). The subject field has a fairly long history (e.g., [21]), but we shall focus on the modern practice LMM, which are based on ML and REML. The ML was first implemented in mixed model analysis by Hartley and Rao [22] and the REML was first proposed by Thompson [23]. Major theoretical breakthroughes, in terms of asymptotics, include establishments of the asymptotic properties of MLE [24], and asymptotic superiority of REML over ML [25], [26]. Most of the asymptotic results are associated with parameter estimation, such as consistency and asymptotic distribution of the estimators, and asymptotic covariance matrix of the estimator. However, asymptotics in other types of inference, such as prediction, model diagnostics and model selection, have also been studied. These topics will be covered in consecutive sections in Chapter 2.

A natural topic after LMM would be GLMM. In earlier time, due to the computational difficulties in computing the MLE, approximations were used in inference about these models, following the work of Breslow and Clayton [14]. It should be noted that, here, the approximation is not in terms of a large sample, but rather in terms of numerical approximation to integrals that cannot be computed analytically. In fact, it is the large-sample asymptotics that discovered limitations of the numerical approximations. The stories will be entertained in the first section of Chapter 3. After the opening section, we shall focus on another alternative to the ML method based on estimating equations. Unlike the approaches covered in Section 3.1, the estimating equation approach typically leads to consistent estimators, although there may be a lack of efficiency. Approaches to improving asymptotic efficiency of the estimating equations, including the GEE method (see Section 1.1), are discussed. The ML method is covered in Section 3.3. Before getting involved in the asymptotic results, we first discuss a recently developed numerical procedure, call _data cloning_

(DC; [27], [28]), that has to do with the asymptotics. The DC is based on Laplace approximation to integrals; however, unlike [14], large sample theory plays a role in DC so that the approximation is asymptotically accurate. Regarding asymptotic properties of MLE, we first consider a case that is fairly straightforward, where the random effects are clustered, such as in Example 1.2. A much more difficult case, in which the random effects are crossed, as in the salamander problem, is discussed next, where the breakthrough of Jiang [20] is presented. This includes an interesting method, called *subset argument*, that was developed to prove consistency of MLE in GLMM with crossed random effects. The method has potentially broader applicability. The last section of this chapter is concerned about conditional inference, including (joint) estimation of fixed and random effects in GLMM.

After systematically discussed asymptotics in LMM and GLMM, it is time to consider some applications, before introducing other topics. A rich source of applications of mixed effects model asymptotics is small area estimation (SAE). A comprehensive converage of SAE can be found in [5], but here we shall focus on the asymptotic methods, which have played a major role in SAE. A problem of important practical interest is estimation of the mean squared prediction error (MSPE) of empirical predictors, such as the empirical best linear unbiased predictor (EBLUP). The problem is difficult because these predictors often involve nonlinear estimators of the variance components, and that second-order unbiased estimator of the MSPE is often desired. Following a seminal paper of Prasad and Rao [29], MSPE estimation in SAE has seen major advances, including those in resampling methods in the context of mixed effects models (e.g., [30], [31]).

The last chapter is devoted to topics related to asymptotics in non-linear mixed effects models, semi- and non-parametric mixed effects models. These models have been developed as extensions of the LMM, with applications in areas such as pharmacokinetics and analysis of longitudinal data.

We conclude each chapter with a section of bibliographical notes and open problems. As is typical for a monograph, exercises are not provided.

Chapter 2

Asymptotic Analysis of Linear Mixed Models

Linear regression is widely known as a useful tool for statistical analysis, but its role has been gradually replaced by LMM due to the flexibility of the latter as well as its ability to incorporate correlations in the observations. C. R. Henderson has been credited for the early developments of LMM, although the origination of the latter can be traced back to some time much earlier (e.g., [21], Chapter 2). Among many of Henderson's contributions are analysis of variance (ANOVA) methods for mixed model analysis, known as Henderson's methods I, II, and III [32]. The ANOVA methods are computationally simpler, compared to some of the methods used today. This fitted well with the earlier time, when computational resources and tools were limited. However, the ANOVA estimators are inefficient in that they may have large variation; they also suffer from possibilities of negative values for quantities that are supposed to be nonnegative, such as variances of the random effects. For such reasons, the ANOVA methods have given way to more sophisticated methods, such as ML and REML, once the latter became computationally feasible.

A general LMM may be expressed as

$$y = X\beta + Z\alpha + \varepsilon, \tag{2.1}$$

where y is an $n \times 1$ vector of observations, X is an $n \times p$ matrix of known covariates, β is a $p \times 1$ vector of *fixed effects*, Z is a $n \times m$ known matrix, α is a $m \times 1$ vector of random effects, and ε is a $n \times 1$ vector of errors. Typically, it is assumed that α, ε satisfy $E(\alpha) = 0$, $Var(\alpha) = G$, $E(\varepsilon) = 0$, $Var(\varepsilon) = R$, and $Cov(\alpha, \varepsilon) = 0$. Several special cases of the general LMM are described below.

Example 2.1 (Gaussian mixed model). If normality is assumed, that is, $\alpha \sim N(0, G)$, $\varepsilon \sim N(0, R)$, and α, ε are independent, the LMM is called a Gaussian LMM, or simply Gaussian mixed model (GMIM). In fact, this is how LMM is usually defined, and ML and REML estimators are derived (see below).

Example 2.2 (Gaussian marginal model). It is seen that, under the normality assumption, one has

$$y \sim N(X\beta, V), \tag{2.2}$$

where $V = ZGZ' + R$. In fact, as long as ML and REML analyses are concerned,

11

the GMM of Example 2.1 is equivalent to (2.2), which is called Gaussian *marginal model* (GMAM). More generally, a GMAM can be defined as (2.2), where V can have any (parametric) structure not necessarily expressed as $ZGZ' + R$. Note that, unlike GMIM, random effects are not present in a GMAM; in particular, one cannot make inference about the random effects under a GMAM.

Example 2.3 (mixed ANOVA model). Suppose that, in (2.1), one has $Z = [Z_1 \, Z_2 \, \cdots \, Z_s]$ (partitioned matrix) and, correspondingly, $\alpha = (\alpha_1', \alpha_2', \ldots, \alpha_s')'$ so that

$$Z\alpha \;=\; Z_1\alpha_1 + \cdots + Z_s\alpha_s, \qquad (2.3)$$

where $\alpha_1, \ldots, \alpha_s$ are subvectors such that $\alpha_r \sim N(0, \sigma_r^2)$, $1 \leq r \leq s$. Furthermore, supposed that $\varepsilon \sim N(0, \tau^2)$, and that $\alpha_1, \ldots, \alpha_s, \varepsilon$ are independent. Then, the LMM is called a Gaussian mixed ANOVA model. It is seen that, in this case, one has $V = \mathrm{Var}(y) = \tau^2 I_n + \sum_{r=1}^{s} \sigma_r^2 Z_r Z_r'$, where I_n is the n-dimensional identity matrix.

A non-Gaussian mixed ANOVA model is defined the same way as the Gaussian one except without the normality assumption. In other words, the components of α_r are i.i.d. with mean 0 and variance σ_r^2, $1 \leq r \leq s$, and the components of ε are i.i.d. with mean 0 and variance τ^2.

Example 2.4 (longitudinal model). Another type of LMM can be expressed as

$$y_i \;=\; X_i\beta + Z_i\alpha_i + \varepsilon_i, \quad i = 1, \ldots, m, \qquad (2.4)$$

where y_i is an $n_i \times 1$ vector of responses, X_i is an $n_i \times p$ matrix of known covariates, Z_i is an $n_i \times b$ known matrix, α_i is a $b \times 1$ vector of random effects, and ε_i is an $n_i \times 1$ vector of errors. It is assumed that $\alpha_i \sim N(0, G_i)$, $\varepsilon_i \sim N(0, R_i)$, and that $\alpha_i, \varepsilon_i, i = 1, \ldots, m$ are independent. This is called a Gaussian longitudinal model. Typically, the covariance matrices $G_i, R_i, 1 \leq i \leq m$ have certain parametric structures that depend on ψ, a vector of variance components.

A non-Gaussian longitudinal model is defined the same way as the Gaussian one except that the normality assumption is replaced by $\mathrm{E}(\alpha_i) = 0$, $\mathrm{Var}(\alpha_i) = G_i$, $\mathrm{E}(\varepsilon_i) = 0$, $\mathrm{Var}(\varepsilon_i) = R_i$, $1 \leq i \leq m$.

See, for example, [2] (Section 1.2) for further discussion of different types of LMMs.

2.1   Maximum likelihood

Suppose that the covariance matrices, G and R, or, more generally, V, depend on a vector of variance components, $\theta = (\theta_k)_{1 \leq k \leq q}$. Then, under the normality assumption, namely, (2.2), the log-likelihood function for estimating β and θ can be expressed as

$$l \;=\; c - \frac{1}{2}\left\{ \log(|V|) + (y - X\beta)'V^{-1}(y - X\beta) \right\}, \qquad (2.5)$$

where $|V|$ denotes the determinant of V. By differentiating with respect to β and components of θ, it can be shown (e.g., [2], p. 10) that the likelihood equation, or

ML equation, is equivalent to

$$\beta = (X'V^{-1}X)^{-1}X'V^{-1}y, \tag{2.6}$$

$$(y - X\beta)'V^{-1}\frac{\partial V}{\partial \theta_k}V^{-1}(y - X\beta) = \mathrm{tr}\left(V^{-1}\frac{\partial V}{\partial \theta_k}\right), \quad 1 \le k \le q. \tag{2.7}$$

Here, for simplicity, we assume that X is of full rank, p. Given a specific parametric form of V, as a function of θ, the partial derivatives $\partial V/\partial \theta_k$ can be specified. For example, under the mixed ANOVA model, a standard vector of variance components is $\theta = (\tau^2, \sigma_1^2, \dots, \sigma_s^2)'$, and one has $V = \tau^2 I_n + \sum_{r=1}^{s} \sigma_r^2 Z_r Z_r'$. Alternatively, one may use the Hartley-Rao form of variance components, defined as $\theta = (\tau^2, \gamma_1, \dots, \gamma_s)'$, where $\gamma_r = \sigma_r^2/\tau^2, 1 \le r \le s$. The MLE of $\psi = (\beta', \theta')'$ is, by definition, the maximier of l over the parameter space of ψ. Under regularity conditions, the MLE is a solution to the ML equation, but not every solution to the ML equation is the MLE. We shall come back to this, but would like to point out first that there is a difference.

Although asymptotic properties of the MLE, such as consistency and asymptotic normality, are well known in the classical case of i.i.d. observations (e.g., [11], Chapter 6), it is not straightforward, as it turned out, to extend these results to LMM. One complication is what is meant by "sample size going to infinity." We use an example to illustrate.

Example 2.5. Consider an LMM defined as $y_{ij} = x_{ij}'\beta + u_i + v_j + e_{ij}$, $i = 1, \dots, m_1$, $j = 1, \dots, m_2$, where x_{ij} is a vector of known covariates, β is a vector of unknown regression coefficients (the fixed effects), u_i, v_j are random effects, and e_{ij} is an additional error. It is assumed that u_i's, $v_j's$ and e_{ij}'s are independent, and that, for the moment, $u_i \sim N(0, \sigma_u^2)$, $v_j \sim N(0, \sigma_v^2)$, $e_{ij} \sim N(0, \sigma_e^2)$. It is well known (e.g., [33], [24]) that, in this case, the effective sample size for estimating σ_u^2 and σ_v^2 is not the total sample size $n = m_1 m_2$, but m_1 and m_2, respectively, for σ_u^2 and σ_v^2.

Hartley and Rao [22] were the first to have studied asymptotic properties of the MLE in LMM, and realized the complexity in the limiting process. Here an LMM with a particular data structure is called a "design," which includes the overall sample size, n, and other numbers that may increase as n increases. For example, under the mixed ANOVA model, the number of random effects for the rth random effect factor involved in (2.3), m_r, which is the dimension of α_r, may increase with n, $1 \le r \le s$. One may denote m_r as $m_r(n)$ to indicate its dependence on n, but such a detail is often suppressed to simplify the notation. As n increases, one considers the design as a sequence, and studies the asymptotic behavior of the MLE along the sequence. Hartley and Rao imposed a condition on how the design sequence grows that is somewhat restrictive. Namely, as n increases, the number of observations at any level of any random effect factor was required to be bounded by some constant. This sometimes limits the scope of application. For example, consider the last example. At any given level, i, of u, the number of related observations is m_2; similarly, at any given level, j, of v, the number of related observations is m_1. If both m_1 and m_2 are bounded, the design sequence cannot increase to infinity. For the most part, the Hartley-Rao condition applies to LMM with clustered random effects, such as the following.

Example 2.6 (nested-error regression). Battese et al. [34] used a form of the so-called nested-error regression (NER) model to estimate mean hectares of crops per segment for some counties in the state of Iowa. In general, let y_{ij} denote the jth observation from the ith small area (e.g., county). Suppose that

$$y_{ij} = x'_{ij}\beta + v_i + e_{ij}, \tag{2.8}$$

$i = 1, \ldots, m, j = 1, \ldots, n_i$, where x_{ij} is a known vector of auxiliary variables, β is an unknown vector of regression coefficients, v_i is an area-specific random effect, and e_{ij} is a sampling error. It is assumed that the v_i's and e_{ij}'s are independent and normally distributed with mean zero, $\text{var}(v_i) = \sigma_v^2$ and $\text{var}(e_{ij}) = \sigma_e^2$, where σ_v^2 and σ_e^2 are unknown variances. The NER model is broadly used in SAE (see [5]), where the sample sizes, n_i, from the small areas (e.g., counties) are typically small or moderate. Note that n_i is the number of observations corresponding to the ith level of the random effect factor, v (which is the only random effect factor involved in the NER model). Therefore, the Hartley-Rao condition is reasonable in this case.

Miller [24] relaxed the Hartley-Rao condition by allowing the number of observations at any level of any random effect factor to grow with the sample size, under the mixed ANOVA model. More specifically, it is assumed that $m_r \to \infty, 1 \le r \le s$. Miller noted the following point that has influenced subsequent research in LMM asymptotics: Estimation of different parameters in a LMM may require different "effective sample sizes," or normalizing constants, in measuring the convergence rates. Miller used the following simple example to illustrate. Suppose that ψ_1, ψ_2 are two parameters and their MLEs are $\hat{\psi}_1, \hat{\psi}_2$, respectively. There may be no single function of n, $v(n)$, such that $\sqrt{v(n)}(\hat{\psi}_1 - \psi_1, \hat{\psi}_2 - \psi_2)'$ converges in distribution to a bivariate normal distribution, as $n \to \infty$. In fact, it may be necessary to use two such functions, $v_1(n)$ and $v_2(n)$, such that $[\sqrt{v_1(n)}(\hat{\psi}_1 - \psi_1), \sqrt{v_2(n)}(\hat{\psi}_2 - \psi_2)]'$ convergens to a bivariate normal distribution, but $v_1(n)/v_2(n)$ may go to 0 or ∞.

Another note for the notation. Strictly speaking, when consider asymptotic properties of estimators, such as in the above example, an estimators, $\hat{\psi}$, should be denoted by $\hat{\psi}_n$ to indicate its dependence on n, the sample size. We suppress such an indication not only for notation simplicity but also because of the complexity of the limiting process, as discussed above, so that the overall sample size n may not represent the full design sequence.

Going back to Miller's influential paper, the author went on to derive exact expressions for the normalizing constants. As it turns out, these normalizing constants are closely related to the degrees of freedom (d.f.) in the analysis of variance (ANOVA) table (e.g., [35]). We use some examples to illustrate.

Example 2.6 (continued). Consider a special case with $x'_{ij}\beta = \mu$, an unknown mean, and write $m = I$. Also suppose that $n_i = J, 1 \le i \le I$. (This example is adopted from [24]; thus, we keep the notation consistent.) Miller showed that, in this simple case, the normalizing constants for obtaining the asymptotic normality result is $\sqrt{I}, \sqrt{I}, \sqrt{IJ}$ for estimating the parameters $\mu, \sigma_v^2, \sigma_e^2$, respectively.

Example 2.5 (continued). Again consider a special case with $x'_{ij}\beta = \mu$. Miller considered an example which is the same except with an interaction. Thus, again, we write $m_1 = I$ and $m_2 = J$ to keep the notation consistent. It can be shown (see [24]) that, in this case, the normalizing constants are $\sqrt{I}, \sqrt{J}, \sqrt{IJ}$ for estimating

$\sigma_u^2, \sigma_v^2, \sigma_e^2$, respectively. The normalizing constant for estimating μ is a bit complicated; however, provided that I/J is asymptotically bounded from above as well as away from zero, it can be simply chosen as $\sqrt{I}$ [24].

To prove asymptotic properties of the MLE, Miller noted that the case is different from the i.i.d. situation, for which the classical limit theory (e.g., [1], Chapter 6) can be used. Earlier, Weiss [36], [37] gave conditions for consistency and asymptotic normality of the MLE in some nonstandard cases. Miller's approach to the proof was therefore to verify these conditions. On the one hand, Weiss' conditions are very general that they apply, in principle, to almost every case, including the one that Miller was concerned about. On the other hand, because the conditions are too general (see below) that they essentially state what are needed to prove the asymptotic properties of the MLE. As a result, it took some major efforts (for Miller) to verify Weiss' conditions. Nevertheless, Weiss' approach to establish the asymptotic properties of the MLE is neat, which we outline below.

Let $l = l(\psi)$ denote an object function, which may be the log-likelihood. By the second-order Taylor expansion, one can write

$$l(\psi) - l(\psi_0) \approx \sum_{j=1}^{a} \left.\frac{\partial l}{\partial \psi_j}\right|_{\psi_0} (\psi_j - \psi_{0j})$$
$$+ \frac{1}{2}\sum_{j,k=1}^{a} \left.\frac{\partial^2 l}{\partial \psi_j \partial \psi_k}\right|_{\psi_0} (\psi_j - \psi_{0j})(\psi_k - \psi_{0k}), \qquad (2.9)$$

where ψ_0 is some "point of interest," say, the true parameter vector, and a is the dimension of ψ. Now suppose that one wants to determine the normalizing constants, v_j for ϕ_j, such that $\sqrt{v_j}(\hat\psi_j - \psi_{0j})$ is bounded in probability, $1 \le j \le a$ (note that this is not yet asymptotic normality, but is an important first-step), one can write the right side of (2.9) as

$$\sum_{j=1}^{a} \frac{1}{\sqrt{v_j}} \left.\frac{\partial l}{\partial \psi_j}\right|_{\psi_0} \sqrt{v_j}(\psi_j - \psi_{0j})$$
$$+ \frac{1}{2}\sum_{j,k=1}^{a} \frac{1}{\sqrt{v_j v_k}} \left.\frac{\partial^2 l}{\partial \psi_j \partial \psi_k}\right|_{\psi_0} \sqrt{v_j}(\psi_j - \psi_{0j})\sqrt{v_k}(\psi_k - \psi_{0k})$$
$$= v^{-1/2} \left.\frac{\partial l}{\partial \psi'}\right|_{\psi_0} \{v^{1/2}(\psi - \psi_0)\}$$
$$+ \frac{1}{2}\{v^{1/2}(\psi - \psi_0)\}' v^{-1/2} \left.\frac{\partial^2 l}{\partial \psi \partial \psi'}\right|_{\psi_0} v^{-1/2}\{v^{1/2}(\psi - \psi_0)\}, \quad (2.10)$$

where $v^h = \text{diag}(v_j^h, 1 \le j \le a)$ for any real number h, and $\partial l/\partial \psi, \partial^2 l/\partial \psi \partial \psi'$ are the vector and matrix of first, second derivatives, respectively. The point is to consider the boundary of an ellipse of ψ satisfying $|v^{1/2}(\psi - \psi_0)| = \rho$, where ρ is a positive number. If conditions are set such that $v^{-1/2}(\partial l/\partial \psi|_{\psi_0})$ is bounded in probability (e.g., [1], Section 3.4) and that

$$v^{-1/2} \left.\frac{\partial^2 l}{\partial \psi \partial \psi'}\right|_{\psi_0} v^{-1/2} \xrightarrow{P} A, \qquad (2.11)$$

where A is a positive definite (constant) matrix, then the first term on the right side of (2.10) is bounded by $O_P(1)\rho$ (again, see, e.g., [1], Section 3.4 for the notation O_P) while the second term is $\geq (1/2)\delta\rho^2$ for some positive constant δ. By choosing ρ sufficiently large, arguments can be put together to show that, with probability tending to one (w. p. $\to$ 1), the right side of (2.10) is positive. Thus, combined with (2.9), it can be shown that, w. p. $\to$ 1, $l(\psi)$ is greater than $l(\psi_0)$ for any ψ on the boundary of the ellipse. It follows that, w. p. $\to$ 1, there is a root, $\hat{\psi}$, to $\partial l/\partial \psi = 0$ that satisfies $|v^{1/2}(\hat{\psi} - \psi_0)| < \rho$.

It has now become clear how the normalizing constants, $v_j, 1 \leq j \leq a$, should be chosen: They are chosen such that (2.11) holds. Under suitable conditions, Miller proved the following results:

(i) w. p. $\to$ 1, there is a root, $\hat{\psi} = (\hat{\beta}', \hat{\theta}')'$, to the ML equation that satisfies $\sqrt{v_{q+1}}(\hat{\beta} - \beta_0) = O_P(1)$ and $v^{1/2}(\hat{\theta} - \theta_0) = O_P(1)$, where $v_k, 1 \leq k \leq q+1$ are sequences of positive constants, $v^{1/2} = \text{diag}(\sqrt{v_1}, \ldots, \sqrt{v_q})$, q is the dimension of θ, and $\psi_0 = (\beta_0', \theta_0')'$ is the true parameter vector.

(ii) The same $\hat{\psi}$ satisfies $[\sqrt{v_{q+1}}(\hat{\beta} - \beta_0)', v^{1/2}(\hat{\theta} - \theta_0)']' \xrightarrow{d} N(0, J^{-1})$ for some positive definite matrix J.

It should be noted that the results proved by Miller are consistency of Cramér-type [38], which states that a root to the likelihood equation is consistent. However, it does not always imply that the MLE, which by definition is the (global) maximizer of the likelihood function, is consistent. A stronger result is called Wald-type consistency ([39]; also see [40]), which states that the MLE is consistent. Although Wald-type consistency results are more desirable, they are often more difficult to prove. As a result, most of the consistency results for the MLE in the literature (not just on LMM but in general) are of Cramér-type.

Another difference between the asymptotic theory of MLE in LMM and that in the i.i.d. case is that, unlike the latter, there is no Fisher information matrix associated with a single observation that is identical for all of the observations. Instead, the Fisher information matrix is defined as

$$I(\psi) = \text{Var}\left(\frac{\partial l}{\partial \psi}\right) = -\text{E}\left(\frac{\partial^2 l}{\partial \psi \partial \psi'}\right), \tag{2.12}$$

provided that certain regularity conditions hold. It can be shown that, under suitable conditions, one has

$$I(\psi)^{1/2}(\hat{\psi} - \psi_0) \xrightarrow{d} N(0, I_a), \tag{2.13}$$

where $\xrightarrow{d}$ means convergence in distribution, a is the dimension of ψ, and the square root of a nonnegative matrix, A, is defined as follows. Let T be the orthogonal matrix that diagonalizes A such that $A = T\text{diag}(\lambda_1, \ldots, \lambda_a)T'$, where $\lambda_j, 1 \leq j \leq a$ are the eigenvalues of A, which are nonnegative. Then, $A^{1/2}$ is defined as $T\text{diag}(\lambda_1^{1/2}, \ldots, \lambda_a^{1/2})T'$. (Alternatively, $A^{1/2}$ may be defined via the Cholesky decomposition, yielding the same asymptotic result.) (2.12) may be interpreted as that $\hat{\psi}$ is asymptotically normal with mean ψ_0 and covariance matrix $I(\psi)^{-1}$, although the latter should be understood as a sequence of matrices. It is in this sense that the MLE is (asymptotically) efficient ([24]; [2], p. 11).

Our discussion on the asymptotic properties of MLE shall continue in the next section in terms of comparison with the REML estimator.

2.2   Restricted maximum likelihood

REML estimation was introduced earlier in Section 1.2 in several, simple, special cases. We now introduce the method under a general setting. Let A be an $n \times (n-p)$ matrix, where $p = \text{rank}(X)$ (again, X is assumed to be full-rank), such that

$$\text{rank}(A) = n - p \quad \text{and} \quad A'X = 0. \tag{2.14}$$

Then, under the normality assumption, the distribution of $z = A'y$ is $N(0, A'VA)$. Here z is viewed as a vector of data under the transformation, A'. With the transformation, known as the REML transformation, the fixed effects have been eliminated from the distribution of z. One can now estimate the variance components, θ, using ML based on z, and this is REML estimation. See, for example, [2] (Section 1.3.2) for details. Let $\hat{\theta}$ denote the REML estimator of θ. Then, under regularity conditions, $\hat{\theta}$ is a root to the following REML equation:

$$y'P\frac{\partial V}{\partial \theta_k}Py = \text{tr}\left(P\frac{\partial V}{\partial \theta_k}\right), \quad 1 \le k \le q, \tag{2.15}$$

where q is the dimension of θ, and

$$P = A(A'VA)^{-1}A' = V^{-1} - V^{-1}X(X'V^{-1}X)^{-1}X'V^{-1} \tag{2.16}$$

(e.g., Appendix B of [2]). It is seen that P, hence the REML equation, do not depend on any particular choice of A. Once the REML estimator of θ is obtained, REML estimator of β is obtained the same way as the MLE of β except where θ is replaced by its REML estimator.

2.2.1   Asymptotic behavior of REML: Earlier studies

The first published study on asymptotic properties of REML estimators was, apparently, due to Das [41] in 1979, two years after Miller's work on the MLE was published. In fact, Das obtained similar results as Miller's for the REML estimator under conditions slightly stronger than those of Miller. Furthermore, Das showed that, under certain conditions, the ML and REML estimators are asymptotically equivalent in the sense that their suitably normalized difference converges to zero in prabil-ity. More specifically, let $\hat{\psi}$ be the REML estimator of ψ. Das showed that, under suitable conditions, one has $[\sqrt{v_j}(\hat{\psi}_j - \psi_{0j})]_{1 \le j \le a} \xrightarrow{d} N(0, J^{-1})$ for some positive definite matrix J, and $[\sqrt{v_j}(\hat{\psi}_j - \tilde{\psi})]_{1 \le j \le a} \xrightarrow{P} 0$, where $\hat{\psi}, \tilde{\psi}$ are the REML and ML estimators of ψ, respectively, ψ_0 is the true parameter, and $v_j, 1 \le j \le a$ are the normalizing constants. It follows that the REML, ML estimator has the same asymptotic covariance matrix, J^{-1}, and therefore equally efficient.

An important assumption that was imposed for obtaining Das's results is that the number of fixed effects, p, is fixed. Note that, without this assumption, the MLE of θ may not even be consistent, as shown by the Neyman-Scott example; on the other hand, it is known that the REML estimator of σ^2 is is consistent in the latter example (Example 1.1). Thus, at least for the Neyman-Scott problem, the REML and ML estimators are *not* asymptotically equivalent.

Somehow, Das' earlier work on REML asymptotics remained largely unknown for the next fifteen years. A couple of studies on the same topic emerged in the first half of 1990s, and neither seemed to be aware of Das' work. Cressie and Lahiri [42] derived general results on asymptotic normality of the REML estimator, using a result of Sweeting [43]. The latter author claimed to have derived "a very general result" on the existence of a local maximum of the likelihood function and the asymptotic normality of such a local maximum. However, the conditions are so general that they essentially state what are needed for the proof. Such a result is not very useful, from a practical standpoint. For example, a practitioner may want to know if the conditions of the theorem are satisfied in some simple, standard situations, such as a two-way random effects model; the answer does not follow immediately fromm the theorem. In fact, this was noted in [42]. Namely, when the latter authors tried to specialize their general result to the mixed ANOVA model, the conditions were not met for some simple cases, such as the two-way random effects model. So, at least for the latter case, the asymptotic behavior of the REML estimator is unclear, despite the simplicity of the model.

In another work that has impacted future studies, Richardson and Welsh [44] considered asymptotic properties of the REML estimator in hierarchical LMMs, that is, LMM in which the random effects have a nested structure. For example, consider $y_{ijk} = x'_{ijk}\beta + u_i + v_{ij} + e_{ijk}$, where x_{ijk} is a vector of known covariates, β is a vector of unknown fixed effects, u_i, v_{ij} are random effects, and e_{ijk} is an error. Here the random effect v_{ij} is nested within the random effect u_i; thus, the model is a special case of the hierarchical LMM. Of course, not every LMM is hierarchical. For example, the random effects u_i and v_j in Example 2.5 are crossed, rather than nested; thus, the corresponding LMM is not a hierarchical one. Although the type of LMM considered by Richardson and Welsh might seem a bit more restrictive compared to those by Cressie and Lahiri [42] (actually, not necessarily, as it is not very clear what types of LMMs are covered by [42]), a major advance in [44] is that, for the first time, it studied asymptotic behavior of the REML estimator, which is derived under the normality assumption, in non-normal situations. This is important, especially from a practical point of view, because the normality assumption about the random effects is likely to be violated in practice. To be clear, in the sequel we shall use the term Gaussian REML (ML) estimator for the REML (ML) estimator derived under normality in a non-normal situation, or a situation in which one is not sure about the normality. The results of [44] ensures that, at least for the hierarchical LMMs, the Gaussian REML estimator is consistent and asymptotically normal, even if the normality fails [but provided that some (much) milder conditions are satisfied; see the next subsection].

Still, the hierarchical LMM assumption limits the scope of applications, as many

LMMs that are used in practice (e.g., in genetics) do not have the hierarchical or nested structure that is assumed by [44].

In all of the studies so far discussed, the number of fixed effects, p, was assumed fixed. As shown by [41], in such a case, the REML and ML estimators are asymptotically equivalent. Of course, finite-sample performances of the REML and ML estimators can be different; see, for example, [50–53]. However, at least in large sample, there seemed to be little difference between the two procedures. Thus, an important question regarding the potential asymptotic superiority of REML over ML, as suggested by the Neyman-Scott example (Example 1.1) as well as numerous other studies, such as those cited above, had not been answered.

2.2.2 Asymptotic behavior of REML: Asymptotic superiority

In his Ph.D. dissertation, Jiang [25], [26] studied asymptotic behavior of the REML estimator under the following basic scenarios: (i) the number of fixed effects, p, is increasing with the sample size; (ii) the random effects and errors are not necessarily normally distributed. Scenario (i) is what distinguished Jiang's work from all of the previous studies, and the case where REML is shown to be asymptotically superior over the ML. Scenario (ii) followed Richardson and Welsh's approach by considering Gaussian REML estimation without assuming normality; in addition, asymptotic behavior of Gaussian ML estimator was also studied, and compared to that of the Gaussian REML estimator.

Consider a non-Gaussian mixed ANOVA model, defined in Example 2.3. The model is called unconfounded if (i) the fixed effects are not confounded with the random effects and errors in the sense that $\text{rank}(X, Z_r) > p$, $\forall r$ and $X \neq I$; and (ii) the random effects and errors are not confounded in the sense that the matrices I and $Z_r Z_r'$, $1 \leq r \leq s$ are linearly independent (e.g., [24]). The model is called non-degenerate if $\text{var}(\alpha_{r1}^2)$, $0 \leq r \leq s$ are bounded away from zero, where α_{r1} is the first component of α_r, $1 \leq r \leq s$ and $\alpha_0 = \varepsilon$. Note that if $\text{var}(\alpha_{r1}^2) = 0$, then, with probability one, $\alpha_{r1} = -c$ or c for some constant c.

Consider the Hartley-Rao form of variance components, defined below (2.7). For a non-Gaussian mixed ANOVA model, the REML estimator of $\theta = (\tau^2, \gamma_1, \ldots, \gamma_s)'$ is defined as a solution to the following REML equation, derived under normality (see [2], p. 16):

$$
\begin{cases}
y'Py &= n - p, \\
y'PZ_r Z_r'Py &= \text{tr}(Z_r'PZ_r), \quad 1 \leq r \leq s,
\end{cases}
\tag{2.17}
$$

where P is defined by (2.15). Similarly, the MLE of θ is defined as a solution to the following ML equation, also derived under normality (again, see [2], p. 16):

$$
\begin{cases}
y'Py &= n, \\
y'PZ_r Z_r'Py &= \text{tr}(Z_r'V^{-1}Z_r), \quad 1 \leq i \leq s.
\end{cases}
\tag{2.18}
$$

We call the components of the REML estimator (MLE) REML estimators (MLEs). Sequences of estimators, $\hat{\tau}^2, \hat{\gamma}_1, \ldots, \hat{\gamma}_s$, are said to be asymptotically normal if there are sequences of positive constants $p_r \to \infty$, $0 \leq r \leq s$ and a sequence of matrices $\mathcal{M}$ satisfying $\limsup(\|\mathcal{M}^{-1}\| \vee \|\mathcal{M}\|) < \infty$, where the (spectral) norm of a matrix, M, is defined as $\|M\| = \sqrt{\lambda_{\max}(M'M)}$ and $\lambda_{\max}$ denotes the largest eigenvalue, such that

$$\mathcal{M}[p_0(\hat{\tau}^2 - \tau^2), p_1(\hat{\gamma}_1 - \gamma_1), \ldots, p_s(\hat{\gamma}_s - \gamma_s)]' \xrightarrow{d} N(0, I_{s+1}).$$

Note that the sequences $p_r, 0 \leq r \leq s$ and matrices $\mathcal{M}$ all depend on n, the sample size or, more generally, the design (see Section 2.1), but the dependence is suppressed in the notation for the sake of simplicity. Two sequences of positive constants, a and b, are of the same order, denoted by $a \sim b$, if $\liminf(a/b) > 0$ and $\limsup(a/b) < \infty$.

We now introduce a few matrices that are associated with the asymptotic covariance matrix of the REML estimator. Let A be as in (2.13). Define $V(A, \gamma) = A'A + \sum_{r=1}^{s} \gamma_r A'Z_r Z_r' A$, and $V(\gamma) = AV(A, \gamma)^{-1}A'$. Note that $V(\gamma)$ does not depend on the choice of A. Let $V_0(\gamma) = b(\gamma)V(\gamma)b(\gamma)'$, where $b(\gamma) = (I, \sqrt{\gamma_1}Z_1, \ldots, \sqrt{\gamma_s}Z_s)'$, and $V_r(\gamma) = b(\gamma)V(\gamma)Z_r Z_r' V(\gamma)b(\gamma)'$, $1 \leq r \leq s$. Furthermore, define $V_0 = I_{n-p}/\tau^2$, $V_r = V(A, \gamma)^{-1/2}A'Z_r Z_r' AV(A, \gamma)^{-1/2}$, $1 \leq r \leq s$. Let $\mathscr{I}$ be the matrix whose (r,t) element is $\mathrm{tr}(V_r V_t)/p_r p_t$, $0 \leq r, t \leq s$, where p_r, $0 \leq r \leq s$ are given sequences of positive constants, and $\mathscr{K}$ the matrix whose (r,t) element is

$$\sum_{l=1}^{m+n} (\mathrm{E}\omega_l^4 - 3)V_{r,ll}(\gamma)V_{t,ll}(\gamma)/p_r p_t (\tau^2)^{1(r=0)+1(t=0)},$$

where $m = \sum_{r=1}^{s} m_r$ and

$$\omega_l = \begin{cases} \varepsilon_l/\tau, & 1 \leq l \leq n, \\ \alpha_{r,l-n-\sum_{t<r} m_t}/\tau\sqrt{\gamma_r}, & n + \sum_{t<r} m_t + 1 \leq l \leq n + \sum_{t\leq r} m_t, 1 \leq r \leq s, \end{cases}$$

where $V_{r,kl}(\gamma)$ is the (k,l) element of $V_r(\gamma)$, $0 \leq r \leq s$, and m_r is the dimension of α_r, $1 \leq r \leq s$. Hereafter, $\theta = (\tau^2, \gamma_1, \ldots, \gamma_s)'$ denotes the true parameter vector, whenever the notation appears in a theorem.

Before presenting the general results, let us first take a look at a special case, namely, the balanced mixed ANOVA model, or balanced data [21]. A mixed ANOVA model is balanced if X and Z_r, $1 \leq r \leq s$ in (2.1) and (2.3) can be expressed as Kronecker products such that $X = \otimes_{l=1}^{w+1} 1_{n_l}^{a_l}$, and $Z_r = \otimes_{l=1}^{w+1} 1_{n_l}^{b_{r,l}}$, where $(a_1, \ldots, a_{w+1}) \in S_{w+1} = \{0,1\}^{w+1}$, $(b_{r,1}, \ldots, b_{r,w+1}) \in S \subset S_{w+1}$, $1 \leq r \leq s$. In other words, there are w factors in the model; n_l represents the number of levels for factor l ($1 \leq l \leq w$); and the $(w+1)$st factor corresponds to "repetition within cells." Thus, we have $a_{s+1} = 1$ and $b_{r,s+1} = 1$ for all r. We consider some examples.

Example 2.7 (One-way random effects model). Consider a special case of Example 2.6 with $x_{ij}'\beta = \mu$, an unknown mean, and $n_i = J, 1 \leq i \leq I$. This is a special case of a balanced mixed ANOVA model with $w = 1, n_1 = I, n_2 = J$, and $S = \{(0,1)\}$.

Example 2.8 (Two-way random effects model). Consider a special case of Example 2.5 with $x_{ij}'\beta = \mu$, an unknown mean. This is a spacial case of a balanced mixed ANOVA model with $w = 2, n_1 = m_1, n_2 = m_2, n_3 = 1$, and $S = \{(0,1,1), (1,0,1)\}$.

If $u, v \in S_{w+1}$, define $u \vee v = (u_1 \vee v_1, \ldots, u_{w+1} \vee v_{w+1})$, $S_u = \{v \in S : v \leq u\}$, $m_u = \prod_{l:u_l=0} n_l$, $m_{u,S} = \min_{v \in S_u} m_v$ if $S_u \neq \emptyset$, and $m_{u,S} = 1$ if $S_u = \emptyset$. Note that, in the case of balanced data, it is more convenience to use the multiple index, $u \in S_{w+1}$, than the single index r. Thus, for example, $m_r, 1 \leq r \leq s$ will be denoted by $m_u, u \in S \subset S_{w+1}$. The following results were proved in [25].

Theorem 2.1. Let the balanced mixed ANOVA model be unconfounded, and $\tau^2, \gamma_r, 1 \leq r \leq s$ be positive. As $n \to \infty$ and $m_r \to \infty$, $1 \leq r \leq s$, the following hold:

(i) There exist with probability tending to one REML estimators $\hat{\tau}^2, \hat{\gamma}_r, 1 \leq r \leq s$ that are consistent, and the sequences $\sqrt{n-p}(\hat{\tau}^2 - \tau^2)$, $\sqrt{m_r}(\hat{\gamma}_r - \gamma_r), 1 \leq r \leq s$ are bounded in probability.

(ii) If, moreover, the model is nondegenerate, the REML estimators in (i) are asymptotically normal with $p_0 = \sqrt{n-p}$, $p_r = \sqrt{m_r}$, $1 \leq r \leq s$, and $\mathcal{M} = \mathcal{J}^{-1/2}\mathcal{I}$, where $\mathcal{J} = 2\mathcal{I} + \mathcal{K}$.

Theorem 2.2. Let the balanced mixed ANOVA model be unconfounded, and $\tau^2, \gamma_r, 1 \leq r \leq s$ be positive. As $n \to \infty$ and $m_r \to \infty$, $1 \leq r \leq s$, the following hold:

(i) There exist with probability tending to one MLEs of $\tau^2, \gamma_r, 1 \leq r \leq s$ that are consistent if and only if

$$\frac{p}{n} \to 0, \qquad \frac{m_{b_r \vee a} m_{b_r \vee a,S}}{m_{b_r}^2} \to 0, \qquad 1 \leq r \leq s.$$

(ii) If, moreover, the model is nondegenerate, then there exist with probability tending to one MLEs of $\tau^2, \gamma_r, 1 \leq r \leq s$ that are asymptotically normal if and only if

$$p_0 \sim \sqrt{n-p}, \qquad p_r \sim \sqrt{m_r}, \qquad 1 \leq r \leq s,$$

and

$$\frac{p}{\sqrt{n}} \to 0, \qquad \frac{m_{b_r \vee a} m_{b_r \vee a,S}}{m_{b_r}^{3/2}} \to 0, \qquad 1 \leq r \leq s.$$

When these conditions are met, the MLEs are asymptotically normal with the same $p_r, 0 \leq r \leq s$ and $\mathcal{M}$ as for the REML estimators.

Theorem 2.1 and 2.2 make it very clear what the difference is between REML and ML in terms of asymptotic properties. For consistency, the REML estimators are consistent under virtually no conditions, that is, conditions that are almost necessary; on the other hand, the MLEs require some additional conditions that control, essentially, how fast p grows with n in order to be consistent, and the additional conditions are also necessary. A similar two-side story is told for the asymptotic normality.

Now let us consider asymptotic comparison between REML and ML under a general mixed ANOVA model. Of course, every theoretical result holds under certain conditions. Instead of listing the mathematical conditions that are required, [25] focuses on finding and interpreting the conditions in such a way that they are intuitive by the "common sense." This is attractive, especially, for the practitioners because, although the latter may not have a sophisticated training in mathematics, or mathematical statistics, they often have a very good common sense due to their subject-matter training/experience. Practically speaking, an asymptotic theory is not very useful unless it can be fully understood by a practitioner.

There are two basic common senses for a parameter estimator to be consistent. The first is that the true parameter has to be unique; the second is that the knowledge about the truth must increase without a limit. Let us further explain these points. First, suppose that there are two parameters, $\theta_1 \neq \theta_2$, so that the distribution of the data under θ_1 is the same as that under θ_2. Then, both θ_1 and θ_2 are true parameters. Now suppose $\hat{\theta}$ is a consistent estimator. Then, as $n \to \infty$, $\hat{\theta}$ has to go to, on the one hand, θ_1, and, on the other hand, θ_2. But, in the end, there can be only one limit. Therefore, $\hat{\theta}$ cannot be consistent in the first place. In mathematical statistics, such a uniqueness of the true parameter (vector) is termed as identifiability of parameters.

Now consider the second point. A popular term in the asymptotic theory is "sample size going to infinity", but what does it mean? Take a look at, perhaps, the simplest example: Estimating the population mean, μ, by the sample mean, $\bar{X}$, based on i.i.d. observations. It is known that the mean squared error (MSE) of this estimation is σ^2/n, where n is the sample size. As the sample size increases to infinity, the estimation error goes away. This means that, as more and more data become available, knowledge about the truth, μ, learned from the data, becomes "complete." It is clear that there can be no limit on the knowledge in order to fully understand μ. In other words, if there is a limit in knowing about a certain aspect of μ, reflected by that the MSE is bounded away from zero, the estimator, or "learner", about μ cannot be consistent. The story of this simple can be generalized to estimating a general parameter vector, θ. If the observations, $Y_1, \ldots, Y_n$, are i.i.d., a measure of their overall knowledge about θ is called *information*, or Fisher information, which is a matrix if θ is multi-dimensional. The latter is equal to $nI(\theta)$ in the i.i.d. case, where $I(\theta)$ is the Fisher information based on a single observation, defined by

$$I(\theta) = \mathrm{Var}\left\{ \frac{\partial}{\partial \theta} \log f(Y_1|\theta) \right\}, \qquad (2.19)$$

where $f(\cdot|\theta)$ is the probability densisty function (pdf), or probability mass function (pmf), of Y_1 given that θ is the true parameter vector, assumed continuously differentiable with respect to θ. Under regularity conditions, the Fisher information has a second expression:

$$I(\theta) = -\mathrm{E}\left\{ \frac{\partial^2}{\partial \theta \partial \theta'} \log f(Y_1|\theta) \right\}. \qquad (2.20)$$

It is seen that, as long as $I(\theta) > 0$ (positive definite in the case of multivariate θ), the information, or knowledge, about θ will increase without a limit. This is the story for the i.i.d. case. Under a general LMM, the situation is more complicated so that the overall information about θ is not as simple as $nI(\theta)$; but, a similar principle applies.

Turning into some details, here we are considering, as the sample size increases, a sequence of LMMs, each having its own design (e.g., [24]). It is possible that, for a fixed design, the parameters are unique, or identifiable but, as n increases, the identifiability diminishes to none. The theory needs to rule out such a case by requiring

that the design is *asymptotically identifiable*. Furthermore, because the REML estimator is based on error contrast, that is, $A'y$, where A is as in (2.14), only the error
contrasts, known as the *invariant class*, is relevant to the identifiability. Thus, what
one actually needs is asymptotic identifiability under the invariant class, or AI^2 using the notation of [25]. Similarly, the requirement on the unlimited information, as
discussed in the last paragraph, was termed as *infinitely informative*, or I^2, using the
notation of [25]. Again, the I^2 is also considered under the invariant class. When the
two conditions are combined, it was called AI^4 (asymptotically identifiable and infinitely informative under the invariant class). The detailed, technical definition can
be found in [25].

Let $\|M\|_2$ denote the Euclidean norm, or L^2-norm of a matrix M defined by
$\|M\|_2 = \sqrt{\mathrm{tr}(M'M)}$. Write $C_r = Z'_r\{V_\gamma^{-1} - V(\gamma)\}Z_r, 1 \leq r \leq s$, where $V_\gamma = I_n + \sum_{r=1}^{s} \gamma_r Z_r Z'_r$. Also let $m^* = \max_{1 \leq r \leq s} m_r$. The following results were proved in [25].

Theorem 2.3. Consider a general non-Gaussian mixed ANOVA model with positive variance components $\tau^2, \gamma_r, 1 \leq r \leq s$.

(i) If the model is AI^4, then there exist with probability tending to one REML
estimators $\hat{\tau}^2, \hat{\gamma}_r, 1 \leq r \leq s$ that are consistent, and the sequences $\sqrt{n-p}(\hat{\tau}^2 - \tau^2), \|V_r\|_2(\hat{\gamma}_r - \gamma_r), 1 \leq r \leq s$ are bounded in probability.

(ii) If, moreover, the model is nondegenerate, the REML estimators in (i) are
asymptotically normal with $p_0 = \sqrt{n-p}, p_r$ being any sequence $\sim \|V_r\|_2, 1 \leq r \leq s$,
and the same $\mathcal{M}$ as in Theorem 2.1.

Recall the matrix $\mathcal{I}$ defined earlier, which depend on the sequences $p_r, 0 \leq r \leq s$.
Also let $\lambda_{\min}$ ($\lambda_{\max}$) denote the smallest (largest) eigenvalue.

Theorem 2.4. Consider a general non-Gaussian mixed ANOVA model with positive variance components $\tau^2, \gamma_r, 1 \leq r \leq s$.

(i) For the MLE to exist with probability tending to one, and be consistent, it is
necessary that $p/n \to 0, \mathrm{tr}(C_r)/m^* \to 0, 1 \leq r \leq s$.

(ii) If, moreover, the model is nondegenerate, then the following are equivalent:

(a) there exist with probability tending to one MLE of $\tau^2, \gamma_r, 1 \leq r \leq s$ that are
asymptotically normal for some $p_r, 0 \leq r \leq s$ satisfying $\liminf \lambda_{\min}(\mathcal{I}) > 0$ and
$\limsup \lambda_{\max}(\mathcal{I}) < \infty$;

(b) The model is AI^4, and $p/\sqrt{n} \to 0, \mathrm{tr}(C_r)/\|V_r\|_2 \to 0, 1 \leq r \leq s$.
In either (a) or (b), the REML estimator and MLE are equivalent in the sense that
they are asymptotically normal for the same $p_r, 0 \leq r \leq s$ and $\mathcal{M}$ as in Theorem 2.3.

Again, the results show a clear difference in the asymptotic behaviors of REML
and ML estimators. For example, for the REML estimator to be asymptotically normal, all one needs is that the model is AI^4 and nondegenerate (note that the latter implies that the variance components are positive). However, for the MLE to be asymptotically normal, one needs, in addition to AI^4 and nondegeneracy, some conditions
that control the rate at which p increases with the sample size.

In proving Theorems 2.1–2.4, [25] used a new technique that had never been used
before in mixed effects model asymptotics, the martingale limit theory (e.g., [45]).
To see how a martingale appears naturally in REML or ML, denote, for example, the
left side of the second equation in (2.17) by Q and note that it can be expressed as
$Q = (y - X\beta)'PZ_rZ'_rP(y - X\beta) = \xi'Z'_rPZ_rZ'_rPZ\xi$, where β is the true vector of fixed

effects, $Z = [I_n \ Z_1 \cdots Z_s]$, and $\xi = (\varepsilon', \alpha'_1, \ldots, \alpha'_s)'$. It follows that Q is a quadratic form in independent random variables. In fact, under a non-Gaussian mixed ANOVA model, both the REML and ML estimators are functions of quadratic forms in independent random variables. This led [25] to study central limit theorem (CLT) for quadratic forms in independent random variables expressed in the general form of $Q_n = \xi'_n A_n \xi_n$, where $\xi_n = (\xi_{ni})_{1 \leq i \leq k_n}$ is a vector of independent random variables with mean 0, and $A_n = (a_{nij})_{1 \leq i, j \leq k_n}$ is a sequence of nonrandom sysmetric matrices. There had been results on similar topics prior to [25]. Some of these apply only to a special kind of random variables (e.g., [46]) or to A_n with a special structure (e.g., [47]). Rao and Kleffe [48] derived a more general form of CLT for quadratic forms in independent random variables, extending an earlier result of [49]. However, as noted by [48](p. 51), "the applications (of the theorem) might be limited as it is essentially based on the assumption that the off diagonal blocks of A_n tend to zero." Such restrictions were removed by [25], whose approach turned out to be a classical application of the martingale CLT (e.g., [45], Theorem 3.2; also see [1], Section 8.5).

To see how the quadratic form is associated with a martingale, note that $E(Q_n) = \sum_{i=1}^{k_n} a_{nii} E(\xi_{ni}^2)$. Thus, we have

$$
\begin{aligned}
Q_n - E(Q_n) &= \sum_{1 \leq i, j \leq k_n} a_{nij} \xi_{ni} \xi_{nj} - \sum_{i=1}^{k_n} a_{nii} E(\xi_{ni}^2) \\
&= \sum_{i=1}^{k_n} a_{nii} \{\xi_{ni}^2 - E(\xi_{ni}^2)\} + \sum_{i \neq j} a_{nij} \xi_{ni} \xi_{nj} \\
&= \sum_{i=1}^{k_n} a_{nii} \{\xi_{ni}^2 - E(\xi_{ni}^2)\} + 2 \sum_{i=1}^{k_n} \left(\sum_{j<i} a_{nij} \xi_{nj} \right) \xi_{ni} \\
&= \sum_{i=1}^{k_n} X_{ni},
\end{aligned}
$$

where $X_{ni} = a_{nii}\{\xi_{ni}^2 - E(\xi_{ni}^2)\} + 2(\sum_{j<i} a_{nij} \xi_{nj})\xi_{ni}$. Let $\mathscr{F}_{ni} = \sigma(\xi_{nj}, 1 \leq j \leq i)$, $1 \leq i \leq k_n$, where σ denotes σ-field (e.g., [1], p. 558). It is easy to verify that $X_{ni}, \mathscr{F}_{ni}$, $1 \leq i \leq k_n$, is an array of martingale differences satisfying $X_{ni} \in \mathscr{F}_{ni}$ (i.e., X_{ni} is measurable with respect to $\mathscr{F}_{ni}$) and $E(X_{ni}|\mathscr{F}_{ni-1}) = 0$; hence, $M_k = \sum_{i=1}^{k} X_{ni}, \mathscr{F}_{nk}, 1 \leq k \leq k_n$ is a martingale (e.g., [1], p. 240).

The following theorem was proved in [25] as a consequence of a broader result. Let $\mathscr{S}_n = \{1 \leq i \leq k_n, a_{nii} \neq 0\}$.

Theorem 2.5. Suppse that $\xi_{ni}, 1 \leq i \leq k_n$ are independent such that

$$
\inf_n \left\{ \min_{1 \leq i \leq k_n} \mathrm{var}(\xi_{ni}) \right\} \wedge \left\{ \min_{i \in \mathscr{S}_n} \mathrm{var}(\xi_{ni}^2) \right\} > 0,
$$

$$
\sup_n \left[\max_{1 \leq i \leq k_n} E\{\xi_{ni}^2 1_{(|\xi_{ni}|>x)}\} \right] \vee \left[\max_{i \in \mathscr{S}_n} E\{\xi_{ni}^4 1_{(|\xi_{ni}|>x)}\} \right] \to 0,
$$

as $x \to \infty$. Then $\|A_n\|/\|A_n\|_2 \to 0$ implies

$$
\frac{\xi'_n A_n \xi_n - E(\xi'_n A_n \xi_n)}{\text{s.d.}(\xi'_n A_n \xi_n)} \xrightarrow{d} N(0,1), \tag{2.21}
$$

where s.d.$(\eta) = \sqrt{\mathrm{var}(\eta)}$ for any random variable η.

The condition $\|A_n\|/\|A_n\|_2 \to 0$ is almost necessary for the asymptotic normality. To see this, first recall that $\|A_n\|^2 = \lambda_{\max}(A_n'A_n)$ and $\|A_n\|_2^2 = \mathrm{tr}(A_n'A_n)$. Consider a special case, in which $k_n = n$ and $\xi_n \sim N(0, I_n)$. Let $A_n = T_n D_n T_n'$ be the eigenvalue decomposition, where T_n is an orthogonal matrix and $D_n = \mathrm{diag}(\lambda_{n,1}, \ldots, \lambda_{n,n})$, where $\lambda_{n,i}, 1 \le i \le n$ are the eigenvalues of A_n. Then, we have $Q_n = \xi_n' A_n \xi_n = \eta_n' D_n \eta_n = \sum_{i=1}^n \lambda_{n,i} \eta_{ni}^2$, where $\eta_n = T_n' \xi_n = (\eta_{ni})_{1 \le i \le n} \sim N(0, I_n)$. Thus, the distribution of Q_n is the same as $\sum_{i=1}^n \lambda_{n,i} \eta_i^2$, where $\eta_i, 1 \le i \le n$ are independent $N(0, 1)$ random variables (that do not depend on n). If the $\lambda_{n,i}$'s are relatively homogeneous in their magnitude, say, $\lambda_{n,i} = 1, 1 \le i \le n$ for an extreme case, then the left side of (2.21) has the same distribution as

$$\frac{1}{\sqrt{n}} \sum_{i=1}^n \frac{\eta_i^2 - 1}{\sqrt{2}} \qquad (2.22)$$

By the classical CLT (e.g., [1], Chapter 6), (2.22) converges in distribution to $N(0, 1)$. Note that, in this case, we have $\|A_n\|^2 = 1$ and $\|A_n\|_2^2 = n$, hence $\|A_n\|/\|A_n\|_2 = 1/\sqrt{n} \to 0$. Now assume that the $\lambda_{n,i}$'s are very inhomogeneous, say, $\lambda_{n,1} = 1$ and $\lambda_{n,i} = 0, 2 \le i \le n$. In this case, the left side of (2.21) has the same distribution as $(\eta_1^2 - 1)/\sqrt{2}$, which is not normal. Note that, in this case, we have $\|A_2\|^2 = 1 = \|A_n\|_2^2$. In general, $\sum_{i=1}^n \lambda_{n,i} \eta_i^2$ may be viewed as a weighted sum of i.i.d. random variables. The two extreme cases we have seen are, in fact, special cases of the following "theory": Asymptotic normality is the result of (standardized) sum of many terms with similar contributions, rather than one dominated by a few terms. The condition $\|A_n\|/\|A_n\|_2 \to 0$ makes sure that the theory is in good hands. However, the condition alone does not necessarily imply (2.21). Jiang [25] (p. 267) gives a simple counterexample.

Finally, a key lemma that allowed Jiang [25] to remove the unpleasant condition that "the off diagonal blocks of A_n tend to zero", as noted by Rao and Kleffe [48], is the following result of linear algebra, proved in [25]:

Lemma 2.1. Let $B = [b_{ij} 1_{(i>j)}]$ be a lower triangular matrix. Then

$$\|B'B\|_2 \le \sqrt{2}\|B' + B\| \cdot \|B\|_2.$$

2.2.3 Asymptotic behavior of REML: Wald consistency

So far, all of the consistency results, either for REML or for ML, are in terms of Cramér consistency [38]. As noted, such results ensure existence of a local maximum, or a root to the likelihood equation (REML equation, etc.), that is consistent. However, in case the root is not unique, the result usually gives no clue on which root is consistent. Some strategies have been suggested in the literature to deal with the issue of multiple roots. For example, suppose that a consistent estimator, $\tilde{\theta}$, of θ is available. Then, if one can find all of the roots to the likelihood equation, one may simply choose the one that is closet to $\tilde{\theta}$, and this root is going to be consistent, by the Cramér consistency. In view of the well-known asymptotic efficiency of the MLE (e.g., [11], Chapter 6), there have also been strategies to obtain a root to the likelihood equation that is asymptotically efficient. For example, consider the case of univariate parameter and let $\tilde{\theta}$ be a consistent estimator of θ. One can make a one-step Newton-Raphson correction on $\tilde{\theta}$:

$$\hat{\theta} = \tilde{\theta} - \frac{l'(\tilde{\theta})}{l''(\tilde{\theta})},$$

where $l(\cdot)$ is the log-likelihood function. It can be shown that $\hat{\theta}$ is asymptotically efficient under some regularity conditions; moreover, if the procedure is iterated, then the limit is a root to the likelihood equation, and asymptotically efficient. See, for example, [11] (Section 6.4). Furthermore, Gan and Jiang [50] developed a statistical test for a root to the likehihood equation to be an asymptotically efficient estimator.

In spite of all of these considerations, it is desirable to answer one basic question: Does the global maximum of the likelihood function correspond to a consistent estimator? Note that the global maximum of the likelihood, not a local one, is what the MLE is supposed to mean. Therefore, an answer to the question is fundamental. A positive answer may also help to determine which root to the likelihood equation is consistent, if there are multiple roots. Suppose that one can identify all of the roots to the likelihood equation. Then, because, typically, the global maximum corresponds to one of the roots, by comparing the values of the likelihood at those roots, and choosing the one with the maximum value, one would be able to identify a consistent estimator.

The discussions have left little doubt that Wald consistency, which simply means that the estimator corresponding to the global maximum of the likelihood function is consistent [39], is among the most elegant properties, asymptotic or not, of the maximum likelihood. On the other hand, Wald consistency results are usually more difficult to prove, compared to Cramér consistency results. In fact, in the context of LMM, the only Wald-type consistency results were proved by Jiang [26] for REML estimation under a non-Gaussian mixed ANOVA model.

The Gaussian restricted log-likelihood function for estimating the variance components, $\tau^2, \gamma_r, 1 \leq r \leq s$ [that is, the log-likelihood function based on $z = A'y$, where A satisfies (2.14)] is given by $l_R(\theta) =$

$$-\frac{1}{2}\left\{(n-p)\log(2\pi) + (n-p)\log(\tau^2) + \log(|A'V_\gamma A|) + \frac{1}{\tau^2}y'V(\gamma)y\right\}, \quad (2.23)$$

using the notation from the previous subsection, where $\theta = (\tau^2, \gamma_1, \ldots, \gamma_s)'$, and $|M|$ denotes the determinant of matrix M. Under a non-Gaussian mixed ANOVA model, by wald consistency, it means that the global maximizer of $l_R(\cdot)$, $\hat{\theta} = (\hat{\tau}^2, \hat{\gamma}_1, \ldots, \hat{\gamma}_s)'$, is consistent, that is, $\hat{\theta}$ converges in probability to the true θ. As in [25], first consider the case of balanced data. The following result was proved in [26].

Theorem 2.6. Let the balanced mixed ANOVA model be unconfounded. As $n \to \infty$ and $m_r \to \infty, 1 \leq r \leq s$, the following hold:

(i) $\hat{\theta}$ is consistent, and the sequences $\sqrt{n-p}(\hat{\tau}^2 - \tau^2), \sqrt{m_r}(\hat{\gamma}_r - \gamma_r), 1 \leq r \leq s$ are bounded in probability.

(ii) If, moreover, the variance components $\tau^2, \gamma_r, 1 \leq r \leq s$ are positive and the model is nondegenerate, then $\hat{\theta}$ is asymptotically normal with the same $p_r, 0 \leq r \leq s$ and $\mathcal{M}$ as in Theorem 2.1.

Note that part (i) of Theorem 2.6 does not require positiveness of the variance components, while the latter was required by part (i) of Theorem 2.1. Thus, at least for part (i), the conditions (actually, there is only one condition, that is, the model is unconfounded) of Theorem 2.6 are even weaker than those of Theorem 2.1. The reason is that, in part (i) of Theorem 2.1, the result was shown by arguing that a local maximum of $l_R(\cdot)$ occurs in the interior of a neighborhood of the true θ; thus, one needs a neighborhood of the true θ that is entirely within the parameter space. However, the global maximum need not be in the interior of the parameter space.

Regarding the proof of Theorem 2.6, personally speaking (and this applies to any mathematical proof, not just that of Theorem 2.6), there is nothing called "the method of proof." Instead, there are only two things: One is called *idea*; the other *language*. Let us explain a little more about these two terms. Idea is the most important part of a proof, although it may not count for much of the "actual work". For example, it may take 5% or less of the space to write down the idea, compared to the entire proof. The most difficult thing about the idea is not just to come with it but also to convince oneself that the idea is going to work. The latter turns out to be the key in many cases–believe or not, it is *not* so easy to make oneself believe (in this regard, the 2008 American animated film *Kung Fu Panda* is definitely worth watching). On the other hand, language refers to a strategy, or plan, to execute the idea. This usually consists most of the "actual work," and one may have to overcome various technical difficulties en route to carry out the plan. In terms of the writing, the language part may count 95% or more of the entire proof.

The idea for the proof of Theorem 2.6 is actually quite simple. Write the true vector of variance components as θ_0. For any θ in the parameter space, write the difference in the restricted log-likelihood as

$$l_R(\theta) - l_R(\theta_0) \quad = \quad \mathcal{E}(\theta) - \mathcal{D}(\theta), \tag{2.24}$$

where $\mathcal{E}(\theta) = \mathrm{E}\{l_R(\theta) - l_R(\theta_0)\}$ and $\mathcal{D}(\theta) = l_R(\theta) - l_R(\theta_0) - \mathrm{E}\{l_R(\theta) - l_R(\theta_0)\}$. Essentially, what one is going to show is that, with probability tending to one, and uniformly for θ outside a small neighborhood of θ_0, the first term on the right side of (2.24) is negative, while the second is negligible compared to the first term. Turning to the language, [26] showed that, under the balanced mixed ANOVA model, both terms on the right side of (2.24) have nice expressions, which allows one to execute the idea (see Lemma 3.1 of [26]).

However, the language has its limitation that it only holds together in the balanced case. In fact, [26] gives a counterexample (see below) showing that in some unbalanced cases Wald consistency fails, but the so-called sieve-Wald consistency holds, defined as follows. The idea was first introduced by Grenander [51]. Consider a sequence of expanding subspaces of the parameter vector, θ, say, $\Theta_n \subset \Theta$, such that, as $n \to \infty$, Θ_n eventually covers every parameter vector. Let $\hat{\theta}_n$ be the MLE of θ restricted to Θ_n. The sequence $\hat{\theta}_n$ is called sieve MLE. By sieve-Wald consistency, it means that $\hat{\theta}_n$, understood as the global maximizer of the likelihood function over $\theta \in \Theta_n$, converges in probability to the true θ. Jiang [26] extends the idea to REML estimation under an unbalanced mixed ANOVA model. It is more convenience to construct the sieve under an alternative parameterization. Let $\phi = (\tau^2, \sigma_1^2, \ldots, \sigma_s^2)'$, where $\sigma_r^2 = \mathrm{var}(\alpha_{r1}), 1 \le r \le s$. Let a_n, b_n be two sequences of positive numbers such that $a_n \to 0, b_n \to \infty$, as $n \to \infty$. The full parameter space for ϕ is $\Phi = [0,\infty)^{s+1}$. Let $\Phi_n = \{\phi \in \Phi : a_n \le \tau^2, \sigma_1^2, \ldots, \sigma_s^2 \le b_n\}$. The Gaussian restricted log-likelihood for estimating ϕ is the same as (2.23) except with $\gamma_r = \sigma_r^2 / \tau^2, 1 \le r \le s$. The sequences a_n, b_n are constructed carefully in [26]. Let $\hat{\phi}_n = (\hat{\tau}^2, \hat{\sigma}_{n,1}^2, \ldots, \hat{\sigma}_{n,s}^2)'$ denote the corresponding sieve REML estimator. Recall the term AI^4 condition explained in the previous subsection. Also, define $w_{n,r} = \|V_r\|_2, 0 \le r \le s$ (defined in the previous subsection) with $\tau^2 = \gamma_1 = \cdots = \gamma_s = 1$.

Theorem 2.7. Consider a general non-Gaussian mixed ANOVA model with positive variance components $\tau^2, \sigma_r^2, 1 \le r \le s$, and let a_n, b_n be constructed as in [26]:

(i) If the model is AI^4, then the sieve-Wald consistency holds, and the sequences $w_{n,0}(\hat{\tau}_n^2 - \tau^2), w_{n,r}(\hat{\sigma}_{n,r}^2 - \sigma_r^2), 1 \le r \le s$ are bounded in probability.

(ii) If, moreover, the model is degenerate, then $\hat{\phi}_n$ is asymptotically normal with $p_0 = \sqrt{n-p}$, p_r being any sequence $\sim w_{n,r}, 1 \le r \le s$, and $\mathcal{M}$ as in Theorem 2.1.

The proof of Theorem 2.7, given in [26], followed the same idea based on (2.24), but the language is modified to take advantage of the sieves.

We conclude this subsection with a counterexample showing that, in some cases of unbalanced non-Gaussian mixed ANOVA model, there is sieve-Wald consistency but no Wald consistency.

Example 2.9 ([26], p. 1800). Let $\lambda_1 = 0; \lambda_2, \ldots, \lambda_n$ be positive numbers such that

$$\limsup \frac{\sum_{i=2}^n \lambda_i/(1 + u\lambda_i)}{[n \sum_{i=2}^n \{\lambda_i/(1 + u\lambda_i)\}^2]^{1/2}} < 1,$$

$$\lim \sum_{i=2}^n \left(\frac{\lambda_i}{1 + u\lambda_i} \right)^2 = \infty$$

for all $u \ge 0$. To see that there are such λ_i's, consider, for example, $\lambda_1 = 0, \lambda_2 = \cdots = \lambda_{[n/2]} = a, \lambda_{[n/2]+1} = \cdots = \lambda_n = b$, where $a, b > 0$ and $a \ne b$. Consider $y_i = \sqrt{\lambda_i}\alpha_i + \varepsilon_i, i = 1, \ldots, n$, where the α_i's are independent random effects with mean zero and variance σ^2, and ε_i's are independent errors with mean zero, variance τ^2, and $\mathrm{P}(\varepsilon_1 = 0) > 0$. It is easy to show that the model satisfies the AI^4 condition. Therefore, by Theorem 2.7, the sieve-Wald consistency holds for the REML estimator, which is the same as the MLE in this case. On the other hand, denote the true $\phi = (\tau^2, \sigma^2)'$ by $\phi_0 = (\tau_0^2, \sigma_0^2)'$, and assume that $\phi_0 \in \Phi^\circ$, the interior of Φ, the parameter space for ϕ. It is easy to derive

$$\tilde{l}_R(\phi) - \tilde{l}_R(\phi_0) = \frac{1}{2} \left\{ \sum_{i=1}^n \log \left(\frac{\tau_0^2 + \sigma_0^2 \lambda_i}{\tau^2 + \sigma^2 \lambda_i} \right) + \sum_{i=1}^n \left(1 - \frac{\tau_0^2 + \sigma_0^2 \lambda_i}{\tau^2 + \sigma^2 \lambda_i} \right) w_i^2 \right\}, \quad (2.25)$$

where $w_i = (\varepsilon_i + \sqrt{\lambda_i}\alpha_i)/\sqrt{\tau_0^2 + \sigma_0^2 \lambda_i}$. Let $B = \{\phi : , \tau^2 \ge \tau_0^2/2, \sigma^2 \ge \sigma_0^2/2\}$, and

$\phi_* = (\tau_*^2, \sigma_0^2)'$ with

$$0 < \tau_*^2 < \tau_0^2 \exp\left\{ -\frac{3}{\delta}\left(n + \frac{\tau_0^2}{\sigma_0^2} \sum_{i=2}^{n} \frac{1}{\lambda_i} \right) - n\log(2) \right\},$$

where $\delta = P(\varepsilon_1 = 0)$. Define $\xi_n = (1/n)\sum_{i=1}^{n} w_i^2$, $\eta_n = \sum_{i=2}^{n} \lambda_i^{-1} w_i^2 / \sum_{i=2}^{n} \lambda_i^{-1}$, and $S = \{\varepsilon_1 = 0, \xi_n \vee \eta_n \leq 3/\delta\}$. Then, from (2.25), it follows that on S, we have

$$\begin{aligned}
\tilde{l}_R(\phi) &\leq \tilde{l}_R(\phi_0) + (n/2)\{\log(2) + \xi_n\} \\
&< \tilde{l}_R(\phi_0) + \frac{1}{2}\left\{ \log\left(\frac{\tau_0^2}{\tau_*^2}\right) - \left(\frac{\tau_0^2}{\sigma_0^2}\right)\left(\sum_{i=2}^{n} \frac{1}{\lambda_i}\right)\eta_n \right\} \\
&< \tilde{l}_R(\phi_*), \quad \forall \phi \in B.
\end{aligned}$$

Therefore, we have $P\{\sup_{\phi \in B} \tilde{l}_R(\phi) < \tilde{l}_R(\phi_*)\} \geq P(S) \geq \delta/3$. In other words, with probability greater than some positive constant, the global maximizer, $\hat{\phi}$, of $\tilde{l}_R(\cdot)$ will not fall into B no matter how large n is; hence, $\hat{\phi}$ cannot be consistent.

2.3   Asymptotic covariance matrix in non-Gaussian LMM

Statistics is not just about getting numbers, for example, for estimation or prediction; more importantly, it is also about assessing uncertainties associated with the numbers. (Otherwise, several fields in applied mathematics have seen plenty of numbers.) The same is true, in particular, for mixed model analysis.

Under the normality assumption, the uncertainties about the REML or ML estimators are typically evaluated using the asymptotic covariance matrices (ACMs) associated with these estimators. The latter are equal to the inverses of the corresponding Fisher information matrices. See, for example, (2.12) and (2.13), for the MLE. The problem is complicated, however, when normality does not hold.

According to the previous section, it is known that even without the normality assumption, the REML estimator is still consistent and asymptotically normal under mild conditions; the same can be said about the MLE, under more restrictive conditions. But, what can we say about the ACMs? To be specific, let us focus on REML, as the same idea that we develop here can be easily extended to ML. It can be shown that, without the normality assumption, the ACM of the REML estimator, $\hat{\theta}$, of the variance components, has the following "sandwich" expression [compare to (1.3)]:

$$\Sigma_R = \left\{ E\left(\frac{\partial^2 l_R}{\partial \theta \partial \theta'}\right) \right\}^{-1} Var\left(\frac{\partial l_R}{\partial \theta}\right) \left\{ E\left(\frac{\partial^2 l_R}{\partial \theta \partial \theta'}\right) \right\}^{-1}, \qquad (2.26)$$

where l_R is the Gaussian restricted log-likelihood, which can be expressed as (2.23) under the mixed ANOVA model. When normality indeed holds, one has

$$E\left(\frac{\partial^2 l_R}{\partial \theta \partial \theta'}\right) = -Var\left(\frac{\partial l_R}{\partial \theta}\right), \qquad (2.27)$$

so (2.26) reduces to the inversed Fisher information matrix:

$$I_R^{-1}(\theta) = \left\{ \text{Var}\left(\frac{\partial l_R}{\partial \theta} \right) \right\}^{-1} = -\left\{ E\left(\frac{\partial^2 l_R}{\partial \theta \partial \theta'} \right) \right\}^{-1}. \tag{2.28}$$

An attractive feature of (2.28) is that it only depends on the variance components, θ [this is more easily seen using the second expression in (2.28)], of which one already has the (REML) estimator. There may not be such a luxury, if (2.27) does not hold, and therefore one has to use (2.26). To see where exactly the problem is, note that $\mathscr{I}_2 = E(\partial^2 l_R / \partial \theta \partial \theta')$ depends only on θ, even without normality, so this part does not cause any "trouble". What may be troublesome is $\mathscr{I}_1 = \text{Var}(\partial l_R / \partial \theta)$, which may involve higher (than 2nd) moments of the random effects and errors, which are not part of θ. We illustrate with an example.

Example 2.8 (continued). Consider the balanced two-way random effects model. To avoid potential confusion in the notation, write $a = I$ and $b = J$. Let $I_a, 1_a$ denote the $a \times a$ identity matrix and $a \times 1$ vector of 1's, respectively, and $J_a = 1_a 1_a'$. It can be shown that, in this case, one has

$$\frac{\partial l_R}{\partial \sigma_e^2} = \frac{\xi' H \xi - (ab - 1)\sigma_e^2}{2\sigma_e^4}, \tag{2.29}$$

where $H = I_a \otimes I_b + \lambda_1 I_a \otimes J_b + \lambda_2 J_a \otimes I_b + \lambda_3 J_a \otimes J_b$, where $\otimes$ denote the Kronecker product, $\lambda_1 = -b^{-1}\{1 - (1 + \gamma_1 b)^{-1}\}$, $\lambda_2 = -a^{-1}\{1 - (1 + \gamma_2 a)^{-1}\}$, and $\lambda_3 = (ab)^{-1}\{1 - (1 + \gamma_1 b)^{-1} - (1 + \gamma_2 a)^{-1}\}$ with $\gamma_1 = \sigma_u^2 / \sigma_e^2$ and $\gamma_2 = \sigma_v^2 / \sigma_e^2$, and $\xi = y - \mu 1_a \otimes 1_b$. It is clear that, by differentiating (2.29) again, with respective to the variance components, one still ends up with quadratic forms in ξ. Thus, it is easy to see that $E(\partial^2 l_R / \partial \sigma_e^4)$, etc. are functions of $\theta = (\sigma_e^2, \sigma_u^2, \sigma_v^2)'$ regardless of normality. Similarly, it can be shown that other elements of $\mathscr{I}_2$ are functions of θ.

On the other hand, it is easy to see from (2.29) that $\text{var}(\partial l_R / \partial \sigma_e^2)$ involves fourth moments of the random effects and errors (but no third moments; and this has nothing to do with normality, or symmetry). Such quantities are not functions of θ, and therefore should be considered as new parameters, unless normality holds [in which case one has, for example, $E(e_{ij}^4) = 3\sigma_e^4$]. Thus, in case of non-Gaussian random effects and errors, $\mathscr{I}_1$ involves $E(e_{ij}^4), E(u_i^4), E(v_j^4)$, in addition to θ.

Therefore, In order to make use of the ACM, one has to find a way to evaluate $\mathscr{I}_1$. Below we consider two approaches.

2.3.1   Empirical method of moments

A simple-minded approach would be to estimate the higher moments involved in $\mathscr{I}_1$. Note that estimates of the higher moments are usually not provided in standard packages of mixed model analysis, such as those in SAS or R. Jiang [52] used an approach, called *empirical method of moments* (EMM) to estimate the higher moments. Let θ be a vector of parameters. Suppose that a consistent estimator of θ, $\hat{\theta}$, is available. Let φ be a vector of additional parameters about which knowledge is needed. Let $\psi = (\theta', \varphi')'$, and $M(\psi, y) = M(\theta, \varphi, y)$ be a vector-valued function of the same dimension as φ that depends on ψ and y, a vector of observations. Suppose that $E\{M(\psi,y)\}=0$ when ψ is the true parameter vector. Then, if θ were known, a method of moments estimator of φ would be obtained by solving

$$M(\theta,\varphi,y) \;=\; 0 \qquad\qquad (2.30)$$

for φ. Note that this is more general than the classical method of moments, in which the function M is a vector of sample moments minus their expected values. In econometric literature, this is referred to as generalized method of moments (e.g., [53]). Because θ is unknown, we replace it in (2.30) by $\hat{\theta}$. The result is an EMM estimator of φ, denoted by $\hat{\varphi}$, which is obtained by solving

$$M(\hat{\theta},\varphi,y) \;=\; 0. \qquad\qquad (2.31)$$

Note that here we use the words "an EMM estimator" instead of "the EMM estimator", because sometimes there may be more than one consistent estimators of θ, and each may result in a different EMM estimator of φ. Jiang [52] shows that, under suitable conditions, the EMM estimator is consistent for estimating φ.

To apply the EMM to non-Gaussian mixed ANOVA model, Jiang [52] assumed that the third moments of the random effects and errors are zero, that is,

$$E(\varepsilon_1^3) = 0, \;\; E(\alpha_{r1}^3) = 0, \;\; 1 \le r \le s, \qquad\qquad (2.32)$$

where ε_1 (α_{r1}) is the first component of ε (α_r) [recall the components of ε (α_r) are i.i.d.] (2.32) is satisfied if, in particular, the distributions of the random effects and errors are symmetric. Under such an assumption, it can be shown that the ACM of REML estimator of $\psi = (\beta',\phi')'$ (ϕ is defined, e.g., above Theorem 2.7) depends on ϕ as well as the kurtoses: $\kappa_0 = E(\varepsilon_1^4) - 3\tau^2$, $\kappa_r = E(\alpha_{r1}^4) - 3\sigma_r^4$, $1 \le r \le s$.

For any matrix $H = (h_{ij})$, define $\|H\|_4 = (\sum_{i,j} h_{ij}^4)^{1/4}$; similarly, if $h = (h_i)$ is a vector, define $\|a\|_4 = (\sum_i h_i^4)^{1/4}$. Let $\mathscr{L}$ be a linear space. Define $L^\perp$ as the linear space $\{v : v'u = 0, \forall u \in \mathscr{L}\}$. If $\mathscr{L}_j, j = 1,2$ are linear spaces such that $\mathscr{L}_1 \subset \mathscr{L}_2$, then $\mathscr{L}_2 \ominus \mathscr{L}_1$ represents the linear space $\{v : v \in \mathscr{L}_2, v'u = 0, \forall u \in \mathscr{L}_1\}$. If $M_1,\ldots,M_k$ are matrices with same number of rows, then $\mathscr{L}(M_1,\ldots,M_k)$ represents the linear space spanned by the columns of $M_1,\ldots,M_k$. Suppose that the matrices $Z_1,\ldots,Z_s$ in (2.3) have been suitably ordered such that

$$\mathscr{L}_r \;\ne\; \{0\}, \quad 0 \le r \le s, \qquad\qquad (2.33)$$

where $\mathscr{L}_0 = \mathscr{L}(Z_1,\ldots,Z_s)^\perp$, $\mathscr{L}_r = \mathscr{L}(Z_r,\ldots,Z_s) \ominus \mathscr{L}(Z_{r+1},\ldots,Z_s)$, $1 \le r \le s-1$, and $\mathscr{L}_s = \mathscr{L}(Z_s)$. Let C_r be a matrix whose columns constitute a base of $\mathscr{L}_r$, $0 \le r \le s$. Define $a_{rq} = \|Z_q'C_r\|_4^4$, $0 \le q \le r \le s$. It is easy to see that, under (2.33), one has $a_{rr} > 0$, $0 \le r \le s$. Let n_r be the number of columns of C_r, and c_{rk} the kth column of C_r, $1 \le k \le n_r$, $0 \le r \le s$. Define

$$b_r(\phi) \;=\; 3 \sum_{k=1}^{n_r} \left(\sum_{q=0}^{r} |Z_q' c_{rk}|^2 \phi_v \right)^2, \quad 0 \le r \le s.$$

where $\phi = (\phi_r)_{0 \leq r \leq s}$ with $\phi_0 = \tau^2$ and $\phi_r = \sigma_r^2, 1 \leq r \leq s$. Let $\kappa = (\kappa_r)_{0 \leq r \leq s}$. Consider $M(\beta, \phi, \kappa, y) = [M_r(\beta, \phi, \kappa, y)]_{0 \leq r \leq s}$, where

$$M_r(\beta, \phi, \kappa, y) = \|C_r'(y - X\beta)\|_4^4 - \sum_{q=0}^{r} a_{rq} \kappa_q - b_r(\phi) , \quad 0 \leq r \leq s .$$

Then, by Lemma 2.2 below and the definition of the C_r's, it is easy to show that $E\{M(\beta, \phi, \kappa, y)\} = 0$ when β, ϕ, κ are the true parameter vectors. Thus, a set of EMM estimators can be obtained by solving $M(\hat{\beta}, \hat{\phi}, \kappa, y) = 0$, where $\hat{\beta}, \hat{\phi}$ are the REML estimators of β, ϕ, respectively [see the note below (2.16)]. The EMM estimators can be computed recursively, as follows:

$$\hat{\kappa}_0 = a_{00}^{-1} \hat{d}_0 ,$$

$$\hat{\kappa}_r = a_{rr}^{-1} \hat{d}_r - \sum_{q=0}^{r-1} \left(\frac{a_{rq}}{a_{rr}} \right) \hat{\kappa}_q , \quad 1 \leq r \leq s , \tag{2.34}$$

where $\hat{d}_r = \|C_r'(y - X\hat{\beta})\|_4^4 - b_r(\hat{\phi}), 0 \leq r \leq s$.

Lemma 2.2. Let $\xi_1, \ldots, \xi_n$ be independent random variables such that $E(\xi_i) = 0$ and $E(\xi_i^4) < \infty$, and $\lambda_1, \ldots, \lambda_n$ be constants. Then, we have

$$E\left(\sum_{i=1}^{n} \lambda_i \xi_i \right)^4 = 3\left\{ \sum_{i=1}^{n} \lambda_i^2 \text{var}(\xi_i) \right\}^2 + \sum_{i=1}^{n} \lambda_i^4 \left[E(\xi_i^4) - 3\{\text{var}(\xi_i)\}^2 \right].$$

We illustrate the EMM with an example.

Example 2.7 (continued). Again, replace the notation, I and J, by a and b, respectively. It is easy to show that $C_0 = I_a \otimes K_b$ and $C_1 = Z_1 = I_a \otimes 1_b$, where

$$K_b = \begin{pmatrix} 1 & \cdots & 1 \\ -1 & \cdots & 0 \\ \vdots & \ddots & \vdots \\ 0 & \cdots & -1 \end{pmatrix}_{b \times (b-1)} .$$

It follows from (2.34) that, in closed-form,

$$\hat{\kappa}_0 = \frac{1}{2a(b-1)} \sum_{i=1}^{a} \sum_{j=2}^{b} (y_{i1} - y_{ij})^4 - 6\hat{\sigma}_e^4 ,$$

$$\hat{\kappa}_1 = \frac{1}{ab^4} \sum_{i=1}^{a} (y_{i \cdot} - b\hat{\mu})^4 - \frac{1}{2ab^3(b-1)} \sum_{i=1}^{a} \sum_{j=2}^{b} (y_{i1} - y_{ij})^4$$
$$- \frac{3}{b^2} \left(1 - \frac{2}{b} \right) \hat{\sigma}_e^4 - \frac{6}{b} \hat{\sigma}_e^2 \hat{\sigma}_v^2 - 3\hat{\sigma}_v^4 ,$$

where $y_{i \cdot} = \sum_{j=1}^{b} y_{ij}$, and $\hat{\mu}$, $\hat{\sigma}_e^2$, and $\hat{\sigma}_v^2$ are the REML estimators of μ, σ_e^2, and σ_v^2, respectively. It is easy to show, either by verifying the conditions of Theorem 1 in [52], or by arguing directly, that the EMM estimators are consistent provided only that $a \to \infty$. The only requirement for b is that $b \geq 2$, but the latter is reasonable, because, otherwise, one cannot separate α and ε.

2.3.2   *Partially observed information*

Another idea of estimating the ACM has to do with the GEE method of estimating the covariance matrix of $\hat{\beta}$, discussed in Section 1.1. Recall that in (1.4) and (1.5), we expressed both $E(\partial g/\partial \beta')$ and $\text{Var}\{g(\beta)\}$ as expectations of sums of random variables. We then dropped the expectation signs and used the expressions insided the expectations to approximate the expectations. This is a useful technique; although the strategy should not be "abused." Jiang [1] (Example 2 in the Preface) used the following example to illustrate when the strategy works, and when it does not.

Example 2.10. It might be thought that, with a large sample, one could always approximate the mean of a random quantity by the quantity itself. Before we get into this, let us first be clear on what is meant by a good approximation. An approximation is good if the error of the approximation is of lower order than the approximation itself. In notation, this means that, suppose one wishes to approximate A by B; the approximation is good if $A - B = o(B)$, so that $A = B + A - B = B + o(B) = B\{1 + o(1)\}$. In other words, B is the "main part" of A.

Once the concept is clear, it turns out that, in some cases, the approximation technique mentioned above works, while in some other cases, it doesn't work. For example, suppose that $Y_1,\ldots,Y_n$ are i.i.d. observations such that $\mu = E(Y_1) \neq 0$. Then one can approximate $E(\sum_{i=1}^{n} Y_i) = n\mu$ by simply removing the expectation sign, that is, by $\sum_{i=1}^{n} Y_i$. This is because the difference $\sum_{i=1}^{n} Y_i - n\mu = \sum_{i=1}^{n}(Y_i - \mu)$ is of the order $O_P(\sqrt{n})$ [provided that $\text{var}(Y_1) < \infty$], which is lower than the order of $\sum_{i=1}^{n} Y_i$, which is $O_P(n)$.

However, the technique completely fails if one considers, for example, approximating $E(\sum_{i=1}^{n} Y_i)^2$, where the Y_i's are i.i.d. with mean 0. For simplicity, assume that $Y_i \sim N(0,1)$. Then, we have $E(\sum_{i=1}^{n} Y_i)^2 = n$. On the other hand, $(\sum_{i=1}^{n} Y_i)^2 = n\chi_1^2$, where χ_1^2 is a random variable with the χ^2 distribution with one degree of freedom. Therefore, $(\sum_{i=1}^{n} Y_i)^2 - E(\sum_{i=1}^{n} Y_i)^2 = n(\chi_1^2 - 1)$, which is of the same order as $(\sum_{i=1}^{n} Y_i)^2$. Thus, $(\sum_{i=1}^{n} Y_i)^2$ is not a good approximation to its mean.

It can be seen that a key to the success of this approximation technique is that the random quantity inside the expectation has to be a sum of random variables with "good behavior" (say, independence). In the context of likelihood inference, the technique is associated with a term called *observed information*. Consider the case of i.i.d. observations. In this case, the information contained in the entire data, $Y_1,\ldots,Y_n$, about the parameter vector, θ, is $\mathscr{I}(\theta) = nI(\theta)$, where $I(\theta)$ has either of the expressions, (2.19) or (2.20), under regularity conditions. If one uses the second expression, (2.19), one can write

$$\mathscr{I}(\theta) = -\sum_{i=1}^{n} E\left\{\frac{\partial^2}{\partial\theta\partial\theta'}f(Y_i|\theta)\right\} = -E\left\{\sum_{i=1}^{n}\frac{\partial^2}{\partial\theta\partial\theta'}f(Y_i|\theta)\right\}.$$

Using the approximation technique, one removes the expectation sign in the last expression to get an approximation to $\mathscr{I}(\theta)$, which is not yet an estimate, because θ is unknown. Thus, as a last step, one replaced θ in the latest approximation by $\hat{\theta}$, the MLE, to get

$$\widehat{\mathscr{I}(\theta)} \;=\; -\sum_{i=1}^{n} \frac{\partial^2}{\partial\theta\partial\theta'} f(Y_i|\hat{\theta}). \tag{2.35}$$

The estimator (2.35) is called observed information (matrix). Alternatively, especially if $I(\theta)$ has an analytic expression, one can replace θ in this expression by, say, the MLE $\hat{\theta}$ to get

$$\widetilde{\mathscr{I}(\theta)} \;=\; nI(\hat{\theta}). \tag{2.36}$$

The estimator (2.36) is called estimated information (matrix). See, for example, [54], for a discussion and comparison of the two estimators in the i.i.d. case.

However, neither the estimated nor the observed information methods work for REML estimation. First, as noted, $I_R(\theta)$ is not a function of θ, the vector of variance components, but involves additional parameters, namely the higher moments. Secondly, as it turns out, one cannot express $I_R(\theta)$ as expectation of a sum of random variables that are observable and have good behavior. Note that, under a LMM, the observations are not independent, so the above derivation for the i.i.d. case does not carry through. Nevertheless, it was found that a combination of these two techniques would work just fine. We first use an example to illustrate.

Example 2.8 (continued). It can be shown that the variance of (2.29) can be expressed as $\mathrm{var}(\partial l_R/\partial\sigma_e^2) = S_1 + S_2$, where

$$S_1 \;=\; \mathrm{E}\left\{ a.\sum_{i,j}\xi_{ij}^4 - a_1\sum_i\left(\sum_j\xi_{ij}\right)^4 - a_2\sum_j\left(\sum_i\xi_{ij}\right)^4 \right\},$$

$\xi_{ij} = y_{ij} - \mu$, $a. = a_0 + a_1 + a_2$, and $a_j, j = 0,1,2$, S_2 are functions of θ. Thus, S_2 can be estimated by replacing θ by $\hat{\theta}$, the REML estimator. As for S_1, it can be estimated using the observed information idea, that is, by removing the expectation sign, and then replacing μ and θ involved in the expression inside the expectaion by their REML estimators.

To describe the method in general, denote the vector of variance components under the mixed ANOVA model by $\theta = (\theta_r)_{0 \le r \le s}$, where $\theta_0 = \tau^2$ and $\theta_r = \gamma_r, 1 \le r \le s$. Then, it can be shown that $\partial l_R/\partial\theta_r = \xi'H_r\xi - h_r, 0 \le r \le s$, where $\xi = y - X\beta$, $H_0 = P/2\theta_0$, with P given by (2.15), and $H_r = (\theta_0/2)PZ_rZ_r'P, 1 \le r \le s$; $h_0 = (n - p)/2\theta_0$, and $h_r = (\theta_0/2)\mathrm{tr}(PZ_rZ_r'), 1 \le r \le s$. Also, with a slight abuse of the notation, let z_{ir}' and z_{rl} be the ith row and lth column of Z_r, respectively, $0 \le r \mathit{leqs}$, where $Z_0 = I$. Define $\Gamma(i_1,i_2) = \sum_{r=0}^{s}\gamma_r(z_{i_1r} \cdot z_{i_2r})$. Here, the dot product of vectors $a_1,\ldots,a_k$ of the same dimension is defined as $a_1 \cdot a_2\cdots a_k = \sum_l a_{1l}a_{2l}\cdots a_{kl}$. Also let $\alpha_0 = \varepsilon$ and recall that m_r is the dimension of $\alpha_r, 0 \le r \le s$ (so, in particular, $m_0 = n$), and that $V = \mathrm{Var}(y) = \theta_0(I_n + \sum_{r=1}^{s}\theta_rZ_rZ_r')$.

We begin with some expressions for $\mathrm{cov}(\xi_{i_1}\xi_{i_2},\xi_{i_3}\xi_{i_4})$, where ξ_i is the ith component of ξ, and $\mathrm{cov}(\partial l_R/\partial\theta_q,\partial l_R/\partial\theta_r)$, the the (q,r) element of $\mathscr{I}_1$ (see [55]).

Lemma 2.3. We have, for any $1 \leq i_j \leq n, j = 1,2,3,4,$

$$\text{cov}(\xi_{i_1}\xi_{i_2}, \xi_{i_3}\xi_{i_4}) = \lambda^2\{\Gamma(i_1,i_3)\Gamma(i_2,i_4) + \Gamma(i_1,i_4)\Gamma(i_2,i_3)\}$$
$$+ \sum_{r=0}^{s} \kappa_r z_{i_1 r} \cdots z_{i_4 r}, \quad (2.37)$$

where $z_{i_1 r} \cdots z_{i_4 r} = z_{i_1 r} \cdot z_{i_2 r} \cdot z_{i_3 r} \cdot z_{i_4 r}$; and, for any $0 \leq q, r \leq s$,

$$\text{cov}\left(\frac{\partial l_R}{\partial \theta_q}, \frac{\partial l_R}{\partial \theta_r}\right) = 2\text{tr}(H_r V H_r V) + \sum_{t=0}^{s} \kappa_t \sum_{l=1}^{m_t} (z_{tl}' H_q z_{tl})(z_{tl}' H_r z_{tl}). \quad (2.38)$$

Let $f_1, \ldots, f_L$ be the different nonzero functional values of

$$f(i_1, \ldots, i_4) = \sum_{r=0}^{s} \kappa_r z_{i_1 r} \cdots z_{i_4 r}. \quad (2.39)$$

Note that this is the second term on the right side of (2.37). Here by functional value it means $f(i_1, \ldots, i_4)$ as a function of $\kappa = (\kappa_r)_{0 \leq r \leq s}$. For example, $\kappa_0 + \kappa_1$ and $\kappa_2 + \kappa_3$ are different functions (even if their values may be the same for some κ). Also, let 0 denote the zero function (of κ). Also, let $H_{r,i,j}$ denote the (i,j) element of H_r. Define $\mathscr{A}_l = \{(i_1, \ldots, i_4) : f(i_1, \ldots, i_4) = f_l\}, 1 \leq l \leq L$, and

$$c_{q,r,l} = \frac{1}{|\mathscr{A}_l|} \sum_{(i_1, \ldots, i_4) \in \mathscr{A}_l} H_{q,i_1,i_2} H_{r,i_3,i_4}, \quad (2.40)$$

where $|\cdot|$ denotes cardinality. Note that $c_{q,r,l}$ depends only on θ. Define $c_{q,r}(i_1, \ldots, i_4) = c_{q,r,l}$, if $f(i_1, \ldots, i_4) = f_l, 1 \leq l \leq L$. The following result was proved in [55]. Let $\mathscr{I}_{1,qr}$ denote the (q,r) element of $\mathscr{I}_1$.

Theorem 2.8. For any non-Gaussian mixed ANOVA model, we have, for any $0 \leq q, r \leq s, \mathscr{I}_{1,qr} = \mathscr{I}_{1,1,qr} + \mathscr{I}_{1,2,qr}$, where

$$\mathscr{I}_{1,1,qr} = E\left\{\sum_{f(i_1, \ldots, i_4) \neq 0} c_{q,r}(i_1, \ldots, i_4)\xi_{i_1} \cdots \xi_{i_4}\right\}$$

$$\mathscr{I}_{1,2,qr} = 2\text{tr}(H_q V H_r V) - 3\theta_0^2 \sum_{f(i_1, \ldots, i_4) \neq 0} c_{q,r}(i_1, \ldots, i_4)\Gamma(i_1, i_3)\Gamma(i_2, i_4).$$

Theorem 2.8 shows that any element of $\mathscr{I}_1$ can be expressed as the sum of two terms. The first term is expressed as the expectation of a sum with the summands being products of four ξ_i's and a function of θ; the second term is a function of θ only. The first term, $\mathscr{I}_{1,1,qr}$, can be estimated the same way as the observed information, that is, by removing the expectation sign, and replacing β (in the ξ's) and θ by $\hat{\beta}$ and $\hat{\theta}$, the REML estimators, respectively. Denote the estimator by $\hat{\mathscr{I}}_{1,1,qr}$. The second term, $\mathscr{I}_{1,2,qr}$, can be estimated by replacing θ by $\hat{\theta}$. Denote this estimator by $\hat{\mathscr{I}}_{1,2,qr}$. An estimator of $\mathscr{I}_{1,qr}$ is then given by $\hat{\mathscr{I}}_{1,qr} = \hat{\mathscr{I}}_{1,1,qr} + \hat{\mathscr{I}}_{1,2,qr}$. Because the estimator consists partially of an "observed" term and partially an estimated term, it is called *partially observed information*, or POI. Jiang [55] gives sufficient conditions under which the POI matrix, $\hat{\mathscr{I}}_1 = (\hat{\mathscr{I}}_{1,qr})_{0 \leq q,r \leq s}$, consistently estimate $\mathscr{I}_1$; hence $\hat{\Sigma}_R = \hat{\mathscr{I}}_2^{-1} \hat{\mathscr{I}}_1 \hat{\mathscr{I}}_2^{-1}$ consistently estimate Σ_R of (2.26), where $\hat{\mathscr{I}}_2$ is $\mathscr{I}_2$ with θ replaced by $\hat{\theta}$, the REML estimator. We consider an example.

Example 2.7 (continued). As in the previous subsection, denote I and J by a and b, respectively. It is easy to show that $f(i_1 j_1, \ldots, i_4 j_4) = 0$, if not $i_1 = \cdots = i_4$; κ_1, if $i_1 = \cdots = i_4$ but not $j_1 = \cdots = j_4$; and $\kappa_0 + \kappa_1$, if $i_1 = \cdots = i_4$ and $j_1 = \cdots j_4$. Thus, $L = 2$ [note that L is the number of of different functional values of $f(i_1 j_1, \ldots, i_4 j_4)$]. Define the following functions of $\theta = (\sigma_e^2, \gamma)'$ and $\gamma = \sigma_v^2 / \sigma_e^2$: $t_0 = 1 - \gamma/(1 + \gamma b) - 1/\{(1 + \gamma b)ab\}$, $t_1 = (a-1)b/\{a(1 + \gamma b)\}$, and $t_3 = \{b(1 + \gamma b)^2 - (1 + \gamma)^2\}/(b^3 - 1)$. Then, the POIs are given by $\hat{\mathscr{I}}_{1,qr} = \hat{\mathscr{I}}_{1,1,qr} + \hat{\mathscr{I}}_{1,2,qr}$, $q,r = 0,1$, where

$$\hat{\mathscr{I}}_{1,1,00} = \frac{\hat{t}_1^2 - \hat{t}_0^2 b}{4\hat{\sigma}_e^8 b(b^3-1)} \left\{ \sum_i \left(\sum_j \hat{\xi}_{ij} \right)^4 - \sum_{i,j} \hat{\xi}_{ij}^4 \right\} + \frac{\hat{t}_0^2}{4\hat{\sigma}_e^8} \sum_{i,j} \hat{\xi}_{ij}^4,$$

$$\hat{\mathscr{I}}_{1,1,01} = \frac{(a-1)(\hat{t}_1 b - \hat{t}_0)}{4\hat{\sigma}_e^6 (1 + \hat{\gamma}b)^2 a(b^3-1)} \left\{ \sum_i \left(\sum_j \hat{\xi}_{ij} \right)^4 - \sum_{i,j} \hat{\xi}_{ij}^4 \right\}$$
$$+ \frac{(a-1)\hat{t}_0}{4\hat{\sigma}_e^6 (1 + \hat{\gamma}b)^2 a} \sum_{i,j} \hat{\xi}_{ij}^4,$$

$$\hat{\mathscr{I}}_{1,1,11} = \frac{(a-1)^2}{4\hat{\sigma}_e^4 (1 + \hat{\gamma}b)^4 a^2} \sum_i \left(\sum_j \hat{\xi}_{ij} \right)^4 ;$$

$$\hat{\mathscr{I}}_{1,2,00} = \frac{1}{2\hat{\sigma}_e^4} \left[ab - 1 - \frac{3}{2} ab\hat{t}_0^2 \{(1 + \hat{\gamma})^2 - \hat{t}_3\} - \frac{3}{2} a\hat{t}_1^2 \hat{t}_3 \right],$$

$$\hat{\mathscr{I}}_{1,2,01} = \frac{(a-1)b}{2\hat{\sigma}_e^2 (1 + \hat{\gamma}b)} \left\{ 1 - \left(\frac{3}{2} \right) \frac{(\hat{t}_1 b - \hat{t}_0)\hat{t}_3 + (1 + \hat{\gamma})^2 \hat{t}_0}{1 + \hat{\gamma}b} \right\},$$

$$\hat{\mathscr{I}}_{1,2,11} = -\frac{(a-1)(a-3)b^2}{4a(1 + \hat{\gamma}b)^2},$$

$\hat{\xi}_{ij} = y_{ij} - \bar{y}_{..}$, and the $\hat{t}$s are the ts with θ replaced by $\hat{\theta}$, the REML estimator.

As an application, we consider using the POI in testing a hypothesis regarding the variance components, or dispersion parameters, in a non-Gaussian mixed ANOVA model. The test is considered robust in the sense that it does not require normality.

Example 2.7 (continued). Suppose that one wishes to test the hypothesis H_0: $\gamma = 1$, i.e., the variance contribution due to the random effects is the same as that due to the errors. For example, the null hypothesis is equivalent to $H_0 : h^2 = 2$, where $h^2 = 4\gamma/(1 + \gamma)$ is a quantity of genetic interest, called *heritability*. The null hypothesis can be expressed as $H_0 : K'\theta = 0$ with $K = (0,1)'$. Furthermore, we have $K'\Sigma_R K = \Sigma_{R,11}$, which is the asymptotic variance of $\hat{\gamma}$, the REML of γ. Thus, a test statistic is

$\hat{\chi}^2 = (\hat{\gamma} - 1)^2 / \hat{\Sigma}_{R,11}$, where $\hat{\Sigma}_{R,11}$ is the POI of $\Sigma_{R,11}$. It is easy to show that

$$\hat{\Sigma}_{R,11} = \frac{\hat{\mathscr{I}}_{1,11}\hat{\mathscr{I}}_{2,00}^2 - 2\hat{\mathscr{I}}_{1,01}\hat{\mathscr{I}}_{2,00}\hat{\mathscr{I}}_{2,01} + \hat{\mathscr{I}}_{1,00}\hat{\mathscr{I}}_{2,01}^2}{(\hat{\mathscr{I}}_{2,00}\hat{\mathscr{I}}_{2,11} - \hat{\mathscr{I}}_{2,01}^2)^2}, \qquad (2.41)$$

where $\hat{\mathscr{I}}_{1,qr} = \hat{\mathscr{I}}_{1,1,qr} + \hat{\mathscr{I}}_{1,2,qr}$, $q, r = 0, 1$, and $\hat{\mathscr{I}}_{1,j,qr}$, $j = 1, 2$ are given earlier, but with $\hat{\gamma}$ replaced by 1, its value under H_0. Furthermore, we have $\hat{\mathscr{I}}_{2,00} = -(ab - 1)/2\hat{\sigma}_e^4$, $\hat{\mathscr{I}}_{2,01} = -(a-1)b/2\hat{\sigma}_e^2(1 + \hat{\gamma}b)$, $\hat{\mathscr{I}}_{2,11} = -(a-1)b^2/2(1 + \hat{\gamma}b)^2$, again with $\hat{\gamma}$ replaced by 1, where $\hat{\sigma}_e^2$ is the REML estimator of σ_e^2 under the null, given by

$$\hat{\sigma}_e^2 = \frac{1}{ab-1}\left(\text{SSE} + \frac{\text{SSA}}{b+1}\right), \qquad (2.42)$$

where $\text{SSE} = \sum_{i=1}^{a}\sum_{j=1}^{b}(y_{ij} - \bar{y}_{i\cdot})^2$, $\text{SSA} = b\sum_{i=1}^{a}(\bar{y}_{i\cdot} - \bar{y}_{\cdot\cdot})^2$ with $\bar{y}_{i\cdot} = b^{-1}\sum_{j=1}^{b}y_{ij}$ and $\bar{y}_{\cdot\cdot} = (ab)^{-1}\sum_{i=1}^{a}\sum_{j=1}^{b}y_{ij}$. The asymptotic null distribution is χ_1^2.

Jiang [55] carried out a simulation study on the performance of the dispersion test proposed above, which we call robust dispersion test (RDT), and compared it with an alternative delete-group jackknife method proposed by Arvesen [56]. The latter applies to cases where data can be divided into i.i.d. groups, such as the current situation. We refer to this test as jackknife. More specifically, Arvesen and Schmitz [57] proposed to use the jackknife estimator with a logarithm transformation, and this is the method that was compared.

We are interested in the situation when a is increasing while b remains fixed. Therefore, the following sample size configurations are considered: (I) $a = 50, b = 2$; (II) $a = 400, b = 2$. (I) represents a case of moderate sample size while (II), a case of large sample. In addition, we would like to investigate different cases in which normality and symmetry may or may not hold. Therefore, the following combinations of distributions for the random effects and errors are considered: (i) Normal-Normal; (ii) DE-NM$(-2, 2, 0.5)$, where DE represents the double exponential distribution and NM(μ_1, μ_2, ρ) the mixture of two normal distributions with means μ_1, μ_2, variance one, and mixing probability ρ [i.e., the probabilities $1 - \rho$ and ρ correspond to $N(\mu_1, 1)$ and $N(\mu_2, 1)$, respectively]; and (iii) CE-NM$(-4, 1, 0.2)$, where CE represents the centralized exponential distribution, i.e., the distribution of $X - 1$, where $X \sim \text{Exponential}(1)$. Note that in case (ii) the distributions are not normal but symmetric, while in case (iii) the distributions are not even symmetric, a further departure from normality. Also note that all these distributions have mean zero. They are standardized so that the distributions of the random effects and errors have variances σ_v^2 and σ_e^2, respectively. The true value of μ is set to 1.0. The true value of σ_e^2 is also chosen as 1.0. Tables 2.1 and 2.2 are taken from [55], in which the simulated size and power, with the nominal level of $\alpha = 0.05$, are reported for testing the null hypothesis against the alternative $H_1 : \gamma \neq 1$.

Overall, the jackknife appears to be more accurate in terms of the size, especially when a is relatively small (Case I). On the other hand, the simulated powers for RDT are higher at all alternatives, especially when a is relatively small (Case I). Also note that the jackknife with the logarithmic transformation is specifically designed for this

Table 2.1 *RDT versus Jackknife—Size*

Nominal Level	Method	Simulated Size					
		I-i	I-ii	I-iii	II-i	II-ii	II-iii
0.01	RDT	0.022	0.026	0.028	0.011	0.013	0.015
	Jackknife	0.010	0.014	0.020	0.009	0.011	0.013
0.05	RDT	0.070	0.078	0.091	0.054	0.057	0.063
	Jackknife	0.052	0.053	0.068	0.053	0.053	0.060
0.10	RDT	0.123	0.132	0.151	0.106	0.108	0.114
	Jackknife	0.099	0.103	0.122	0.104	0.103	0.109

Table 2.2 *RDT versus Jackknife—Power (Nominal Level 0.05)*

Alternative	Method	Simulated Power					
		I-i	I-ii	I-iii	II-i	II-ii	II-iii
$\gamma = 0.2$	RDT	0.747	0.807	0.745	1.000	1.000	1.000
	Jackknife	0.728	0.709	0.668	1.000	1.000	1.000
$\gamma = 0.5$	RDT	0.283	0.336	0.286	0.980	0.966	0.917
	Jackknife	0.277	0.271	0.275	0.981	0.958	0.912
$\gamma = 2.0$	RDT	0.532	0.424	0.369	0.999	0.993	0.973
	Jackknife	0.411	0.317	0.223	0.999	0.993	0.970
$\gamma = 5.0$	RDT	0.997	0.984	0.956	1.000	1.000	1.000
	Jackknife	0.991	0.971	0.903	1.000	1.000	1.000

kind of model where the observations are divided into independent groups, while the POI method is for a much richer class of LMM where the observations may or may not be divided into independent groups. See, for example, Section 6.2 of [55] for another simulation study under a LMM that involved crossed random effects; hence, the observations cannot be divided into independent groups in this case.

2.4 Mixed model prediction

Mixed model prediction (MMP) has a fairly long history starting with Henderson's early work in animal breeding ([58]). The field has since flourished thanks to its broad applications in various fields. The traditional fields of applications include genetics, agriculture, education, and surveys. This is a field where frequentist and Bayesian approaches found common grounds. See, for example, [59], and [2] (Section 2.3).

As noted in [2], there are two types of prediction problems associated with the mixed effects models. The first type, which is encountered, by far, more often in practice is prediction of mixed effects; the second type is prediction of future observation. We illustrate the difference between the two using a hypothetical example.

Example 2.11. An online shopper has made a few recent purchases at a commercial website. Naturally, the website has saved the shopper's records in doing business with the website. Suppose that the website wishes to predict the shopper's average purchase for the next month, it is a first type of prediction problem. Here, the average, or mean, is subject-specific, where the subject is shopper, in this case. Typically, the subject mean is consider a random variable. One reason is that the website has no idea which shopper is going to show up. There is another reason, or advantage, for treating the subject mean as a random variable, because there are many such means, one for each shopper. If all of these means are considered fixed parameters, there will be too many parameters to estimate. Note that the number of purchases by each shopper is usually not very large, at least within a month. As a result, the website does not have sufficient information to estimate the mean monthly purchase of the shopper. On the other hand, by treating the subject-specific mean as a random effect, the problem reduces to estimation the distribution of the random effects, which, under a parametric model (e.g., normal), may depend only on a few parameters. Therefore, we often use the term "prediction," rather than "estimation," for the subject-specific mean. Now suppose that, instead, the website wishes to predict the shopper's next purchase spending. Then, it is a second type of prediction problem.

For the most part, methods of MMP are centered around an idea, or concept, called best prediction (BP). Jiang (2007, p. 75) used a hypothetical example (from [60]) to explain the main point. Suppose that we are interested in predicting a person's intelligence quotient, widely known as IQ. Suppose that the person's IQ test score is 130. The test score may be viewed as an observed IQ; however, it is not necessarily the true IQ, because it is subject to variation (on a different day, the test score may be different). But, without any additional information, the best prediction of the person's IQ is 130, because this is all we know about. Now suppose, in addition, that we know the IQ's have a population mean of 100, with a standard deviation (s.d.) of 15. Then, all of a sudden, the best prediction of the person's IQ is no longer 130. The reason is simple: On average, the IQ is about 100, plus/minus 15. How come the person's test score is so high? This makes us think that the person might be a bit lucky to get the high score in the IQ test; his/her true IQ is likely to be a little lower than his/her test score. In fact, with some further information about the variation in the person's test scores (in repeated IQ tests), a BP for the person's IQ can be calculated (and it is, indeed, lower than 130). A more detailed example is given below.

Example 2.12 (Fay–Herriot model). Fay and Herriot [61] proposed the following model to estimate per-capita income of small places with population size less than 1,000: $y_i = x_i'\beta + v_i + e_i$, $i = 1, \ldots, m$, where y_i is a direct estimate (sample mean) for the ith area, x_i is a vector of known covariates, β is a vector of unknown regression coefficients, v_i's are area-specific random effects and e_i's are sampling errors. It is assumed that v_i's, e_i's are independent with $v_i \sim N(0, A)$ and $e_i \sim N(0, D_i)$. The variance A is unknown, but the sampling variances D_i's are assumed known (in practice, the D_i's can be estimated with accuracy, hence can be assumed known, at least approximately). Here, the interest is to estimate the small area mean $\zeta_i = E(y_i|v_i) = x_i'\beta + v_i$ for each i. The quantity of interest is in the form of a linear mixed effect, that is, a linear combination of fixed and random effects. It can be shown that the BP of ζ_i is not the direct estimator, y_i, but a weighted average of y_i and the regression estimator, $x_i'\beta$ (see below).

In general, suppose that, under the LMM (2.1), one is interested in estimate, or predict, a mixed effect in the form of $\zeta = b'\beta + a'\alpha$, where b, a are known vectors. Under normal theory [e.g., Jiang (2007), sec. C.1], the BP of ζ, in the sense of minimal mean squared prediction error (MSPE), is the conditional expectation of ζ given $y = (y_i)_{1\leq i\leq m}$, $E(\zeta|y)$, which is given by

$$\zeta_{BP} = b'\beta + a'GZ'V^{-1}(y - X\beta), \qquad (2.43)$$

where $V = \text{Var}(y) = R + ZGZ'$. Here, in (2.43), β, G and R are understood as the true fixed effects and covariance matrices (for α and ε).

Example 2.12 (continued). For the Fay–Herriot model, consider $\zeta = \zeta_i$ for a fixed i. Here $b = x_i$; a is the m-dimensional vector whose ith component is 1 and other components are 0, and $\alpha = v = (v_i)_{1\leq i\leq m}$. By (2.43), the BP for ζ_i is

$$\tilde{\zeta}_{i,BP} = \frac{A}{A + D_i}y_i + \frac{D_i}{A + D_i}x_i'\beta, \qquad (2.44)$$

where β, A are the true parameters. As indicated earlier, the BP is a weighted average of the direct estimator, y_i, and regression estimator, $x_i'\beta$, and the weights have the following interpretations. First note that the regression estimator relies on the assumed model that links different areas (note that β is common for all the areas). If A is very large compared to D_i, which means there is a huge between-area variation, there is little link between the areas. In other words, there is not much "strength" that the BP can "borrow" from other areas. As a result, the BP is almost the same as the direct estimator, y_i. On the other hand, if A is small compared to D_i, there is a strong link between areas so that the BP can borrow more strength. As a result, more weight is given to the regression estimator, $x_i'\beta$ (in the extreme case of $A = 0$, the BP is the same as $x_i'\beta$, which is identical to ζ_i a.s., because $v_i = 0$ a.s. when $A = 0$).

2.4.1 Best linear unbiased prediction

In practice, the fixed effects, β, and covariance matrices, G, R are usually unknown; hence, the BP is not computable. To make a practical use of the BP, a first step is to replace β by its MLE, assuming that G, R are known. The MLE is given by the right side of (2.6) with $V = R + ZGZ'$. The result of this replacement is called best linear unbiased predictor, or BLUP. In summary, the BLUP of ζ is given by

$$\tilde{\zeta}_{BLUP} = b'\tilde{\beta} + a'GZ'V^{-1}(y - X\tilde{\beta}), \qquad (2.45)$$

where $\tilde{\beta}$, known as the best linear unbiased estimator (BLUE), is given by

$$\tilde{\beta} = (X'V^{-1}X)^{-1}X'V^{-1}y. \qquad (2.46)$$

Example 2.12 (continued). In this case, the BLUE is given by

$$\tilde{\beta} = \left(\sum_{i=1}^{m}\frac{x_ix_i'}{A + D_i}\right)^{-1}\sum_{i=1}^{m}\frac{x_iy_i}{A + D_i}. \qquad (2.47)$$

The BLUP of ζ_i is given by (2.44) with β replaced by $\tilde{\beta}$.

Still, the BLUP is not computable unless G, R are known. Typically under a LMM, G, R depend on a vector, θ, of variance components. It is customary to replace θ in the expression of BLUP by $\hat{\theta}$, a consistent estimator, such as the ML or REML estimator. The result is the so-called empirical best linear unbiased predictor (EBLUP).Thus, the EBLUP of ζ is given by

$$\hat{\zeta}_{\text{EBLUP}} = b'\hat{\beta} + a'\hat{G}Z'\hat{V}^{-1}(y - X\hat{\beta}), \qquad (2.48)$$

where

$$\hat{\beta} = (X'\hat{V}^{-1}X)^{-1}X'\hat{V}^{-1}y \qquad (2.49)$$

and $\hat{G}, \hat{V}$ are G, V with θ replaced by $\hat{\theta}$, respectively. The estimator (2.49) is naturally called empirical BLUE (EBLUE). It is identical to the ML (REML) estimator of β when $\hat{\theta}$ is the ML (REML) estimator.

Note that, unlike BLUP, EBLUP is not a linear function of y. Nevertheless, under mild conditions, the EBLUE is a consistent estimator of β. As for the EBLUP, it is not consistent for estimating the random effects at individual level, unless the number of observations associated with the random effect goes to infinity. We illustrate with a simple example.

Example 2.13. Consider a special case of Example 2.6 with $x'_{ij}\beta = 0$. Suppose that one wishes to estimate v_i, the random effect for the ith area, consistently. It is easy to show that, if the random effects are treated as a fixed parameters, the MLE of v_i is $\bar{y}_{i\cdot} = n_i^{-1}\sum_{j=1}^{n_i} y_{ij}$. Clearly, $\bar{y}_{i\cdot}$ cannot be consistent unless $n_i \to \infty$, and this is true for any i. Thus, if one wants to consistently estimate all the individual random effects, then it is necessary that $n_i \to \infty, 1 \leq i \leq m$.

The asymptotic properties of EBLUE and EBLUP were studied by Jiang [62], who considered, in particular, consistency of EBLUP in the mean square (MS) sense. For example, in the case of Example 2.13, this means $m^{-1}\sum_{i=1}^{m}(\hat{v}_i - v_i)^2 \xrightarrow{\text{P}} 0$. Note that MS consistency is not necessarily weaker, or stronger, than consistency of EBLUP for each individual random effect. In addition, the author considered convergence in probability of the empirical distribution (e.d.) of the EBLUPs to the true distribution of the random effects under the non-Gaussian mixed ANOVA model. For example, in Example 2.13, this means that $m^{-1}\sum_{i=1}^{m} 1_{(\hat{v}_i \leq x)} \xrightarrow{\text{P}} F(x)$ for every x that is a continuity point of F, the true distribution of v_i (recall the v_i's are i.i.d.). The conditions for the MS consistency and convergence of e.d. are slightly weaker than what is required for consistency of EBLUP for each individual random effect. For example, in Example 2.13, the conditions for MS consistency and convergence of e.d. require, roughly speaking, that $n_i \to \infty$ for most of the i's.

An important, and difficult, problem for the EBLUP is assessing its uncertainty. A standard measure of uncertainty is MSPE. However, unlike the BLUP, the MSPE of EBLUP typically does not have an analytic expression. A naive estimator of the MSPE of EBLUP may be obtained by first deriving the MSPE of BLUP, which has a closed-form expression as a function of θ (see below), and then replacing the unknown θ in the expression by $\hat{\theta}$. However, this strategy is likely to under-estimate the true MSPE of EBLUP, as it does not take into account the additional variability, in the EBLUP, due to estimation of θ. In a seminal work, Prasad and Rao [29] use Taylor series linearization to obtain second-order unbiased estimator of the MSPE of EBLUP in the context of small area estimation. An MSPE estimator is second-order unbiased if its bias is of the order $o(m^{-1})$, where m is the number of small areas. The naive estimator of the MSPE, mentioned above, typically has a bias of the order $O(m^{-1})$. The Prasad–Rao method will be revisited later in Chapter 4. Here, we consider an extension, due to [63], of the Prasad–Rao method to (Gaussian) mixed ANOVA model. We begin with a nice decomposition of the MSPE of EBLUP, known as Kackar-Harville identity [64], which holds under a Gaussian LMM:

$$
\begin{aligned}
\mathrm{MSPE}(\hat{\zeta}_{\mathrm{EBLUP}}) &= \mathrm{E}(\hat{\zeta}_{\mathrm{EBLUP}} - \zeta)^2 \\
&= \mathrm{MSPE}(\tilde{\zeta}_{\mathrm{BLUP}}) + \mathrm{E}(\hat{\zeta}_{\mathrm{EBLUP}} - \tilde{\zeta}_{\mathrm{BLUP}})^2. \quad (2.50)
\end{aligned}
$$

Note that, if $\tilde{\zeta}_{\mathrm{BLUP}}$ is replaced by $\tilde{\zeta}_{\mathrm{BP}}$ in (2.50), the decomposition is obvious; however, what is less obvious is that the decomposition still holds, even though $\tilde{\zeta}_{\mathrm{BLUP}}$ and $\tilde{\zeta}_{\mathrm{BP}}$ are different [compare (2.43) and (2.45)], but only under the normality assumption. The idea is obtain second-order unbiased estimators of both terms on the right side of (2.50). The following general result, proved in [63], plays a key role in dealing with the second term. Let $l(\theta) = l(\theta, y)$ be a function, where $\theta = (\theta_i)_{1 \le i \le s}$ is a parameter with the parameter space Θ and y is the vector of observations. For example, l may be the restricted log-likelihood function [see (2.23)] that leads to the REML estimator of θ, the vector of variance components.

Lemma 2.4. Suppose that the following hold:
(i) $l(\theta, y)$ is three-times continuously differentiable with respect to θ;
(ii) the true $\theta \in \Theta^o$, the interior of Θ;
(iii) $-\infty < \limsup_{n \to \infty} \lambda_{\max}(D^{-1}HD^{-1}) < 0$, where $H = \mathrm{E}\{\partial^2 l / \partial \theta^2\}$ with the second derivative evaluated at the true θ and $D = \mathrm{diag}(d_1, \ldots, d_s)$, with d_i's being positive constants satisfying $d_* = \min_{1 \le i \le s} d_i \to \infty$, as $n \to \infty$;
(iv) the gth moments of the following are bounded for some $g > 0$:

$$
\frac{1}{d_i}\left|\frac{\partial l}{\partial \theta_i}\right|, \quad \frac{1}{\sqrt{d_i d_j}}\left|\frac{\partial^2 l}{\partial \theta_i \partial \theta_j} - \mathrm{E}\left(\frac{\partial^2 l}{\partial \theta_i \partial \theta_j}\right)\right|, \quad \frac{d_*}{d_i d_j d_k} M_{ijk}, \ 1 \le i, j, k \le s,
$$

where the first and second derivatives are evaluated at the true θ and

$$
M_{ijk} = \sup_{\tilde{\theta} \in S_\delta(\theta)} \left|\frac{\partial^3 l}{\partial \theta_i \partial \theta_j \partial \theta_k}\right|_{\theta==\tilde{\theta}}
$$

with $S_\delta(\theta) = \{\tilde{\theta} : |\tilde{\theta}_i - \theta_i| \le \delta d_* / d_i, 1 \le i \le s\}$ for some $\delta > 0$. Then, there exists $\hat{\theta}$ such that for any $0 < \rho < 1$, there is an event set $\mathscr{B}$ satisfying for large n and on $\mathscr{B}$, $\hat{\theta} \in \Theta$, $\partial l / \partial \theta|_{\theta=\hat{\theta}} = 0$, $|D(\hat{\theta} - \theta)| < d_*^{1-\rho}$, and

$$
\hat{\theta} = \theta - H^{-1}h + r, \quad (2.51)
$$

where θ is the true θ, $h = \partial l / \partial \theta$, evaluated at the true θ, and $|r| \le d_*^{-2\rho} u$ with $\mathrm{E}(u^g) = O(1)$; and $\mathrm{P}(\mathscr{B}^c) \le c d_*^{\tau g}$ with $\tau = (1/4) \wedge (1 - \rho)$ and c being a constant.

Equation (2.51) was used in the Taylor series expansion of the BLUP, as a function of θ, to approximate the difference $\hat{\zeta}_{\text{EBLUP}} - \tilde{\zeta}_{\text{BLUP}}$, which eventually leads to the following approximation, under suitable conditions [see [63]; note that a minus sign is missing before the trace in Equation (3.4) of the latter reference]:

$$\text{E}\{\hat{\zeta}_{\text{EBLUP}} - \tilde{\zeta}_{\text{BLUP}}\}^2 = -\text{tr}\left(\frac{\partial S'}{\partial \theta} V \frac{\partial S}{\partial \theta'} H^{-1}\right) + o(d_*^{-2}), \qquad (2.52)$$

where $S = V^{-1}ZGa$ (a in the expression of ζ following Example 2.12), $\theta = (\tau, \sigma_1, \ldots, \sigma_s)'$ [see (2.3)], and $H = \text{E}(\partial^2 l_R / \partial \theta)$, l_R being the restricted log-likelihood, given by (2.23), considered as a function of θ. Note that, unlike Lemma 2.4, normality is required for (2.52) to hold.

On the other hand, the first term on the right side of (2.50) has an analytic expression, namely (e.g., [65]),

$$\text{MSPE}(\tilde{\zeta}_{\text{BLUP}}) = g_1(\theta) + g_2(\theta), \qquad (2.53)$$

where $g_1(\theta) = a'(G - GZ'V^{-1}ZG)a$ and $g_2(\theta) = (b - X'S)'(X'V^{-1}X)^{-1}(b - X'S)$. Thus, if we denote the first term on the right side of (2.52) by $g_3(\theta)$, we have

$$\text{MSPE}(\hat{\zeta}_{\text{EBLUP}}) = g_1(\theta) + g_2(\theta) + g_3(\theta) + o(d_*^{-2}). \qquad (2.54)$$

Equation (2.54) is an important step forward (note that all the g functions have analytic expressions), but it is not yet a second-order unbiased estimator. To obtain the latter, a naive approach would be to take take $g_1(\theta) + g_2(\theta) + g_3(\theta)$, and replace θ by $\hat{\theta}$. However, this typically results in a bias of the order $O(d_*^{-2})$. Thus, one needs to do a bias-correction to reduce the bias to $o(d_*^{-2})$. This will be discussed in Chapter 4. After the bias-correction, a second-order unbiased estimator is obtained as

$$\widehat{\text{MSPE}}(\hat{\zeta}_{\text{EBLUP}}) = g_1(\hat{\theta}) + g_2(\hat{\theta}) + 2g_3(\hat{\theta}), \qquad (2.55)$$

where $\hat{\theta}$ is the REML estimator of θ, which is also used in $\hat{\zeta}_{\text{EBLUP}}$. A similar result can be obtained when the MLE of θ is used in the EBLUP. See [63] for details.

2.4.2 Observed best prediction

If one thinks more carefully, the BLUP may be viewed as a hybrid of optimal prediction–BP and optimal estimation–ML. Note that the $\tilde{\beta}$ in (2.45) is the MLE of β, assuming that G, R are known. If prediction of the mixed effect is of main interest, it may be wondered why one has to hybridize. To make it simpler, let us stay, for now, with the assumption that G, R are known. Can we estimate β in a way that is also to the best interest of prediction (of the mixed effect)? Note that the MLE is known to be (asymptotically) optimal in terms of estimation, but not necessarily in terms of prediction (in fact, it is not, as we shall show). This idea leads to a new approach for mixed model prediction, proposed by Jiang et al. [66]. The approach has a surprising bonus, as it turns out, that is, it leads to a predictor that is more robust to model misspecification than the BLUP, or EBLUP.

A key to the new approach is to entertain two models. One is called assumed model, which is (2.1); the other is called broader model, or true model. For simplicity, we focus on the case that the potential model misspecification is only in terms of the mean function; in other words, the true model is

$$y = \mu + Z\alpha + \varepsilon, \tag{2.56}$$

where $\mu = E(y)$. Here, E denotes expectation under the true distribution of y, which may be unknown, but is not model-dependent. Our interest is prediction of a vector of mixed effects, expressed as

$$\zeta = F'\mu + C'\alpha, \tag{2.57}$$

where F, C are known matrices. For example, in Example 2.12, we are interested in predicting $\zeta = (\zeta_i)_{1 \leq i \leq m} = \mu + v$, where $\mu = (\mu_i)_{1 \leq i \leq m}$ with $\mu_i = E(y_i)$, and $v = (v_i)_{1 \leq i \leq m}$. In other words, $\zeta = E(y|v)$, which can be expressed as (2.57) with $F = C = I_m$. Suppose that G, R are known, as in the derivation of BLUP. Then, the BP of ζ, in the sense of minimum MSPE, under the assumed model, is given by

$$\begin{aligned}
E_M(\zeta|y) &= F'\mu + C'E_M(\alpha|y) \\
&= F'X\beta + C'GZ'V^{-1}(y - X\beta) \\
&= F'y - \Gamma(y - X\beta), \tag{2.58}
\end{aligned}$$

where $\Gamma = F' - B$ with $B = C'GZ'V^{-1}$. Here, unlike E, E_M denotes conditional expectation under the assumed model, (2.1), denoted by M. Let $\check{\zeta}$ denote the right side of (2.58) with a fixed, arbitrary β. Then, by (2.56), (2.57), we have $\check{\zeta} - \zeta = H'\alpha + F'\varepsilon - \Gamma(y - X\beta)$, where $H = Z'F - C$. Thus, we have

$$\begin{aligned}
\text{MSPE}(\check{\zeta}) &= E(|\check{\zeta} - \zeta|^2) \\
&= E(|H'\alpha + F'\varepsilon|^2) - 2E\{(\alpha'H + \varepsilon'F)\Gamma(y - X\beta)\} \\
&\quad + E\{(y - X\beta)'\Gamma'\Gamma(y - X\beta)\} \\
&= I_1 - 2I_2 + I_3.
\end{aligned}$$

It is easy to see that I_1 does not depend on β. In fact, I_2 does not depend on β either, because $I_2 = E\{(\alpha'H + \varepsilon'F)\Gamma(y - \mu)\} + \{E(\alpha'H + \varepsilon'F)\}\Gamma(\mu - X\beta) = E\{(\alpha'H + \varepsilon'F)\Gamma(Z\alpha + \varepsilon)\}$, by (2.56). Thus, we can express the MSPE as

$$\text{MSPE}(\check{\zeta}) = E\{(y - X\beta)'\Gamma'\Gamma(y - X\beta) + \cdots\}, \tag{2.59}$$

where $\cdots$ does not depend on β. The idea is to estimate β by minimizing the observed MSPE, which is the expression inside the expectation in (2.59). Then, because $\cdots$ does not depend on β, this is equivalent to minimizing $(y - X\beta)'\Gamma'\Gamma(y - X\beta)$. The solution is what we call best predictive estimator, or BPE, given by

$$\hat{\beta} = (X'\Gamma'\Gamma X)^{-1}X'\Gamma'\Gamma y, \tag{2.60}$$

assuming nonsingularity of $\Gamma'\Gamma$ and that X is full rank. The resulting predictor of ζ

is called *observed best predictor*, or OBP, given by the right side of (2.58) with β replaced by $\hat{\beta}$. The term OBP is due to the fact that the BPE is obtained by minimizing the observed MSPE (if the observed MSPE were the true MSPE, the same procedure would lead to the BP).

Note that $\hat{\beta}$ is different from the MLE (or BLUE) of β, $\tilde{\beta}$, given by (2.46). Correspondingly, the BLUP is obtained by the right side of (2.58) with β replaced by $\tilde{\beta}$. We consider a specific example.

Example 2.12 (continued). We have known that, in this case, the ith component of the BLUP, $\tilde{\zeta}_i$, is given by the right side of (2.44) with β replaced by $\tilde{\beta}$ of (2.47), which is the MLE of β under normality. On the other hand, the ith component of the OBP, $\hat{\zeta}_i$, is given by the right side of (2.44) with β replace by the BPE,

$$\hat{\beta} = \left\{ \sum_{i=1}^{m} \left(\frac{D_i}{A+D_i} \right)^2 x_i x_i' \right\}^{-1} \sum_{i=1}^{m} \left(\frac{D_i}{A+D_i} \right)^2 x_i y_i. \tag{2.61}$$

It is interesting to note that both the MLE and BPE are in the forms of weighted averages; the only difference is the weights. The BPE gives more weights to areas with larger sampling variance, D_i, while the MLE does just the opposite–assigning more weights to areas with smaller sampling variance. The question is: Who is right?

To answer this question, first note that, from the expression of the BP, (2.44), it is evident that the assumed model is involved only through $x_i'\beta$, whose corresponding weight, $D_i/(A+D_i)$, is increasing with D_i. In other words, the model-based BP is more relevant to those areas with larger D_i. Imagine that there is a meeting of representatives from the different small areas to discuss what estimate of β is to be used in the BP. The areas with larger D_i think that their "voice" should be heard more (i.e., they should receive more weights), because the BP is more relevant to their business. Their request is reasonable (although, politically, this may not work out within a democratic voting system).

Not only the OBP has a intuitive explanation, it also has an attractive property that it is more robust to model misspecification than BLUP, in terms of MSPE. This is not surprising, given the way that OBP is derived, but the following theorem, proved in [66], makes this a precise story. Consider an empirical best predictor (EBP), $\check{\zeta}$, which is the BP [i.e., right side of (2.58)] with β replaced by a weighted least squares (WLS) estimator (recall that G, R are assumed known),

$$\check{\beta} = (X'WX)^{-1}X'Wy, \tag{2.62}$$

where W is a positive definite weighting matrix. The BPE and MLE are special cases of the WLS estimator, with $W = \Gamma'\Gamma$ for the former and $W = V^{-1}$ for the latter. Thus, the OBP and BLUP are special cases of the EBP. Also recall that $\mu = \mathrm{E}(y)$.

Theorem 2.9. The MSPE of EBP can be expressed as

$$\mathrm{MSPE}(\check{\zeta}) = a_0 + \mu'A_1(W)\mu + \mathrm{tr}\{A_2(W)\}, \tag{2.63}$$

where a_0 is a nonnegative constant that does not depend on W; $A_j(W), j = 1,2$ are nonnegative definite matrices that depend on W. $\mu'A_1(W)\mu$ is minimized by the OBP, with $W = \Gamma'\Gamma$; $\mathrm{tr}\{A_2(W)\}$ is minimized by the BLUP, with $W = V^{-1}$.

The expressions of $a_0, A_j(W), j = 1, 2$ are given below. Define

$$L = F' - \Gamma\{I - X(X'WX)^{-1}X'W\},$$

where I is the identity matrix, and recall $B = C'GZ'V^{-1}$. Then, we have

$$
\begin{aligned}
a_0 &= \mathrm{tr}\{C'(G - GZ'V^{-1}ZG)C\}, \\
A_1(W) &= (F' - L)'(F' - L), \\
A_2(W) &= (L - B)V(L - B)'.
\end{aligned}
$$

Here are some of the important implications of Theorem 2.9. First note that the second terms disappear when the mean of y is correctly specified, that is, $\mu = X\beta$ for some β. So, when the underlying model is correct, BLUP usually wins the battle of MSPE, because it minimizes the only (remaining) term that depends on W. On the other hand, when the mean is misspecified, the second term is usually of higher order than the third term. Thus, in this situation, OBP is likely to win the battle, because it minimizes a higher order term. We use an example to illustrate.

Example 2.14. We use a very simple example to show that the gain of OBP over BLUP can be substantial, if the underlying model is misspecified. Consider a special case of Example 2.12, in which $x_i'\beta = \beta$, an unknown mean. To make it even simpler, suppose that A is known, so that one can actually compute the BLUP. Furthermore, suppose that $m = 2n$, $D_i = a, 1 \leq i \leq n$, and $D_i = b, n+1 \leq i \leq m$, where a, b are positive known constants. Now suppose that, actually, the underlying model is

$$y_i = c + v_i + e_i, \ 1 \leq i \leq n \text{ and } y_i = d + v_i + e_i, \ n+1 \leq i \leq m,$$

where $c \neq d$; in other words, we have a model misspecification by assuming $c = d$. Consider $\zeta = (\zeta_i)_{1 \leq i \leq m}$, where $\zeta_i = c + v_i, 1 \leq i \leq n$, and $\zeta_i = d + v_i, n+1 \leq i \leq m$.

For this special case, we can actually derive the exact expressions of the MSPEs. It can be shown (see [66], Appendix) that

$$
\begin{aligned}
\mathrm{MSPE}(\hat{\zeta}_{\mathrm{OBP}}) &= \left\{ \left(\frac{a}{A+a} + \frac{b}{A+b}\right)A + \frac{a^2 b^2 (c-d)^2}{a^2(A+b)^2 + b^2(A+a)^2} \right\} n \\
&\quad + \frac{a^4(A+b)^3 + b^4(A+a)^3}{(A+a)(A+b)\{a^2(A+b)^2 + b^2(A+a)^2\}} \\
&= g_1 n + h_1, \hspace{4cm} (2.64) \\
\mathrm{MSPE}(\hat{\zeta}_{\mathrm{BLUP}}) &= \left\{ \left(\frac{a}{A+a} + \frac{b}{A+b}\right)A + \frac{(a^2+b^2)(c-d)^2}{(2A+a+b)^2} \right\} n \\
&\quad + \frac{a^2(A+b)^2 + b^2(A+a)^2}{(A+a)(A+b)(2A+a+b)} \\
&= g_2 n + h_2. \hspace{4cm} (2.65)
\end{aligned}
$$

If $c \neq d$, then $g_1 \leq g_2$ and $h_1 \geq h_2$ with equality holding in both cases if and only if

$$a^2(A+b) = b^2(A+a). \hspace{3cm} (2.66)$$

Now suppose that (2.66) does not hold, then we have $g_1 < g_2$, and $h_1 > h_2$, but the latter is not important when n is large. In fact, we have

$$\lim_{n\to\infty} \frac{\text{MSPE}(\hat{\zeta}_{\text{OBP}})}{\text{MSPE}(\hat{\zeta}_{\text{BLUP}})} = \frac{g_1}{g_2} < 1.$$

For example, suppose that $A/(c-d)^2 \approx 0$, $A/b \approx 0$ and $b/a \approx 0$. Then it is easy to show that $g_1/g_2 \approx 0.5$. Thus, in this case, the MSPE of the OBP is asymptotically about half of that of the BLUP. On the other hand, if $c = d$, that is, if the underlying model is correctly specified, then we have $g_1 = g_2$ while, still, $h_1 \geq h_2$. Therefore, in this case, $\text{MSPE}(\hat{\zeta}_{\text{OBP}}) \geq \text{MSPE}(\hat{\zeta}_{\text{BLUP}})$. However, we have $\lim_{n\to\infty} \text{MSPE}(\hat{\zeta}_{\text{OBP}})/\text{MSPE}(\hat{\zeta}_{\text{BLUP}}) = 1$; hence, in this case, the MSPEs of the OBP and BLUP are asymptotically the same.

In addition to the MSPE property, [66] also studied asymptotic behavior of the BPE, which is a defining element of the OBP, under possible model misspecification. This is similar, in spirit, to an earlier work of White [67], who studied asymptotic behavior of the MLE when the underlying model is misspecified. Note that, under a potentially misspecified model, it is no longer meaningful to talk about consistency, because the true parameters do not exist. However, it was shown that, in this situation, the BPE still converges in probability to "something." To see what the something is, let ψ denote the vector of parameters that one intends to estimate via the BPE. In general, the idea is that the MSPE can be expressed as

$$E\{Q(y, \psi)\} \equiv \text{mspe}(\psi). \tag{2.67}$$

Here, again, E denotes expectation under the true distribution of y, which may be unknown, but not model dependent, and (2.67) is considered a function of ψ. The BPE of ψ, $\hat{\psi}$, is then obtained by minimizing the left side of (2.67) without the expectation sign, that is, $Q(y, \psi)$, over $\psi \in \Psi$, where Ψ is the parameter space. Then, the something to which the BPE converges (in probability) is actually the minimizer of $\text{mspe}(\psi)$ over all $\psi \in \Psi$. Jiang [66] shows that the convergence actually happens under some regularity conditions; in fact, $\hat{\psi}$ is shown to be $\sqrt{m}$ consistent in the sense that $\sqrt{m}(\hat{\psi} - \psi_*) = O_P(1)$ (e.g., [1], Section 3.4), where m is the number of clusters, and ψ_* the minimizer of $\text{mspe}(\psi)$ over $\psi \in \Psi$.

2.4.3 Classified mixed model prediction

Nowadays, new and challenging problems have emerged from such fields as business and health sciences, in addition to the traditional fields of MMP (see the beginning of Section 2.4), to which methods of MMP are potentially applicable, but not without further methodology and computational developments. Some of these problems occur when interest is at subject level (e.g., personalized medicine), or (small) subpopulation level (e.g., county), rather than at large population level (e.g., epidemiology). In such cases, it is possible to make substantial gains in prediction accuracy by identifying a class that a new subject belongs to. As a first step, we need to find a "match" between the class that the new subject belongs to and a class among the existing data, which we call training data.

Suppose that we have a set of training data, $y_{ij}, i = 1, \ldots, m, j = 1, \ldots, n_i$ in the sense that their classifications are known, that is, one knows which group, i, that y_{ij} belongs to. The assumed linear mixed model (LMM) for the training data is

$$y_i = X_i\beta + Z_i\alpha_i + \varepsilon_i, \qquad (2.68)$$

where $y_i = (y_{ij})_{1 \le j \le n_i}$, $X_i = (x'_{ij})_{1 \le j \le n_i}$ is a matrix of known covariates, β is a vector of unknown regression coefficients (the fixed effects), Z_i is a known $n_i \times q$ matrix, α_i is a $q \times 1$ vector of group-specific random effects, and ε_i is an $n_i \times 1$ vector of errors. It is assumed that the α_i's and ε_i's are independent, with $\alpha_i \sim N(0, G)$ and $\varepsilon_i \sim N(0, R_i)$, where the covariance matrices G and R_i depend on a vector ψ of variance components.

Our goal is to make a classified prediction for a mixed effect associated with a set of new observations, $y_{n,j}, 1 \le j \le n_{\text{new}}$ (the subscript n referrs to "new") such that

$$y_{n,j} = x'_n\beta + z'_n\alpha_I + \varepsilon_{n,j}, \quad 1 \le j \le n_{\text{new}}, \qquad (2.69)$$

where x_n, z_n are known vectors, $I \in \{1, \ldots, m\}$ but one does not know which element $i, 1 \le i \le m$, is equal to I. Furthermore, $\varepsilon_{n,j}, 1 \le j \le n_{\text{new}}$ are new errors that are independent with $E(\varepsilon_{n,j}) = 0$ and $\text{var}(\varepsilon_{n,j}) = R_{\text{new}}$, and are independent with the α_is and ε_is. Note that the normality assumption is not always needed for the new errors, unless prediction interval is concerned (see below). Also, the variance R_{new} of the new errors does not have to be the same as the variance of ε_{ij}, the jth component of ε_i associated with the training data. The mixed effect that we wish to predict is given by (2.43), which in this special case can be expressed as

$$\theta = E(y_{n,j}|\alpha_I) = x'_n\beta + z'_n\alpha_I. \qquad (2.70)$$

From the training data, one can estimate the parameters, β and ψ. For example, one can use the standard mixed model analysis to obtain ML or REML estimators (see Sections 2.1 and 2.2). Alternatively, one may use the OBP method (see the previous subsection) to obtain estimators of β, ψ, which is more robust to model misspecifications in terms of the predictive performance. Thus, we can assume that estimators $\hat{\beta}, \hat{\psi}$ are available for β, ψ.

Suppose that $I = i$. Then, the vectors $y_1, \ldots, y_{i-1}, (y'_i, \theta)', y_{i+1}, \ldots, y_m$ are independent. Thus, we have $E(\theta|y_1, \ldots, y_m) = E(\theta|y_i)$. By the normal theory, we have

$$E(\theta|y_i) = x'_n\beta + z'_nGZ'_i(R_i + Z_iGZ'_i)^{-1}(y_i - X_i\beta). \qquad (2.71)$$

The right side of (2.71) is the BP under the assumed LMM, if the true parameters, β and ψ, are known. Because the latter are unknown, we replace them by $\hat{\beta}$ and $\hat{\psi}$, respectively. The result is what we call empirical best predictor (EBP), denoted by $\tilde{\theta}_{(i)}$. In practice, however, I is unknown and treated as a parameter. In order to identify, or estimate, I, we consider the mean squared prediction error (MSPE) of θ by the BP when I is classified as i, that is $\text{MSPE}_i = E\{\tilde{\theta}_{(i)} - \theta\}^2 = E\{\tilde{\theta}^2_{(i)}\} -$

$2E\{\tilde{\theta}_{(i)}\theta\}+E(\theta^2)$. Using the expression $\theta=\bar{y}_n-\bar{\varepsilon}_n$, where $\bar{y}_n=n_{\text{new}}^{-1}\sum_{j=1}^{n_{\text{new}}}y_{n,j}$ and $\bar{\varepsilon}_n$ is defined similarly, we have $E\{\tilde{\theta}_{(i)}\theta\}=E\{\tilde{\theta}_{(i)}\bar{y}_n\}-E\{\tilde{\theta}_{(i)}\bar{\varepsilon}_n\}=E\{\tilde{\theta}_{(i)}\bar{y}_n\}$. Thus, we have the expression:

$$\text{MSPE}_i = E\{\tilde{\theta}_{(i)}^2-2\tilde{\theta}_{(i)}\bar{y}_n+\theta^2\}. \qquad (2.72)$$

It follows that the observed MSPE corresponding to (2.72) is the expression inside the expectation. Therefore, a natural idea is to identify I as the index i that minimizes the observed MSPE. Because θ^2 does not depend on i, the minimizer is given by

$$\hat{I} = \text{argmin}_i\left\{\tilde{\theta}_{(i)}^2-2\tilde{\theta}_{(i)}\bar{y}_n\right\}. \qquad (2.73)$$

The classified mixed-effect predictor (CMEP) of θ is then given by $\hat{\theta}=\tilde{\theta}_{(\hat{I})}$.

Remark. Because $\tilde{\theta}_{(i)}^2-2\tilde{\theta}_{(i)}\bar{y}_n=\{\tilde{\theta}_{(i)}-\bar{y}_n\}^2-\bar{y}_n^2$, (2.73) is equivalent to

$$\hat{I} = \text{argmin}_i\{\tilde{\theta}_{(i)}-\bar{y}_n\}^2. \qquad (2.74)$$

Thus, the class identifier, $\hat{I}$, may be interpreted as finding the index i whose corresponding EBP is closest to the observed mean of the new observations.

Jiang et al. [68] studied theoretical properties of CMMP. The most striking result of their findings is the following. The CMMP procedure essentially matches the new observations to a group among the training data in terms of their corresponding random effects. In practice, however, such a "match" may not exist. What appears to be striking is that, even if the match does not, or exist but mis-matched, CMMP still helps in improving prediction accuracy; in particular, the CMEP $\hat{\theta}$ is consistent in estimating the mixed effect, θ, associated with the new observations, and outperforms the standard regression predictor. We state the result below as a theorem. For simplicity of presentation, we focus on the following nested-error regression (NER) model [34]:

$$y_{ij} = x'_{ij}\beta+\alpha_i+\varepsilon_{ij}, \qquad (2.75)$$

$i=1,\ldots,m,j=1,\ldots,n_i$, where the random effects, α_i, and errors ε_{ij}, are independent such that $\alpha_i\sim N(0,G)$ and $\varepsilon_{ij}\sim N(0,R)$. Suppose that the new observations, $y_{n,1},\ldots,y_{n,n_{\text{new}}}$, satisfy a similar NER model:

$$y_{n,j} = x'_n\beta+\alpha_{\text{new}}+\varepsilon_{\text{new},j}, \quad 1\le j\le n_{\text{new}}, \qquad (2.76)$$

where x_n is a vector of known covariates that does not depend on j; in other words, the new observations have the same (unconditional) mean, and $\varepsilon_{\text{new},j}$ is the new error. As for α_{new}, we assume that it is either (a) a new random effect or (b) identical to one of the $\alpha_i, 1\le i\le m$, but we do not know which case, (a) or (b), is true. We assume that $E(\alpha_{\text{new}})=0, E(\varepsilon_{\text{new},j})=0, 1\le j\le n_{\text{new}}$, and that $\text{var}(\alpha_{\text{new}}),\text{var}(\varepsilon_{\text{new},j}),1\le j\le n_{\text{new}}$ are bounded. However, it is not assumed that $\alpha_{\text{new}},\varepsilon_{\text{new},j},1\le j\le n_{\text{new}}$ are normally distributed; neither do we assume that $\text{var}(\alpha_{\text{new}})=G$ and $\text{var}(\varepsilon_{\text{new},j})=R$.

The following notation will be used: Let $\theta_n=x'_n\beta+\alpha_{\text{new}}$ be the true mixed effect; β, G, R denote the true parameters; and $\hat{\beta}, \hat{G}, \hat{R}$ their consistent estimators; $\theta_{n,i} = x_n'\beta + \alpha_i$, where α_i is the true random effect associated with group i of the training data,

$$\tilde{\theta}_{n,i} = x_n'\beta + \frac{n_i G}{R + n_i G}(\bar{y}_{i\cdot} - \bar{x}_{i\cdot}'\beta),$$

where $\bar{y}_{i\cdot} = n_i^{-1}\sum_{j=1}^{n_i} y_{ij}$, $\bar{x}_{i\cdot} = n_i^{-1}\sum_{j=1}^{n_i} x_{ij}$, and

$$\hat{\theta}_{n,i} = x_n'\hat{\beta} + \frac{n_i \hat{G}}{\hat{R} + n_i \hat{G}}(\bar{y}_{i\cdot} - \bar{x}_{i\cdot}'\hat{\beta}).$$

Then, the CMEP of θ_n (see above) is $\hat{\theta}_n = \hat{\theta}_{n,\hat{I}}$, where

$$\hat{I} = \text{argmin}_{1 \leq i \leq m}\left\{\hat{\theta}_{n,i}^2 - 2\hat{\theta}_{n,i}\bar{y}_n\right\}$$

with $\bar{y}_n = n_{\text{new}}^{-1}\sum_{j=1}^{n_{\text{new}}} y_{n,j}$. Similarly, let $\bar{\varepsilon}_n = n_{\text{new}}^{-1}\sum_{j=1}^{n_{\text{new}}} \varepsilon_{n,j}$. Let $\hat{\theta}_{n,\text{r}} = x_n'\hat{\beta}_{\text{LS}}$ denote the regression predictor (RP) of θ_n, where $\hat{\beta}_{\text{LS}}$ is the least squares (LS) estimator of β. Let $\tilde{G}$ and $\tilde{R}$ denote a certain type of consistent estimators of G and R, respectively. For example, $\tilde{G}, \tilde{R}$ may be the ML, or REML, estimators (e.g., [2]). As in [63], we consider truncated versions of these estimators of variance components. Let

$$\hat{G} = \begin{cases} a(\log m)^{-\nu}, & \text{if } \tilde{G} < a(\log m)^{-\nu}, \\ \tilde{G}, & \text{if } a(\log m)^{-\nu} \leq \tilde{G} \leq b(\log m)^{\nu}, \\ b(\log m)^{\nu}, & \text{if } \tilde{G} > b(\log m)^{\nu}, \end{cases}$$

and define $\hat{R}$ similarly, where a, b, ν are positive constants. The corresponding estimator for β is then the empirical best linear unbiased estimator (EBLUE), defined as

$$\hat{\beta} = (X'\hat{V}^{-1}X)^{-1}X'\hat{V}^{-1}y,$$

where $X = (X_i)_{1 \leq i \leq m}$ with $X_i = (x_{ij}')_{1 \leq j \leq n_i}$, $y = (y_i)_{1 \leq i \leq m}$ with $y_i = (y_{ij})_{1 \leq j \leq n_i}$, and $\hat{V} = \text{diag}(\hat{R}I_{n_i} + \hat{G}J_{n_i}, 1 \leq i \leq m)$ with I_{n_i}, J_{n_i} being the n_i-dimensional identity matrix and matrix of 1's, respectively. We assume the following regularity conditions.

A1. $\alpha_i, 1 \leq i \leq m, \alpha_{\text{new}}$ are independent with the ε_{ij}'s; α_{new} is independent with the training data under case (a) below (2.76), and $\bar{\varepsilon}_n$ is independent with all of the α_i's and ε_{ij}'s.

A2. $|x_{ij}|, 1 \leq i \leq m, 1 \leq j \leq n_i$ and x_n are bounded.

A3. $\hat{\beta} \longrightarrow \beta$ in L^2, where β is the true β.

A4. $m \to \infty$, $(\log m)^{2\nu}/n_{\min} \to 0$, where $n_{\min} = \min_{1 \leq i \leq m} n_i$, and $n_{\text{new}} \to \infty$.

Theorem 2.10. Suppose that A1–A4 hold, and $\text{var}(\alpha_{\text{new}}) > 0$. Then, we have

$$E\{(\hat{\theta}_n - \theta_n)^2\} \to 0 \quad \text{and} \quad \liminf\left\{E(\hat{\theta}_{n,\text{r}} - \theta_n)^2\right\} \geq \delta$$

for some constant $\delta > 0$. It follows that the CMEP, $\hat{\theta}_n$, is consistent and, asymptotically, has smaller MSPE than the RP, $\hat{\theta}_{n,\text{r}}$.

Remarks. Theorem 2.10 implies that, under reasonable conditions, the CMEP of the mixed effect is consistent even if there is no match, or mis-match (see below). It should be noted that the consistency is not about estimating the class membership, I. In fact, as m, the number of classes increases, the probability of identifying the true index I decreases, even if there is a match; thus, there is no consistency in terms of estimating I, even in the matched case. In the unmatched case, of course, there is also no consistency. But, in spite of these, the CMEP of the mixed effect of interest is consistent, which is all we care about here. To see why this can ever happen, note that the way we identify I is to find the index i whose corresponding EBP for the mixed effect, θ, is closest to the "observed" θ, which is $\bar{y}_n$ [see the remark following (2.73)]. So, even if the class membership is estimated wrong, the corresponding EBP is still the closest to θ, in a certain sense, which is most important according to our current interest. Furthermore, Theorem 2.10 shows that CMMP is outperforming RP in terms of MSPE even if there is no match. In the following we support the theory with some empirical results.

Example 2.15. (Simulation study) we consider a situation where there may or may not be matches between the group of the new observation(s) and one of the groups in the training data, and compare these two cases in terms of prediction of the mixed effects associated with the new observations. The simulation study was carried out under the following model:

$$y_{ij} \;=\; 1 + 2x_{1,ij} + 3x_{2,ij} + \alpha_i + \varepsilon_{ij},$$

$i = 1, \ldots, m$, $j = 1, \ldots, n$, with $n = 5$, $\alpha_i \sim N(0, G)$, $\varepsilon_{ij} \sim N(0, 1)$, and α_i's, ε_{ij}'s are independent. The $x_{k,ij}, k = 1, 2$ were generated from the $N(0, 1)$ distribution, then fixed throughout the simulation. There are $K = 10$ new observations, generated under two scenarios. Scenario I: The new observations have the same α_i as the first K groups in the training data ($K \leq m$), but independent ε's; that is, they have "matches". Scenario II: The new observations have independent α's and ε's; that is, they are "unmatched." Note that there are K different mixed effects. We use CMMP (and RP) to predict each of the K mixed effects, and report the average of the simulated MSPEs, obtained based on $T = 1000$ simulation runs. We considered $m = 10$ and $m = 50$. The results are presented in Table 2.3.

It appears that, regardless of whether the new observations actually have matches or not, CMMP match them anyway. More importantly, the results show that even a "fake" match still helps. At first, this might sound a little surprising, but it actually makes sense, both practically and theoretically. Think about a business situation. Even if one cannot find a perfect match for a customer, but if one can find a group that is kind of similar, one can still gain in terms of prediction accuracy. This is, in fact, how business decisions are often made. In the simulation study, even if there is no match in terms of the individual random effects, there is at least a "match" in terms of the random effects distribution, that is, the new random effect is generated from the same distribution that has generated the (previous) training-data random effects; therefore, it is not surprising that one can find one among the latter that is close to the new random effect. Comparing RP with CMMP, PR assumes that the mixed effect is $x_i'\beta$ with nothing extra. On the other hand, CMMP assumes that the mixed effect

Table 2.3 *Average MSPE for Prediction of Mixed Effects.*

Scenario		σ_α^2	0.1	1	2	3
$m = 10$	I	RP	0.157	1.002	1.940	2.878
	I	CMMP	0.206	0.653	0.774	0.836
	I	%MATCH	91.5	94.6	93.6	93.2
	II	RP	0.176	1.189	2.314	3.439
	II	CMMP	0.225	0.765	0.992	1.147
	II	%MATCH	91.2	94.1	92.6	92.5
$m = 50$	I	RP	0.112	1.013	2.014	3.016
	I	CMMP	0.193	0.799	0.897	0.930
	I	%MATCH	98.7	98.5	98.6	98.2
	II	RP	0.113	1.025	2.038	3.050
	II	CMMP	0.195	0.800	0.909	0.954
	II	%MATCH	98.8	98.7	98.4	98.4

%MATCH $=$% of times that the new observations were matched to some of the groups in the training data.

is $x_i'\beta$ plus something extra. For the new observation, there is, for sure, something extra, so CMMP is right, at least, in that the extra is non-zero; it then selects the best extra from a number of choices, some of which are better than the zero extra that PR is using. Therefore, it is not surprising that CMMP is doing better than PR, regardless of the actual match (which may or may not exist), because one is comparing "doing something" with "doing nothing."

2.5 Linear mixed model diagnostics

So far we have studied asymptotic analysis of LMM, sometimes with the normality assumption (e.g., Section 2.1 and part of Section 2.2.1) and sometimes without (e.g., Sections 2.2.2, 2.2.3). This makes one wonder, perhaps, about the role that normality plays in the analysis of LMM. More specifically, exactly for what types of analyses the normality assumption is important, and for what types it is not? Unfortunately, there is no simple answer to this question (this is one of those things that one has to accept: sometimes, life is easier; and sometimes it is not). For example, without the normality assumption, one can still use the REML estimators of the fixed effects and variance components as point estimators, and expect them to be accurate when the sample size is large (in a meaningful way; see, e.g., Theorem 2.1). In fact, some of the large-sample tests are still valid (see the next section). On the other hand, measures of uncertainty, such as the standard errors (s.e.), may be affected (see Section 2.3). As another example, the Kackar-Harville decomposition, (2.50), for the EBLUP and expression (2.52) depend heavily on the normality assumption. So, if the latter fails, the MSPE estimator (2.55) may not be second-order unbiased.

This brings up an issue about checking the normality assumption. For example, if one is confident about the assumption, one would feel comfortable using the Gaussian s.e. for the variance component estimator, which is much simpler (see Section 2.3), or the MSPE estimator (2.55) for the EBLUP. Checking the (normality) assumption may be done informally, or formally, as discussed in [6] (Chapter 12). In the context of LMM, Lange and Ryan [69] proposed to use the EBLUP of the random effects for informal checking of the normality of the random effects. Also see [70]. This approach may be supported by the work of [62], under some rather restrictive conditions (see the discussion below Example 2.13 in Section 2.4.1). As for the formal checking, one way to do this is via a goodness-of-fit test.

We can formulate the testing problem under a more general setting. Consider a non-Gaussian mixed ANOVA model, as (2.1) and (2.3). Let $F_r(\cdot|\sigma_r)$ denote the true distribution of the components of α_r (which are assumed i.i.d.), which depends on the standard deviation (s.d.) σ_r, $1 \leq r \leq s$. Similarly, let $G(\cdot|\tau)$ denote the true distribution of the components of ε, which depends on the s.d. τ. Here the F_r's and G are assumed to be standardized distributions with mean equal to 0 and variance equal to 1. The null hypothesis may be expressed as

$$H_0 : \quad F_r(\cdot|\sigma_r) = F_{0r}(\cdot|\sigma_r), \ 1 \leq r \leq s \ \text{ and } \ G(\cdot|\tau) = G_0(\cdot|\tau), \qquad (2.77)$$

where F_{0r}, $1 \leq r \leq s$ and G_0 are known distributions. In the special case of testing for normality, all of the F_{0r}'s and G are the standard normal distribution.

There is a rich history on goodness-of-fit tests in the statistical literature. Perhaps, the most famous of all, is Pearson's χ^2-test. Let $Y_1, \ldots, Y_n$ be i.i.d. observations from an unknown distribution, F. Let $E_1, \ldots, E_K$ be partition of the real line, that is, disjoint intervals. To test the hypothesis $H_0 : F = F_0$, where F_0 is a specified distribution, Pearson suggested to use

$$\hat{\chi}_P^2 \quad = \quad \sum_{k=1}^{K} \frac{\{N_k - E_0(N_k)\}^2}{E_0(N_k)} \qquad (2.78)$$

as the test-statistic, where $N_k = \sum_{i=1}^{n} 1_{(Y_i \in E_k)} = \#\{1 \leq i \leq n : Y_i \in E_k\}$ is the observed count, or frequency, for the cell E_k, and $E_0(N_k) = nP_0(Y_1 \in E_k)$ is the expected frequency under H_0. Here it is assumed that the cell probabilities, $p_k = P_0(Y_1 \in E_k)$, $1 \leq k \leq K$ are completely specified under H_0. In such a case, the asymptotic null distribution of $\hat{\chi}_P^2$ is χ_{K-1}^2, the χ^2 distribution with $K - 1$ degrees of freedom (d.f.). In many cases, however, the cell probabilities are not completely specified under the null; instead, some unknown parameters are involved. In such a case, it is customary to replace the unknown parameters by their estimators. The result is Pearson's χ^2-test with estimated cell probabilities. However, this is the case where the asymptotic theory gets complicated.

In a simple problem of assessing the goodness-of-fit to a Poisson or multinomial distribution, it is known that, when the unknown parameters are estimated by the MLE, the asymptotic null distribution of $\hat{\chi}_P^2$ is χ_{K-q-1}^2, where q is the number of estimated (free) parameters. This is the well-known "subtract one degree of freedom for each parameter estimated" rule written in many statistics books (e.g., [71], p. 242). However, an easy mistake could be made here when one tries to extend the "rule", as it does not even hold in the normal case, that is, testing for goodness-of-fit to a normal distribution with unknown mean and variance. R. A. Fisher was the first to note that the asymptotic null distribution of Pearson's χ^2-statistic, $\hat{\chi}_P^2$, with estimated cell probabilities, is not necessarily χ^2 (see [72]). He showed that, if the unknown parameters are estimated by what he called minimum chi-square method, the χ^2_{K-q-1} rule holds, but it may fail if other methods of estimation are used, such as maximum likelihood. In fact, Chernoff and Lehmann [73] showed that, if the MLEs based on the original observations, rather than the cell frequencies, are used, the asymptotic null distribution of $\hat{\chi}_P^2$ is not χ^2; instead, it is a weighted χ^2, in which the weights are eigenvalues of some nonnegative definite matrix. Note that there is no conflict between the latest story and the well-known results about Poisson and multinomial distributions, because in the latter cases the MLEs are, indeed, based on the cell frequencies. All of these stories, and much more, can be found in a nice historical review by Moore [72].

Given the rich history of χ^2 tests, it would be natural to consider a similar approach to testing the null hypothesis (2.79) for LMM. The extension, however, is not straightforward. The first attempt was made by [74], which considered a special case of the mixed ANOVA model, namely, the NER model (see Example 2.6) , and used a data-splitting technique. Namely, the data is divided into two parts; one part is used to estimate the parameters and the other part in the χ^2 test. One concern is that this may reduce the power of the test, because the data may be used less efficiently.

To introduce a unified approach without data-splitting, first note that the denominator in the χ^2-statistic, $E_0(N_k)$ was used in Pearson's original χ^2 test to normalizing the numerator so that the result has a χ^2 asymptotic null distribution. However, as discussed earlier, the situations where the χ^2-statistic indeed has a χ^2 asymptotic null distribution is rather limited. In fact, it can be shown that, in general, the asymptotic null distribution may never be χ^2, no matter what normalizing constants may be used in place of $E_0(N_k)$ [including $E_0(N_k)$]. Due to this consideration, it would be simpler to use something simpler, as long as it has the right order. Note that $E_0(N_k) = nP_0(Y_1 \in E_k)$ in the i.i.d. case. Thus, for simplicity, we consider

$$\hat{\chi}^2 = \frac{1}{N} \sum_{k=1}^{K} \{N_k - E_{\hat{\psi}}(N_k)\}^2\}^2, \qquad (2.79)$$

where N is a normalizing sequence of constants, $\hat{\psi}$ is the REML estimator of $\psi = (\beta', \tau, \sigma_1, \ldots, \sigma_s)'$, and E_ψ denotes expectation under the non-Gaussian mixed ANOVA model, and the null hypothesis (2.79), when ψ is the true parameters. Note that the distribution of y is completely specified under (2.1), (2.3), and (2.79).

Before stating a general result, let us first look at a simple case, in which there is no random effects. In other words, only the distribution G appears in (2.79), hence $\psi = (\beta', \tau')'$. The model can be expressed as $y_i = x_i'\beta + \varepsilon_i$, where the ε_i's are i.i.d. $\sim G(\cdot|\tau)$. This is a linear regression model with an unknown error distribution. In this case, the normalizing sequence, N in (2.81), can be chosen as n, the sample size. Write $p_{ik}(\psi) = P_\psi(y_i \in E_k)$, and $p_i(\psi) = [p_{ik}(\psi)]_{1 \leq k \leq K}$. Define

$$h_{n,i} = [1_{(y_i \in E_k)} - p_{ik}(\psi)]_{1 \leq k \leq K} - \left(\sum_{j=1}^{n} \frac{\partial p_j}{\partial \beta'}\right)(X'X)^{-1}x_i\varepsilon_i$$

$$-\frac{1-x_i'(X'X)^{-1}x_i}{n-p}\left(\sum_{j=1}^{n}\frac{\partial p_j}{\partial \tau^2}\right)\varepsilon_i^2.$$

Note that, if $E_1=(-\infty,c_1],E_k=(c_{k-1},c_k],1\le k\le K-1,E_K=(c_{K-1},\infty)$, where $c_1<\cdots<c_{K-1}$, and G has the pdf $\tau^{-1}g(\cdot/\tau)$, then, we have

$$\frac{\partial p_{jk}}{\partial \beta_l}=\frac{x_{jl}}{\tau}\left\{g\left(\frac{c_{k-1}-x_j'\beta}{\tau}\right)-g\left(\frac{c_k-x_j'\beta}{\tau}\right)\right\},\quad 1\le l\le p,$$

$$\frac{\partial p_{jk}}{\partial \tau^2}=\frac{1}{2\tau^3}\left\{(c_{k-1}-x_j'\beta)g\left(\frac{c_{k-1}-x_j'\beta}{\tau}\right)-(c_k-x_j'\beta)g\left(\frac{c_k-x_j'\beta}{\tau}\right)\right\},$$

where β_l (x_{jl}) is the lth component of β (x_j), and p is the dimension of β. Here, for convenience, c_0 and c_K are understood as $-\infty$ and ∞, respectively; and $g(u)=ug(u)=0$ when $u=-\infty$ or ∞. Note that $h_{n,i}$ is a $K\times 1$ random vector. Define $\Sigma_n=n^{-1}\sum_{i=1}^{n}\mathrm{Var}(h_{n,i})$. It can be shown that, under regularity conditions, one has $\Sigma_n\to\Sigma$ as $n\to\infty$, for some constant matrix Σ (which is nonnegative definite). Furthermore, if we let $\lambda_1\ge\cdots\ge\lambda_K$ be the eigenvalues of Σ, then, one has $\hat{\chi}^2\xrightarrow{d}\sum_{k=1}^{K}\lambda_k\xi_k^2$, where $\xi_1,\ldots,\xi_K$ are independent $N(0,1)$ random variables. It is seen that, as in [73], the asymptotic distribution is a weighted χ^2, where the weights are eigenvalues of some nonnegative definite matrix. If, in particular, the latter is idempotent (e.g., [2], p. 235), then the eigenvalues are either 0 or 1; thus, in this case, the asymptotic distribution reduces to χ^2, with the d.f. equal to the number of nonzero eigenvalues.

A practical issue regarding the asymptotic distribution is how to evaluate the λ_k's. Let $\hat{\Sigma}_n$ be Σ_n with ψ replaced by $\hat{\psi}=(\hat{\beta}',\hat{\sigma})'$, the REML estimator. Note that $\hat{\beta}$ is the same as the LS estimator, and $\hat{\tau}^2$ is the same as the standard estimator of the error variance in the regression [same as the $\hat{\sigma}^2$ above (1.9)]. Let $\hat{\lambda}_1\ge\cdots\ge\hat{\lambda}_K$ be the eigenvalues of $\hat{\Sigma}_n$. By Weyl's perturbation theorem (e.g., [75], p. 63), we have

$$\max_{1\le k\le K}|\hat{\lambda}_k-\lambda_k|\le\|\hat{\Sigma}_n-\Sigma\|$$

$$\le\|\hat{\Sigma}_n-\Sigma_n\|+\|\Sigma_n-\Sigma\|,\tag{2.80}$$

where $\|\cdot\|$ denotes the spectral norm of matrix. By consistency of the REML estimator, and convergence of Σ_n to Σ, one may argue that both terms on the right side of (2.82) go to zero; hence, so does the left side. Inequality (2.80) ensures that one can approximate the λ_k's by the corresponding $\hat{\lambda}_k$'s; hence, the null distribution is approximated by that of $\sum_{k=1}^{K}\hat{\lambda}_k\xi_k^2$ (the latter is considered a random variable only in terms of the ξ_k's, because the $\hat{\lambda}_k$'s are realized values given the data). The critical values of the latter can be obtained by Monte-Carlo simulation with $\xi_1,\ldots,\xi_K$ generated independently from the $N(0,1)$ distribution.

To generalize the above result, we need to define the corresponding Σ_n under the (non-Gaussian) mixed ANOVA model (2.1) and (2.3). The rest, including obtaining the corresponding $\hat{\Sigma}_n$, and approximating the eigenvalues and critical values, are essentially the same. Recall the notation $V_r(\gamma),p_r,0\le r\le s,b(\gamma),\omega_l,1\le l\le m+n,\mathcal{J}$ introduced in Section 2.2.2. Let $V_\gamma=I_n+\sum_{r=1}^{s}\gamma_rZ_rZ_r'$, and $P=\mathrm{diag}(p_0,p_1,\ldots,p_s)$.

Suppose that each row of Z_r has exactly one nonzero component, denoted by $z_{r,i,i_{(r)}}$ for the ith row, where $i_{(r)}$ is an index between 1 and m_r, so that the ith row of $Z\alpha == \sum_{r=1}^{s} Z_r \alpha_r$ is $\sum_{r=1}^{s} z_{r,i,i,i_{(r)}} \alpha_{r,i_{(r)}}$. Let $\eta_i = x_i'\beta + \sum_{r=1}^{s} z_{r,i,i,i_{(r)}} \alpha_{r,i_{(r)}}$, where x_i' is the ith row of X, and $\eta_i(u)$ be η_i with $\alpha_{r,i_{(r)}} = u_r, 1 \leq r \leq s$ for $u = (u_r)_{1 \leq r \leq s}$. Define $P_i(\psi)$ as the $K \times (p+s+1)$ matrix whose (k,r) element is $(\partial/\partial \psi_r)P_\psi(y_i \in E_k) = $

$$\frac{\partial}{\partial \psi_r} \int \cdots \int \{G(c_k - \eta_i(u)|\tau) - G(c_{k-1} - \eta_i(u)|\tau)\}dF_1(u_1|\sigma_1)\cdots dF_s(u_s|\sigma_s),$$

where $\psi_r = \beta_r, 1 \leq r \leq p$, $\psi_{p+1} = \tau^2$, and $\psi_{p+1+s} = \gamma_r$ with $\gamma_r = \sigma_r^2/\tau^2, 1 \leq r \leq s$. Again, for convenience, let $c_0 = -\infty, c_K = \infty, G(-\infty|\tau) = 0, G(\infty|\tau) = 1$. Let $P_i[c,d](\psi)$ denote the $c,\ldots,d$ columns of $P_i(\psi)$. Define

$$\Phi = \mathcal{I}^{-1}P^{-1}\sum_{i=1}^{n}\{P_i[p+1,p+s+1](\psi)\}',$$

$$\Pi = \tau b(\gamma)V_\gamma^{-1}X(X'V_\gamma^{-1}X)^{-1}\sum_{i=1}^{n}\{P_i[1,p](\psi)\}'.$$

Next, we assume that the random effect vectors, $\alpha_1,\ldots,\alpha_s$, correspond to either random main effects, or random interactions, or nested random effects (e.g., [76], p. 1151). Let $\alpha_{[1]}$ denote the combined vector of random main effects. Define $d_k(\eta_i) = P_\psi(y_i \in E_k|\alpha) = G(c_k - \eta_i|\tau) - G(c_{k-1} - \eta_i|\tau)$, and $\xi_{k,i} = E\{d_k(\eta_i)|\alpha_{[1]}\} - E\{d_k(\eta_i)\}$. Without loss of generality, let $\alpha_1,\ldots,\alpha_a$ be the α_r's corresponding to the random main effects. Recall that all of the random effects and errors are indexed as a single sequence ω_l (see Section 2.2.2); thus, each components of an α_r, with $1 \leq r \leq a$, corresponds to an index l between $n+1$ and $L = n + \sum_{r=1}^{a} m_r$. Define

$$h_{n,l} = \left[\sum_{i \in S_l}E(\xi_{k,i}|\omega_l)\right]_{1 \leq k \leq K} 1_{(n+1 \leq l \leq L)} - \omega_l\Pi_l - \omega_l^2 \sum_{r=0}^{s} \frac{V_{r,ll}(\gamma)}{\tau^{21}{}_{(r=0)}p_r}\Phi_r,$$

$1 \leq l \leq n+m$, where Φ_r' is the rth row of Φ, Π_l' is the lth row of Π, $V_{r,ll'}(\gamma)$ is the (l,l') element of $V_r(\gamma)$, and S_l is the set of index i that correspond to the same ω_l (recall $m = \sum_{r=1}^{s} m_r$). For notation simplicity, write $h_l = h_{n,l}$, and $H = \sum_{l=1}^{n+m} \text{Var}(h_l)$. Similarly, let $\mathcal{R}$ be the $s \times s$ matrix whose (r,r') element is

$$\left[\tau^{2\{1_{(r=0)}+1_{(r'=0)}\}}p_r p_{r'}\right]^{-1}\sum_{l=1}^{n+m}V_{r,ll}(\gamma)V_{r',ll}(\gamma).$$

Let N be a sequence of normalizing constants such that

$$\Sigma_n = N^{-1}\{H + 2\Phi'(\mathcal{I} - \mathcal{R})\Phi\} \longrightarrow \Sigma. \qquad (2.81)$$

Let $\lambda_1 \geq \cdots \geq \lambda_K$ be the eigenvalues of Σ. Jiang ([76]) showed that, under some regularity conditions, one has $\hat{\chi}^2 \xrightarrow{d} \sum_{k=1}^{K} \lambda_k \xi_k^2$, where $\xi_1,\ldots,\xi_K$ are independent $N(0,1)$ random variables. Furthermore, the critical value of the asymptotic distribution can be approximated the same way as in the previous (simple) case.

Notes: It can be shown that $\mathscr{I} - \mathscr{R}$ is nonnegative definite. Thus, compared to the previous (simple) case, the additional term, $2\Phi'(\mathscr{I} - \mathscr{R})\Phi$, in (2.83) corresponds to the extra variation, or uncertainty, due to the variance components estimation (there was no such an issue in the simple case, because no random effect was involved in that case). As for the choice of N, [76] offers some discussion. In particular, it is suggested that $N = n$ in the case of no random effects (i.e, the simple case), and $N = m$ in the case of a single random-effect factor.

Although the asymptotic distribution of $\hat{\chi}^2$ is obtained, it is not simple for the case of more than one random effect factors; it also depends on the choice of the normalizing constant, N, for which no explicity choice was given except for the two simple cases noted above. For such a reason, [76] suggests an alternative method to compute the critical values via *bootstrapping* . Namely, under the null hypothesis, one can estimate the model parameters, including the fixed effects and variance components. One then treat the estimated parameters as the true parameters, and generate data under the null hypothesis. For each generated data set, one compute the test statistic $\hat{\chi}^2$, and obtain the percentiles of the simulated $\hat{\chi}^2$, which are used as the p-values. It is not uncommon, in the literature, to use bootstrap methods as alternatives to the large sample results, when the latter is not simple to use. This raises a question on the role of the asymptotic results. However, there are aspects of large sample techniques that cannot be replaced by bootstrap. For example, a fundamental question regarding the inference is whether there is a "truth"; in other words, do we expect the result to stabilize when the sample size is large. The question cannot be answered by bootstrapping, which is operated under a fixed sample size. Having asked the question about the truth, the next question is "what is the truth?" The asymptotic result shows that the limiting distribution is not a standard one, such as χ^2. Therefore, it is reasonable to use Monte-Carlo method to compute the critical values. A related issue is that, usually, there is a lack of mathematical expressions in the bootstrap approach. As a result, it is difficult to see how a quantity of interest, e.g., the asymptotic distribution, is affected by different factors. In this regard, large sample techniques are often more useful in suggesting "directions of improvements" ([1], Preface).

More recently, there have been some renewed interests in LMM diagnostics. In particular, Claeskens and Hart ([77]) proposes an alternative approach to the χ^2 test for checking the normality assumption in LMM. The authors showed that, in some cases, the χ^2-test is not sensitive to certain departures from the null distribution; as a result, the test may have low power against certain alternatives. As a new approach, the authors considered a class of distributions that include the normal distribution as a reduced, special case. Then test is based on the likelihood-ratio that compares the "estimated distribution" and the null distribution (i.e., normal). Here the estimated distribution is under an identified member of the class of distributions. For the latter, [77] considers Hermite expansion in the form of

$$f_U(u) \;=\; \phi(u)\{1 + \kappa_3 H_3(u) + \kappa_4 H_4(u) + \cdots\}, \tag{2.82}$$

where $\phi(\cdot)$ is the $N(0,1)$ pdf; $\kappa_3, \kappa_4, \ldots$ are related to the cumulants of U and the Hermite polynomilas, H_j, satisfy $H_j(u)\phi(u) = (-1)^j(d^j\phi/du^j)$. In particular, we have

$H_3(u) = u^3 - 3u$, $H_4(u) = u^4 - 6u^2 + 3$. In practice, a truncated version of (2.84) is used which is a polynomial of order M. [77] suggests to determine the latter using a model selection approach, namely, the information criteria, such as Akaike's information criterion (AIC; [78]) or Bayesian information criterion (BIC; [79]). In addition to the Hermite expansion class, other classes of distributions are also considered in [77]. The U in (2.84) may correspond to either a random effect or an error. Thus, the estimated distribution in the LRT is obtained under the order M chosen by the information criterion. [77] indicates that a suitably normalized LRT statistic has an asymptotic null distribution in the form of the distribution of $\sup_{r \geq 1} \{2Q_r / r(r+3)$, where $Q_r = \sum_{q=1}^{r} \chi_{q+1}^2$, and $\chi_2^2, \chi_3^2, \ldots$ are independent such that χ_j^2 has a χ^2 distribution with j degrees of freedom, $j \geq 2$.

2.6   Misspecified mixed model analysis

Genome-wide association study (GWAS), which typically refers to examination of associations between up to millions of genetic variants in the genome and certain traits of interest among unrelated individuals, has been very successful for detecting genetic variants that affect complex human traits/diseases in the past 11 years. According to the web resource of GWAS catalog ([80]; **http://www.genome.gov/gwastudies**), as of September 2016, more than 240,000 single-nucleotide polymorphisms (SNPs) have been reported to be associated with at least one trait/disease at the genome-wide significance level (p-value$\leq 5 \times 10^{-8}$), many of which have been validated/replicated in further studies. However, these significantly associated SNPs only account for a small portion of the genetic factors underlying complex human traits/diseases [81]. For example, human height is a highly heritable trait with an estimated heritability of around 80%, that is, 80% of the height variation in the population may be attributed to genetic factors [82]. However, genome-wide significant SNPs (p-value $< 5 \times 10^{-8}$ after Bonferroni correction) together can only explain 16% of variation in height ([83]). This "gap" between the total genetic variation and the variation that can be explained by the identified genetic loci is universal among many complex human traits/diseases and is referred to as the "missing heritability" (e.g., [84], [85]). One possible explanation for the missing heritability is that many SNPs jointly affect the phenotype, while the effect of each SNP is too weak to be detected at the genome-wide significance level. To address this issue, Yang et al. [86] used a linear mixed model (LMM)-based approach to estimate the total amount of human height variance that can be explained by all common SNPs assayed in GWAS. They showed that 45% of the human height variance can be explained by those SNPs, providing compelling evidence for this explanation: A large proportion of the heritability is not "missing", but rather hidden among many weak-effect SNPs. These SNPs may require a much larger sample size to be detected. The LMM-based approach was also applied to analyze many other complex human traits/diseases (e.g., metabolic syndrome traits, [87], and psychiatric disorders, [88]) and similar results have been observed.

Statistically, the heritability estimation based on the GWAS data can be cast as the problem of variance component estimation in high dimensional regression, where the response vector is the phenotypic values and the design matrix is the standardized genotype matrix (to be detailed below). One needs to estimate the residual variance and the variance that can be attributed to all of the variables in the design matrix. In a typical GWAS data set, although there may be many weak-effect SNPs (e.g., $\sim 10^3$) that are associated with the phenotype, they are still only a small portion of the total number SNPs (e.g., $10^5 \sim 10^6$). In other words, using a statistical term, the true underlying model is sparse. However, the LMM-based approach used by [86] assumes that the effects of all the SNPs are non-zero. It follows that the assumed LMM is misspecified. We explain more in detail below.

2.6.1 A misspecified LMM and MMMA

Consider a LMM that can be expressed as

$$y = X\beta + \tilde{Z}\alpha + \varepsilon, \tag{2.83}$$

where y is an $n \times 1$ vector of observations; X is an $n \times q$ matrix of known covariates β is a $q \times 1$ vector of unknown regression coefficients (the fixed effects); $\tilde{Z} = p^{-1/2}Z$, where Z is an $n \times p$ matrix whose entries are random variables. Furthermore, α is a $p \times 1$ vector of random effects that is distributed as $N(0, \sigma_\alpha^2 I_p)$, I_p being the p-dimensional identity matrix, and ε is an $n \times 1$ vector of errors that is distributed as $N(0, \sigma_\varepsilon^2)$, and α, ε, and Z are independent. See Section 6 for discussion regarding the normality assumption about the random effects and errors. The estimation of σ_ε^2 is among the main interests. Without loss of generality, assume that X is full rank.

The above LMM is what we call assumed model. In reality, however, only a subset of the random effects are nonzero. More specifically, we have $\alpha = \{\alpha_{(1)}', 0'\}'$, where $\alpha_{(1)}$ is the vector of the first m components of α ($1 \leq m \leq p$), and 0 is the $(p-m) \times 1$ vector of zeros. Correspondingly, we have $\tilde{Z} = [\tilde{Z}_{(1)} \tilde{Z}_{(2)}]$, where $\tilde{Z}_{(j)} = p^{-1/2}Z_{(j)}, j = 1, 2$, $Z_{(1)}$ is $n \times m$, and $Z_{(2)}$ is $n \times (p-m)$. Therefore, the true LMM can be expressed as

$$y = X\beta + \tilde{Z}_{(1)}\alpha_{(1)} + \varepsilon. \tag{2.84}$$

With respect to the true model (2.86), the assumed model is misspecified. We shall call the latter a misspecified LMM. However, this may not be known to the investigator, who would proceed with the standard mixed model analysis to obtain estimates of the model parameters, based on the assumed model. This is what we referred to as misspecified mixed model analysis (MMMA). Let us focus on the REML method (see Section 2.2). Furthermore, following Section 2.3.3, we consider estimation of σ_ε^2 and the ratio $\gamma = \sigma_\alpha^2/\sigma_\varepsilon^2$. From (2.15), it is easily derived that REML estimator of γ, denoted by $\hat{\gamma}$, is the solution to the equation

$$\frac{y'P_\gamma\tilde{Z}\tilde{Z}'P_\gamma y}{\text{tr}(P_\gamma\tilde{Z}\tilde{Z}')} = \frac{y'P_\gamma^2 y}{\text{tr}(P_\gamma)}, \tag{2.85}$$

where $P_\gamma = V_\gamma^{-1} - V_\gamma^{-1}X(X'V_\gamma^{-1}X)^{-1}X'V_\gamma^{-1}$ with $V_\gamma = I_n + \gamma\tilde{Z}\tilde{Z}'$. Equation (2.87) is combined with another REML equation, which can be expressed as

$$\sigma_{\varepsilon}^2 = \frac{y' P_{\gamma}^2 y}{\text{tr}(P_{\gamma})}, \tag{2.86}$$

to obtain the REML estimator of σ_{ε}^2, namely, $\hat{\sigma}_{\varepsilon}^2 = y' P_{\hat{\gamma}}^2 y / \text{tr}(P_{\hat{\gamma}})$.

A difference between the LMM considered here from those considered in the previous sections is that, here, the design matrix, Z, for the random effects, is considered random. Note that the standard LMM, that is, those considered in the previous sections, is a conditional model, on the X and Z; hence, in particular, the matrix Z is nonrandom. However, this difference is relatively trivial. A more important difference is, as noted, that the LMM is misspecified. Nevertheless, what appears to be striking is that the estimator $\hat{\sigma}_{\varepsilon}^2$ is, still, consistent. On the other hand, the estimator $\hat{\gamma}$ converges in probability to a constant limit, although the limit may not be the true γ. In spite of the inconsistency of $\hat{\gamma}$, when it comes to estimating some important quantities of genetic interest, such as the heritability (see below), REML still provides the right answer. Before presenting any theory, we first illustrate with a numerical example that also highlights the practical relevance of our theoretical study.

Example 2.16. (A numerical illustration) in SNPs are high-density bi-allelic genetic markers. Loosely speaking, each SNP can be considered as a binomial random variable with two trials and the probability of "success" is defined as "allele frequency" in genetics. Accordingly, the genotype for each SNP can be coded as either 0, 1 or 2. In a simulation study (see [89]), we first simulate the allele frequencies for p SNPs, $\{f_1, f_2, \ldots, f_p\}$, from the Uniform$[0.05, 0.5]$ distribution, where f_j is the allele frequency of the j-th SNP. We then simulate the genotype matrix $U \in \{0, 1, 2\}^{n \times p}$, with rows corresponding to the samples/individuals and columns the SNPs. Specifically, for the j-th SNP, the genotype value of each individual is sampled from $\{0, 1, 2\}$ according to probabilities $(1 - f_j)^2$, $2f_j(1 - f_j)$, and f_j^2, respectively. After that, each column of U is standardized to have zero mean and unit variance, and the standardized genotype matrix is denoted as Z. Let $\tilde{Z} = p^{-1/2} Z$. In [86], an LMM was used to describe the relationship between a phenotypic vector y and the standardized genotype matrix $\tilde{Z}$: $y = 1_n \mu + \tilde{Z} \alpha + \varepsilon$ with $\alpha \sim N(0, \sigma_{\alpha}^2 I_p)$ and $\varepsilon \sim N(0, \sigma_{\varepsilon}^2 I_n)$, where 1_n is the $n \times 1$ vector of 1's, μ is an intercept, α is the vector of random effects, I_n is the $n \times n$ identity matrix, and ε is the vector of errors. An important quantity in genetics is *heritability*, defined as the proportion of phenotypic variance explained by all genetic factors. For convenience, we assume that all of the genetic factors have been captured by the SNPs in GWAS. Under this assumption, the heritability can be characterized via the variance components:

$$h^2 = \frac{\sigma_{\alpha}^2}{\sigma_{\alpha}^2 + \sigma_{\varepsilon}^2}. \tag{2.87}$$

Note that the definition of heritability by (2.89) assumes that $\alpha_j \sim N(0, \sigma_{\alpha}^2)$ for all $j \in \{0, 1, 2, \ldots, p\}$. However, in reality, only a subset of the SNPs are associated with the phenotype. A correct model is a special case of (2.86), that is, $y = 1_n \mu + \tilde{Z}_{(1)} \alpha_{(1)} + \varepsilon$ with $\alpha^{(1)} \sim N(0, \sigma_{\alpha}^2 I_m)$, where m is the total number of SNPs that are associated with the phenotype, $\alpha_{(1)}$ is the subvector of α corresponding to the nonzero components that are associated with the SNPs, and $\tilde{Z}_{(1)} = p^{-1/2}Z_{(1)}$, $Z_{(1)}$ being the submatrix of Z corresponding to the associated SNPs. In this case, the heritability should instead be given by

$$h_{\text{true}}^2 = \frac{(m/p)\sigma_\alpha^2}{(m/p)\sigma_\alpha^2 + \sigma_\varepsilon^2}. \tag{2.88}$$

In practice, it is impossible to identify all of the m SNPs due to the limited sample size. Therefore, we carry out MMMA under the assumed LMM, pretending that we do not know which SNPs are associated with the phenotype. This means that we simply use all the SNPs in Z to estimate the variance components, σ_α^2 and σ_ε^2. The estimated heritability is then obtained as

$$\hat{h}^2 = \frac{\hat{\sigma}_\alpha^2}{\hat{\sigma}_\alpha^2 + \hat{\sigma}_\varepsilon^2}. \tag{2.89}$$

In this illustrative simulation, we fixed $n = 2,000$, $p = 20,000$, $\sigma_\varepsilon^2 = 0.4$ and varied m from 10 to 20,000. We also set the variance component $\sigma_\alpha^2 = 0.6p/m$ so that the proportion of phenotypic variance explained by genetic factors $h_{\text{true}}^2 = 0.6$, based on (2.89). We repeated the simulation 100 times. As shown in Figure 2.1, there is almost no bias in the estimated h^2 regardless of the underlying true model, whether it is sparse (i.e., m/p is close to zero) or dense (i.e., m/p is close to one). This suggests that the REML works well in providing unbiased estimator of the heritability despite the model misspecification. [89] developed an asymptotic theory to support what we observe in Figure 2.1. Before the theory is presented, we need some preliminaries about random matrix theory, which is the main tool used to establish the theory.

2.6.2   Preliminaries

A key preliminary is the following celebrated result in random matrix theory (e.g., [90]). Let Z be an $n \times p$ matrix whose entries are i.i.d., complex-valued random variables with mean 0 and variance 1, where $n \to \infty$ as $p \to \infty$ such that $n/p \to \tau$, as in (2.81). We are interested in the asymptotic behavior of the empirical spectral distribution (ESD) of $S = p^{-1}ZZ'$, defined as

$$F^S(x) = \frac{1}{n}\sum_{k=1}^{n} 1_{(\lambda_k \leq x)}, \; x \in R,$$

where $\lambda_1, \ldots, \lambda_n$ are the eigenvalues of S.
  Lemma 2.5. (Marčenko-Pastur law) Suppose that

$$\frac{n}{p} \longrightarrow \tau, \; \frac{m}{p} \longrightarrow \omega, \tag{2.90}$$

Then, as $p \to \infty$, the ESD of S converges a.s. in distribution to the Marčenko-Pastur (M-P) law, F_τ, whose pdf is given by

$$\varphi_\tau(x) = \frac{1}{2\pi\tau x}\sqrt{\{b_+(\tau) - x\}\{x - b_-(\tau)\}},$$

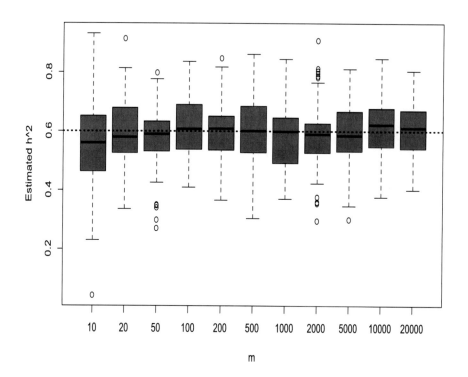

Figure 2.1 *Heritability–REML provides the right answer despite model misspecification.*

if $b_-(\tau) \le x \le b_+(\tau)$, and $\varphi_\tau(x) = 0$ elsewhere, where $b_\pm(\tau) = (1 \pm \sqrt{\tau})^2$.

A result that is frequently referred to is the following which is a consequence of convergence in distribution (e.g., [1], p. 45):

Corollary 2.1. Under the assumptions of Lemma 2,5, we have, for any positive integer l, $n^{-1}\mathrm{tr}(S^l) \xrightarrow{\text{a.s.}} \int_{b_-(\tau)}^{b_+(\tau)} x^l \varphi_\tau(x)dx$ as $p \to \infty$.

The next result ([91], Theorem 2.16) is regarding the extreme eigenvalues of S. Let $\lambda_{\min}(S)$ (respectively, $\lambda_{\max}(S)$) denote the smallest (largest) eigenvalues of S.

Lemma 2.6. Suppose that, in addition to the assumptions of Lemma 2.5, the fourth moment of the entries of Z are finite. Then, we have, as $p \to \infty$, $\lambda_{\min}(S) \xrightarrow{\text{a.s.}} b_-(\tau)$ and $\lambda_{\max}(S) \xrightarrow{\text{a.s.}} b_+(\tau)$.

Let $\xi_1, \dots, \xi_n$ be random variables. We say $\xi = (\xi_i)_{1 \le i \le n}$ is sub-Gaussian if there exists $\sigma > 0$ such that for all $\lambda \in R^n$ we have $E(e^{\lambda' \xi}) \le e^{|\lambda|^2 \sigma^2/2}$. The Gaussian distribution, of course, is a member of the sub-Gaussian class. The following result is due to [92].

Lamma 2.7. A random variable ξ is sub-Gaussian if any of the following equivalent conditions hold:

(I) $E(e^{\xi^2/K_1^2}) < \infty$ for some $0 < K_1 < \infty$;

(II) $\{E(|\xi|^q)\}^{1/q} \leq K_2 \sqrt{q}$ for all $q \geq 1$, for some $0 < K_2 < \infty$.

If, moreover, $E(\xi) = 0$, then the following is equivalent to (I) and (II):

(III) $E(e^{t\xi}) \leq e^{t^2 K_3^2}$ for all $t \in R$, for some $0 < K_3 < \infty$.

Define the sub-Gaussian norm of a random variable ξ as

$$\|\xi\|_{\psi_2} \equiv \sup_{q \geq 1} \left\{ q^{-1/2} (E|\xi|^q)^{1/q} \right\}. \tag{2.91}$$

Clearly, by (II) of Lemma 2.7, ξ is a sub-Gaussian random variable if and only if $|\xi|_{\psi_2} < \infty$. One of the useful characteristics of sub-Gaussianity is that it is preserved under linear combinations. Specifically, we have the following result.

Lemma 2.8. ([92], Lemma 5.9). Suppose that $X_1, \ldots, X_n$ are independent sub-Gaussian random variables, and $b_1, \ldots, b_n \in R$ are nonrandom. Then $\sum_{i=1}^n b_i X_i$ is sub-Gaussian and, for some $C > 0$, we have

$$\left\| \sum_{i=1}^n b_i X_i \right\|_{\psi_2}^2 \leq C \sum_{i=1}^n b_i^2 \|X_i\|_{\psi_2}^2.$$

Lemma 2.8. follows easily from the equivalent characterizations in Lemma 2.7, specifically, by using the moment generating function. The following corollary is very useful for our applications.

Corollary 2.2. Let $X_1, \ldots, X_n$ be independent with $\max_{1 \leq i \leq n} \|X_i\|_{\psi_2} \leq K < \infty$. Then $\sum_{i=1}^n b_i X_i$ is sub-Gaussian and, for some $C > 0$, we have

$$\left\| \sum_{i=1}^n b_i X_i \right\|_{\psi_2}^2 \leq CK^2 \left(\sum_{i=1}^n b_i^2 \right).$$

The following result, due to Rudelson and Vershynin ([93]), is a concentration inequality for quadratic forms involving a random vector with independent sub-Gaussian components. It is referred to as *Hanson-Wright inequality*. For any matrix Q of real entries, the spectral norm of Q is defined as $\|Q\| = \lambda_{\max}^{1/2}(Q'Q)$ and the Euclidean norm is defined as $\|Q\|_2 = \sqrt{\text{tr}(Q'Q)}$.

Proposition 2.1. Let $\xi = (\xi_1, \ldots, \xi_n)'$, where the ξ_i's are independent random variables satisfying $E(\xi_i) = 0$ and $\max_{1 \leq i \leq n} \|\xi_i\|_{\psi_2} \leq K < \infty$. Let Q be an $n \times n$ matrix. Then, for some constant $c > 0$, we have, for any $t > 0$,

$$P\{|\xi'Q\xi - E(\xi'Q\xi)| > t\} \leq 2\exp\left\{ -c\min\left(\frac{t^2}{K^4 \|Q\|_2^2}, \frac{t}{K^2 \|Q\|} \right) \right\}.$$

In the settings that we are interested in, we have $E(\xi_i^2) = 1$ for any i, hence $E(\xi'Q\xi)$ reduces to $\text{tr}(Q)$.

The next result, well known in random matrix theory (e.g., [94], Sections A.5, A.6), is regarding perturbation of the ESD.

Lemma 2.9. For any $n \times p$ matrices A, B we have

(i) $\|F^{AA'} - F^{BB'}\| \leq n^{-1}\text{rank}(A - B)$, where for a real-valued function g on R, $\|g\| = \sup_{x \in R}|g(x)|$;

(ii) $L^4(F^{AA'}, F^{BB'}) \leq 2n^{-2}(\|A\|_2^2 + \|B\|_2^2)\|A - B\|_2^2$, where the Levy distance between two distributions, F and G on R, is defined as $L(F, G) = \inf\{\varepsilon > 0 : F(x - \varepsilon) - \varepsilon \leq G(x) \leq F(x + \varepsilon) + \varepsilon\}$.

The following result is implied by Lemma 2 of Bai and Yin [95].

Lemma 2.10. Suppose that $X_{ij}, i, j = 1, 2, \ldots$ are i.i.d. with $\text{E}(X_{11}^2) < \infty$. Then, we have $\max_{1 \leq j \leq n}|\bar{X}_j - \text{E}(X_{11})| \xrightarrow{\text{a.s.}} 0$, where $\bar{X}_j = n^{-1}\sum_{i=1}^{n} X_{ij}$.

Lemma 2.9 and Lemma 2.10 are used to study the asymptotic ESD of symmetric random matrices involving the standardized design matrix. Note that the standardized design matrix can be expressed as $Z = (U - \bar{u} \otimes 1_n)D_s^{-1}$, where $\bar{u} = (\bar{u}_1, \ldots, \bar{u}_p)$, and $D_s = \text{diag}(s_1, \ldots, s_p)$ (where $\otimes$ denotes the Kronecker product). Let A be the matrix associated with the REML estimation (e.g., [2], p. 13). Consider $\Psi = p^{-1}\zeta\zeta'$, where $\zeta = A'Z$ and A is $n \times (n - q)$ satisfying $A'X = 0$ and $A'A = I_{n-q}$. The following corollary is proved in [89].

Corollary 2.3. Under the assumptions of Lemma 2.5, the ESD of Ψ converges a.s. in distribution to the Marčenko-Pastur law. Furthermore, under the assumptions of Lemma 2.6, $\lambda_{\min}(\Psi)$ and $\lambda_{\max}(\Psi)$ converge a.s. $b_-(\tau)$ and $b_+(\tau)$, respectively.

2.6.3   Consistency and asymptotic distribution

First we state a result regarding the consistency of the misspecified REML estimator of $\sigma_\varepsilon^2, \hat{\sigma}_\varepsilon^2$, and convergence in probability of the misspecified REML estimator of γ, $\hat{\gamma}$. We assume that q, the dimension of β, is fixed, while n, p, and m increase. Furthermore, we consider the case where the design matrix, Z, for the random effects is standardized. Let $U = (u_{ik})_{1 \leq i \leq n, 1 \leq k \leq p}$, whose entries are i.i.d. Define $Z = (z_{ik})_{1 \leq i \leq n, 1 \leq k \leq p}$, where $z_{ik} = (u_{ik} - \bar{u}_k)/s_k$ with $\bar{u}_k = n^{-1}\sum_{i=1}^{n} u_{ik}$ and $s_k^2 = (n-1)\sum_{i=1}^{n}(u_{ik} - \bar{u}_k)^2$. In other words, the new Z matrix has the sample mean equal to 0 and sample variance equal to 1 for each column. We then define $\tilde{Z} = p^{-1}Z$, and proceed as in Section 2.6.1. Also, as noted (see Example 2.16), in GWAS, the entries of U are generated from a discrete distribution which assigns the probabilities $\theta^2, 2\theta(1 - \theta), (1 - \theta)^2$ to the values $0, 1, 2$, where θ is pre-specified so that $\theta \in (0.05, 0.5)$; however, there is also interest in the case where the entries of U are normal. Under the discrete distribution, it makes no difference if we standardize the discrete distribution so that it has mean 0 and variance 1, so, without loss of generality, the entries of U are $u_{ik} = (d_{ik} - \mu)/\sigma$, where d_{ij} has the above discrete distribution, $\mu = \text{E}(d_{ik}) = 2(1 - \theta)$, and $\sigma^2 = \text{var}(u_{ik}) = 2\theta(1 - \theta)$. Both the Gaussian and discrete cases can be treated under the unified framework of sub-Gaussian distribution discussed in the previous subsection.

Theorem 2.11. Suppose that the true $\sigma_\alpha^2, \sigma_\varepsilon^2$ are positive, and (2.81) holds. Then, we have the following results.

(i) With probability tending to one, there is a REML estimator, $\hat{\gamma}$, such that $\hat{\gamma} \xrightarrow{P} \omega\gamma_0$, where γ_0 is the true γ.

(ii) $\hat{\sigma}_\varepsilon^2 \xrightarrow{P} \sigma_{\varepsilon 0}^2$, where $\hat{\sigma}_\varepsilon^2$ is (2.88) with $\gamma = \hat{\gamma}$, as in (i), and $\sigma_{\varepsilon 0}^2$ is the true σ_ε^2.

The proof, given in [89], makes use of the preliminaries in the previous subsection, and an argument that derives a limit using two different methods. Using the first method, the authors showed the existence of the limit using the random matrix theory (see Subsection 2.6.2), but the expression of the limit is too complicated to draw any conclusion needed for the proof. The authors then used a second method to derive the expression of the limit based on convergence of the expectation using the dominated convergence theorem (e.g., [1], p. 32), which resulted in a much simpler result that gave exactly what was needed for the proof of consistency.

Remark. It is interesting to note that the limit of $\hat{\gamma}$ in (i) depends on ω, but not τ. More specifically, the limit is equal to the true γ multiplied by ω, the limiting proportion of the nonzero random effects. The The result seems intuitive. On the other hand, part (ii) of Theorem 2.10 states that the REML estimator of σ_ε^2 is consistent in spite of the model misspecification.

As far as the consistency of $\hat{\sigma}_\varepsilon^2$ is concerned, condition (2.81) can be relaxed. We state this as a corollary.

Corollary 2.4. Suppose that, in Theorem 2.11, condition (2.81) is weakened to

$$\liminf\left(\frac{m \wedge n}{p}\right) > 0, \quad \limsup\left(\frac{m \vee n}{p}\right) \leq 1. \tag{2.92}$$

Then, with probability tending to one, there are REML estimators, $\hat{\gamma}, \hat{\sigma}_\varepsilon^2$, such that
(i) $\hat{\sigma}_\varepsilon^2 \xrightarrow{P} \sigma_{\varepsilon 0}^2$, in other words, the REML estimator of σ_ε^2 is consistent; and
(ii) the adjusted REML estimator of γ is consistent, that is, $(p/m)\hat{\gamma} \xrightarrow{P} \gamma_0$.

The latest asymptotic result may explain what has been observed in Figure 2.1. Note that the estimated heritability, (2.91), can be written as

$$\hat{h}^2 = \frac{(m/p)(p/m)\hat{\gamma}}{1 + (m/p)(p/m)\hat{\gamma}}. \tag{2.93}$$

On the other hand, the true heritability, (2.90), can be written as

$$h_{\text{true}}^2 = \frac{(m/p)\gamma_0}{1 + (m/p)\gamma_0}. \tag{2.94}$$

Because $(p/m)\hat{\gamma}$ converges in probability to γ_0, when we replace the $(p/m)\hat{\gamma}$ in (2.95) by γ_0, the resulting first-order approximation is exactly (2.90). It should also be noted that condition (2.94) requires that the limiting lower bound be positive. This may explain why the bias for $m = 10$ in Figure 2.1 is much more significant compared to the other cases, because the ratio m/p in this case is fairly close to zero.

Another consequence of Theorem 2.11 may be regarded as an extension of the well-known result on consistency of the REML estimator; see Section 2.2. Note that the latter result is based on conditioning on Z, but here the result is without conditioning; in other words, Z is a random matrix.

Corollary 2.5. Suppose that $m = p$, that is, the LMM is correctly specified. Then, as $n, p \to \infty$ such that (2.94) holds with $m \wedge n$ and $m \vee n$ replaced by n, there are REML estimators $\hat{\gamma}$ and $\hat{\sigma}_\varepsilon^2$ such that $\hat{\gamma} \xrightarrow{P} \gamma_0$ and $\hat{\sigma}_\varepsilon^2 \xrightarrow{P} \sigma_{\varepsilon 0}^2$; in other words, the REML estimators are consistent without conditioning on Z.

Given the consistency of $\hat{\sigma}_\varepsilon$, more precise asymptotic behavior of the latter is of interest. The following result establishes asymptotic distribution of the REML estimator of σ_ε^2 as well as that of the adjusted REML estimator of γ. The proof is given by [89].

Theorem 2.12. Suppose that in the assumptions of Theorem 2.11, condition (2.81) is strengthened to

$$\sqrt{n}\left|\frac{n}{p} - \tau\right| \to 0, \quad \sqrt{n}\left|\frac{m}{p} - \omega\right| \to 0, \tag{2.95}$$

and Z has independent sub-Gaussian entries with zero mean, unit variance and bounded sub-Gaussian norm. Then, with $\gamma_* = \omega \gamma_0$, we have

$$\sqrt{n}(\hat{\gamma} - \gamma_*) \xrightarrow{d} N(0, 2\Psi_1), \tag{2.96}$$

$$\sqrt{n}(\hat{\sigma}_\varepsilon^2 - \sigma_{\varepsilon 0}^2) \xrightarrow{d} N(0, 2\sigma_{\varepsilon 0}^4 \Psi_2), \tag{2.97}$$

where Ψ_1, Ψ_2 are functions of γ_*, τ, ω, whose detailed definitions are given in [89].

2.6.4   Further empirical results

To demonstrate our theoretical results numerically, we carry out more comprehensive simulation study following the same procedures as described in Example 2.16. The h^2 was also set at 0.6 ($\sigma_e^2 = 0.4$ and $\gamma = 1.5$). We fix the ratio $\tau = n/p = 0.1$ and varied $\omega = m/p$ from 0.001 to 1. We examine the performance of the REML, under the misspecified LMM, in estimating γ and σ_e^2 as n varies from 1000 to 5000. The performance of the adjusted REML estimator of γ for $\omega = 0.01$ is shown in Figure 2.2. It appears that the adjusted REML always gives nearly unbiased estimate of γ, confirming our observations in Example 2.16 and the theoretical results, namely, part (ii) of Corollary 2.4. More importantly, as both n and p increase (with n/p fixed at 0.1), the standard deviation of the estimate decreases.

Several other methods for high dimensional variance estimation have been proposed recently. As a comparison, we examine the performances of two of these methods, refitted cross validation (c.v.; [96]) and scaled lasso [97], in estimating σ_e^2 under the misspecified LMM. The results for $n = 2000$, $p = 20000$ are shown in Figure 2.3. Again, the REML estimator appears to be unbiased regardless of the value of m. On the other hand, the competing methods tend to have much larger bias, especially when m is large. This is not surprising because the competing methods are largely based on the sparsity assumption that m is relatively small compared to p. Indeed, when $m = 20$, the biases and standard deviations of the competing methods are quite small. In the latter case, the competing method may outperform the REML in terms of mean squared error (MSE). However, the REML performs well consistently across a much broader range of m, as demonstrated by Figure 2.3.

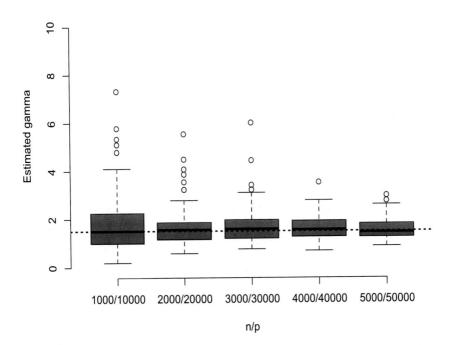

Figure 2.2 *Heritability–REML provides the right answer despite model misspecification.*

We conclude this section with a real-data example. LMM is nowadays commonly used in the genetics community for heritability estimation of complex traits [98], including anthropometric traits [86], metabolic syndrome traits [87], and psychiatric disorders [107–109]. Here we provide a real data example by using LMM to estimate heritability of body mass index (BMI).

We downloaded COGA and SAGE datasets from dbGaP [accession number: phs000125.v1.p1 (COGA) and phs000092.v1.p1 (SAGE)]. First, we remove the duplicated samples in COGA and SAGE. Secondly, we remove samples without height and weight information because BMI is of our interest here. Thirdly, we exclude relatives because these samples can inflate the heritability estimation [86]. As a result, a total of $n = 2294$ individuals from European ancestry remain after these steps. To avoid artifacts from genotyping in our estimation, we apply stringent quality control for the genotype data from these individuals. Specifically, we remove SNPs with a missing rate > 0.01. We test for Hardy-Weinberg Hardy-Weinberg Equilibrium and exclude SNPs with p-value < 0.001. SNPs with minor allele frequency (MAF)

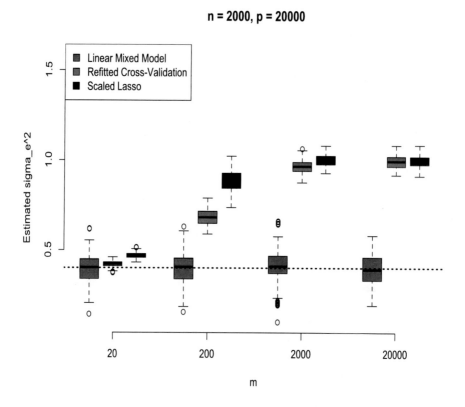

Figure 2.3 *Heritability–REML provides the right answer despite model misspecification.*

$< 5\%$ are also removed to focus on the analysis of common variants. After these quality control steps, $p = 728,000$ SNPs remain for analysis.

After sample and genotype cleaning, we apply the LMM approach to estimate the heritability of BMI. Specifically, we normalize the genotype matrix such that it has zero (sample) mean and unit (sample) variance, for each column, denoted as Z. We then use $\tilde{Z} = p^{-1/2}Z$ as the design matrix for the random effects. As for the matrix X for the fixed effects, we include, in addition to the intercept, the first ten principal component scores computed from $\tilde{Z}\tilde{Z}'$, known as the genetic similarity matrix, to account for the influence of population stratification. It should be noted that, strictly speaking, the X matrix here is not independent with Z. However, this dependence does not affect our asymptotic results (see [89], Remark 3.5), as long as $q = o(\sqrt{n})$. For the current data, $\sqrt{n}$ is about 48 and q is 11, so the condition may be considered satisfied. Another note, from a practical point of view, is regarding the normalization of the genotype matrix so that each SNP had zero mean and unit variance. This is according to the common practice of LMM application to GWAS [86], [99]. Although heritability is not originally defined on the normalized genotypes, heritability estimation based on normalized genotype data explicitly assumes that the genetic variants with lower allele frequencies tend to have larger effect sizes. Speed et al. [100] carefully examined heritability estimation under this assumption and their results suggested that this assumption could give the most stable heritability estimation in the presence of a misspecified distribution of effect sizes. As part of the results, we obtain the REML estimates as $\hat{\sigma}_\alpha^2 = 6.119$ with a standard error (s.e.) of 4.292, and $\hat{\sigma}_\varepsilon^2 = 25.149$ with a s.e. of 4.287, which results in the estimated heritability of $\hat{h}^2 = 19.6\%$ [see (2.91)] with a s.e. of 13.6%.

As a comparison, we use the refitted c.v. method to estimate the residual variance [96]. Specifically, we randomly partition the data into two-groups (with equal sample sizes). We use the first half of the data to select the top $K = \lceil n/\log(n)\rceil = 296$ SNPs, and then estimate the residual variance associated with the second half of the data based on those selected SNPs. We repeat the random partitioning 50 times. The estimated residual variance almost equals to the sample phenotype variance. The result of refitted c.v. suggests that genetics has little contribution to the phenotype, which could further lead to the phenomenon of "missing heritability." However, the result of MMMA suggests that genetic factors can explain a substantial proportion of phenotypic variance. More importantly, the heritability of BMI estimated by LMM (about 16.5%–22.9%) has been replicated based on several independent datasets [99], [101].

2.7   Additional bibliographical notes and open problems

2.7.1   Bibliographical notes

Prior to [44], there had been a few work on asymptotic properties of other types of variance compoments estimators in LMM without assuming normality. In particular, Brown ([102]) proved asymptotic normality of C. R. Rao's MINQUE, as well as the I-MINQUE (e.g., [48], Section 9.1) under replicated error structure (e.g., [103]). Speed [104] proved that for the balanced random effects model (that is $X\beta = \mu 1_n$, where μ is an unknown mean, and the data is balanced), the usual ANOVA estimators of the variance components (e.g., [2], Section 1.5.1) are consistent. Westfall [105] obtained asymptotic normality of the ANOVA estimators of variance components for unbalanced nested mixed ANOVA models.

While most of the literature on asymptotics in LMM has focused on estimation problems, other types of inference have also been considered. Welham and Thompson [106] established asymptotic equivalence of the likelihood ratio, score, and Wald tests in LMM with the normality assumption. On the other hand, Richardson and Welsh [107] considered LRT in LMM without assuming normality. Jiang [108] studied asymptotic distribution of robust versions of the classical tests, that is, the Wald, score and likelihood-ratio tests, with dependent observations, and applied the results to dispersion tests in LMM. Graybill and Wang [109] proposed a method, which they called modified large sample method, that improves Satterthwaite's procedure [110] for constructing an approximate confidence interval for a nonnegative linear combination of the expected sum of squares (SS) corresponding to the fixed or random effects factors in the mixed ANOVA model. Note that many variance components can be expressed in the latter form. See [111] for a review of several approximate procedures for constructing confidence intervals for (not necessarily nonnegative) linear combinations of the expected SS's.

There is an extensive literature on mixed effects model asymptotics associated with small area estimation. For the most part, these involve a special class of mixed effects models with clustered random effects. We shall discuss, in detail, these asymptotics in Chapter 4.

Recently, model selection in LMM has received much attention. See Müller et al. [112] for a review. The latter classifies approaches to LMM selection into three categories: generalized information criteria (GIC), the fence methods, and shrinkage methods. For the most part, there are three types of asymptotic properties associated with a model selection procedure that are usually considered. One is the classical property of consistency in model selection. Namely, the latter means that, as the sample size increases, the probability that the selected model is identical to the true underlying model goes to one. The classical consistency property has been established for GIC (e.g., [113]) and the fence methods (e.g., [114]).

A different type of consistency property, known as *signal-consistency*, was introduced by [115]. Also see [116]. The idea is to consider an "ideal" situation where the "strength" of the parameters–the signals–increases, and show that, in such a case, the probability of selecting the optimal model goes to one. Of course, in real life, one may not be able to increase the signal, but the same can be said regarding increasing the sample size, in real life. The point for signal consistency is as fundamental as consistency in the classical sense, that is, a procedure would work perfectly well, at least, in an ideal situation. The authors argued that signal-consistency is a more reasonable asymptotic property than the classical consistency in a situation where the sample size is (much) smaller than the number of parameters.

Another more modern-type of asymptotic property, called *oracle property* [117], is often considered for shrinkage methods (e.g., [118], [117]).The latter combines model selection with parameter estimation. For a shrinkage procedure, which produces a selected model as well as estimated parameters, the procedure has the oracle property if, with probability tending to one as the sample size increases, the of coefficients that are shrunk to zero is equal to the set of true parameters that are equal to zero; at the same time, the vector of estimated nonzero parameters converges to the corresponding vector of true parameters at the same rate as if the true underlying model is known. Note that the oracle property is usually travial for parameter estimation under the classical setting, where the class of candidate models is finite, which includes a true model. This is because, if a model selection procedure is consistent, then with probability tending to one, the selected model is identical to the true model; thus, from this point on, the parameter estimator based on the selected model is no different from that based on the true model. Therefore, asymptotically, parameter estimator under the selected model behave the same, including rate of convergence, as that under the true model. However, if the dimension of the model increases with the sample size, such as in many of the shrinkage selection problems where the number of parameters is (much) larger than the sample size, the oracle property does not follow trivially from this argument. In the LMM context, the oracle property has been established for joint selection of fixed and random effects in [119], [120], and [121].

2.7.2   Open problems

In [26], Wald consistency was proved for balanced non-Gaussian mixed ANOVA models; an example was also given to show that, in some cases of unbalanced mixed ANOVA models, Wald consistency does not hold (see Section 2.2.3). However, the distribution of the errors in the latter example (see Example 2.9) is not normal; thus, this case is truly non-Gaussian. An interesting question is: Does Wald consistency hold for unbalanced Gaussian mixed ANOVA models? Note that, in the latter case, the REML estimator is what it is supposed to be under the classical definition, that is, the maximizer of the restricted log-likelihood function, rather than the one under the extended definition above (2.17).

The OBP theory developed in Section 2.4.2, namely, Theorem 2.9, is for the case of known covariance matrices, G and R. As indicated in [66] (p. 745), OBP can be derived, in a similar way, if G is unknown, provided that R is known. However, it is not clear if a similar result to Theorem 2.9 holds in the latter case. If R is also unknown, it is not even clear how to derive the OBP under the model-based MSPE [i.e., under the model (2.56)]; however, [122] derived the OBP under a design-based MSPE. Again, a similar result to Theorem 2.9 is not available in the latter case. On the other hand, empirical results have suggested that OBP performs better when the underlying model is misspecified; furthermore, when the underlying model is correctly specified, EBLUP performs better than OBP, but the overall performances of the two are similar in large sample. Thus, there must exist some version of an extension of Theorem 2.9 when the covariance matrices are unknown.

Also not available is a second-order unbiased MSPE estimator similar to (2.55). A nice feature of the latter MSPE estimator is that it is guaranteed nonnegative, which makes sense. Note that all three terms, g_1, g_2, g_3, on the right side of (2.54) are nonnegative for any θ. Although [66] derived a second-order unbiased MSPE estimator, under the Fay–Herriot model (see Example 2.12), the estimator is not guaranteed nonnegative. The main difficulty comes from the potential model misspecification, which is naturally assumed in the context of OBP. See further detail in Chapter 4.

Even under the assumption that the model is correct, the algebraic derivation of (2.55) is tedious. Such a practice requires considerable analytic skills, and a great deal of patience. More importantly, errors often occur in the process of analytic derivations as well as coding, when efforts are made to write computer programs based on the analytic results. Clearly, such errors, if not found, can have serious consequence, both theoretically and practically. The most difficulties are due to (i) the algebraic computation of the partial derivations, especially the second or higher order ones; and (ii) the collection of the coefficients corresponding to the same terms together in order to obtain an "explicit" analytic expression. As it turns out, these tasks, especially (ii), are the major sources of analytic errors. These are also the major sources of the coding errors, because, in spite of the efforts, the explicity (final) expressions as the results of these tedious analytic derivations are usually compli-

cated, based on which the computer codes will be written. One idea to overcome
such difficulties is to let a computer "do the job." For example, many mathematicians
are familiar with the computational software *Mathematica*. It was built on the earlier idea of Cole and Wolfram [123] called *Symbolic Manipulation Program*. Since
then, many versions of *Mathematica* have been developed. Among its many features,
Mathematica can do symbolic derivations that involve partial derivatives, algebraic
reorganizations and simplifications, no matter how tedious. Furthermore, Code Gen-
eration from *Mathematica* can convert compiled functions in *Mathematica* to C codes
that are ready to use for numerical computation. The result of the symbolic derivation
can also be presented with technical word processing (e.g., Tex) for reports and publications. Thus, from all aspects, *Mathematica* seems to be capable of doing exactly
what we have in mind, that is, obtaining the final expressions and converting them
into computational codes, and, more importantly, doing so in an error-free fashion.
For statistical applications, there is also an interest in converting the *Mathematica*
results to popular statistical software (e.g., R) codes.

Regarding statistical software, another complication that occurs with these software packages has to do with obtaining the s.e. in variance component estimation.
The see why there is a problem, note that, for example, the REML estimator of σ_v^2,
say $\hat{\sigma}_v^2$, is supposed to be nonnegative. It follows that $\sqrt{m}(\hat{\sigma}_v^2 - \sigma_v^2)$ cannot be asymptotically normal, if the true σ_v^2 is zero. Thus, the usual concept about the s.e., such
as the plus/minus two s.e. rule as evidence of significance (of $\sigma_v^2 > 0$), does not apply. This is why the s.e. for variance component estimates are usually not provided
in the standard software packages, because the latter (e.g., R) claim that these s.e.'s
are "misleading." Nevertheless, even if $\sqrt{m}(\hat{\sigma}_v^2 - \sigma_v^2)$ is not asymptotically normal,
it is, still, asymptotically "something", although the something is not as simple as
normal. Asymptotic theory can be established, in a way similar to [124] (also see
[125]), to obtain the something, and implement it in the software packages to assess
significance of the variance components.

Regarding the CMMP method, introduced in Section 2.4.3, the method has been
compared to RP, and shown to perform better, asymptotically (Theorem 2.10). It is
noted at the end of Example 2.15 that the superiority of CMMP over RP is due to
"comparing doing something with doing nothing." It is more challenging, however,
to compare doing something with doing something, and a good candidate for the
latter is mixed model prediction (MMP). In order for CMMP to outperform MMP,
one needs to improve the accuracy of matching, that is, identifying the class I. As
noted (see the remark following Theorem 2.10), one does not need much accuracy
in identifying I when comparing with RP, but likely one will need it when it comes
to comparison with MMP. Note that, so far, covariate information has not been used
in the class identification. This is something that is potentially useful for further
advancing the CMMP method.

It has been observed that finite-sample performance of the χ^2-test (see Section
2.5) is sensitive to the choice of the intervals, $E_1, \ldots, E_K$ (e.g., [76], [126]). Typically,
such a problem is approached by choosing the intervals that maximize the power of
the test at the alternative. The difficulty is: What alternative? For example, if the null
hypothesis states that the distribution of the random effects and errors are normal,

there are infinitely many alternatives. One idea to solve the problem is to consider a "unit departure" from the null to an alternative, and maximize the power at the unit-departure alternative (UDA). For example, under the Hermite expansion class of (2.84), the UDA is clearly given by $f_{UDA}(u) = \phi(u)\{1 + \kappa_3 H_3(u)\}$. It would be interesting to see how the χ^2-test performs, if the intervals (say, equiprobability) are chosen by maximizing the power at the UDA.

Finally, regarding MMMA (see Section 2.6), the work of [89] is an important step towards justifying the practice of MMMA in GWAS, obtaining useful inference methods, and potentially extending the methods to other fields. However, the current work has a number of limitations. First, the i.i.d. assumption about the SNPs, which correspond to the columns of the genotype matrix, Z (before the standardization), is too strong to be practical. For example, such an assumption is unlikely to hold in high-dimensional GWAS where the SNPs are dense; in such a case, it is unrealistic to expect that they are in linkage equilibrium. Although some potential extension beyond the i.i.d. case was discussed in the aforementioned work, the problem deserves much more careful consideration, and justification. Second, the expression of the asymptotic variances in (2.96), (2.97) are too complicated to be practically useful, and they involve the unknown parameter ω, whose consistent estimator is not yet available (although τ can be consistently estimated by n/p).

Chapter 3

Asymptotic Analysis of Generalized Linear Mixed Models

Generalized linear mixed models (GLMMs) has become a popular and very useful class of statistical models. See, for example, [2], [3] for some wide-ranging accounts of GLMMs with theory and applications. Suppose that, given a vector of random effects, α, responses $y_1, \ldots, y_N$ are (conditionally) independent such that the conditional distribution of y_i given α is a member of the exponential family with pdf [or probability mass function (pmf)] given by

$$ f_i(y_i|\alpha) = \exp\left\{ \frac{y_i \xi_i - b(\xi_i)}{a_i(\phi)} + c_i(y_i, \phi) \right\}, \tag{3.1} $$

where $b(\cdot)$, $a_i(\cdot)$, $c_i(\cdot, \cdot)$ are known functions, and ϕ is a dispersion parameter, which in some cases is known. Here, ξ_i is the natural parameter of the exponential family [6], which is associated with the conditional mean, $\mu_i = E(y_i|\alpha)$. The latter, in turn, is associated with a linear predictor

$$ \eta_i = x_i'\beta + z_i'\alpha, \tag{3.2} $$

where x_i and z_i are known vectors and β is a vector of unknown parameters (the fixed effects), through a known link function $g(\cdot)$ such that

$$ g(\mu_i) = \eta_i. \tag{3.3} $$

Furthermore, it is assumed that $\alpha \sim N(0, G)$, where the covariance matrix G may depend on a vector θ of unknown variance components.

As usual, a standard approach to inference about GLMMs would be maximum likelihood. However, unlike linear mixed models, the likelihood function under a GLMM typically does not have an analytic expression. In fact, the likelihood function may involve high-dimensional intergrals that are difficult to evaluate even numerically. A infamous example of this type was the salamander data. See Section 1.3. Due to the numerical difficulties of computing the maximum likelihood estimators, some alternative methods of inference have been proposed. One approach is based on Laplace approximation to integrals (e.g., [1], Section 4.6). First note that the likelihood function under the GLMM can be expressed as

$$L_q \;\propto\; |G|^{-1/2} \int \exp\left\{ -\frac{1}{2} \sum_{i=1}^{N} d_i - \frac{1}{2}\alpha' G^{-1}\alpha \right\} d\alpha,$$

where the subscript q indicates quasi-likelihood and

$$d_i \;=\; -2 \int_{y_i}^{\mu_i} \frac{y_i - u}{a_i(\phi)v(u)} du,$$

known as the (quasi-) deviance. What it means is that the method to be described be-low does not require the full specification of the conditional distribution (3.1)—only the first two conditional moments are needed. Here, $v()$ corresponds to the variance function; that is,

$$\mathrm{var}(y_i|\alpha) \;=\; a_i(\phi)v(\mu_i) \tag{3.4}$$

and $\mu_i = \mathrm{E}(y_i|\alpha)$. In particular, if the underlying conditional distribution satisfies (3.1), then L_q is the true likelihood. Using Laplace approximation, one obtains an approximation to the logarithm of L_q:

$$l_q \;\approx\; c - \frac{1}{2}\log|G| - \frac{1}{2}\log|q''(\tilde{\alpha})| - q(\tilde{\alpha}), \tag{3.5}$$

where c does not depend on the parameters,

$$q(\alpha) \;=\; \frac{1}{2}\left(\sum_{i=1}^{N} d_i + \alpha' G^{-1}\alpha \right),$$

and $\tilde{\alpha}$ minimizes $q(\alpha)$. Typically, $\tilde{\alpha}$ is the solution to the equation

$$q'(\alpha) = G^{-1}\alpha - \sum_{i=1}^{N} \frac{y_i - \mu_i}{a_i(\phi)v(\mu_i)g'(\mu_i)} z_i \;=\; 0,$$

where $\mu_i = x_i'\beta + z_i'\alpha$. It can be shown that

$$q''(\alpha) \;=\; G^{-1} + \sum_{i=1}^{N} \frac{z_i z_i'}{a_i(\phi)v(\mu_i)\{g'(\mu_i)\}^2} + r, \tag{3.6}$$

where the remainder term r has expectation 0. If we denote the term in the denomi-nator of (3.6) by w_i^{-1} and ignore the term r, assuming that it is in probability of lower order ([1], Section 3.4) than the leading terms, then we have a further approximation

$$q''(\alpha) \;\approx\; Z'WZ + G^{-1},$$

where Z is the matrix whose ith row is z_i', and $W = \mathrm{diag}(w_1,\dots,w_N)$. Note that the quantity w_i is known as the GLM iterated weights (e.g., [6], Section 2.5). By combining approximations (3.5) and (3.6), one obtains

$$l_q \approx c - \frac{1}{2}\left(\log|I + Z'WZG| + \sum_{i=1}^{N} \tilde{d}_i + \tilde{\alpha}'G^{-1}\tilde{\alpha}\right), \tag{3.7}$$

where $\tilde{d}_i$ is d_i with α replaced by $\tilde{\alpha}$. A further approximation may be obtained by assuming that the GLM iterated weights vary slowly as a function of the mean. Then because the first term inside the $(\cdots)$ in (3.7) depends on β only through W, one may ignore this term and thus approximate l_q by

$$l_{pq} \approx c - \frac{1}{2}\left(\sum_{i=1}^{N} \tilde{d}_i + \tilde{\alpha}'G^{-1}\tilde{\alpha}\right). \tag{3.8}$$

Approximation (3.8) was first derived by [14], who called the procedure penalized quasilikelihood (PQL) by making a connection to the PQL of Green [127].

Let θ denote the vector of variance components. So far in the derivation of PQL we have held θ fixed. Thus, the PQL estimator of β, which is the maximizer of l_{pq}, depends on θ. Breslow and Clayton [14] proposes to substitute this "estimator" back to (3.7) to obtain a profile quasi-log-likelihood function. The authors suggested further approximations that led to a similar form of REML in linear mixed models. See [14] for details.

It is clear that a number of approximations are involved in PQL. If the approximated log-likelihood, l_{pq}, is used in place of the true log-likelihood, one wants to know if the resulting estimator would preserve the desirable properties of the MLE. In particular, a fundamental question is whether the PQL estimator is consistent. The answer is no, as we explain in the next section.

3.1 Asymptotic deficiency of PQL

Consider the following mixed logistic model. Suppose that, given the random effects $\alpha = (\alpha_i)_{1 \le i \le m}$, binary responses $y_{ij}, 1 \le i \le m, 1 \le j \le n_i$ are conditionally independent such that

$$\text{logit}\{P(y_{ij} = 1|\alpha)\} = x_{ij}'\beta + \alpha_i,$$

where $x_{ij} = (x_{ijk})_{1 \le k \le p}$ is a vector of covariates, and $\beta = (\beta_k)_{1 \le k \le p}$, a vector of unknown fixed effects. It is assumed that $\alpha_1, \ldots, \alpha_m$ are independent and distributed as $N(0, \sigma^2)$. For simplicity, let us assume that σ^2 is known and that $n_i, 1 \le i \le m$, are bounded. The x_{ijk}'s are assumed to be bounded as well.

Let $\phi_i(t, \beta)$ denote the unique solution u to the following equation:

$$\frac{u}{\sigma^2} + \sum_{j=1}^{n_i} h(x_{ij}'\beta + u) = t, \tag{3.9}$$

where $h(x) = e^x/(1+e^x)$. Then, the PQL estimator of β is the solution to the following equation:

$$\sum_{i=1}^{m}\sum_{j=1}^{n_i}\{y_{ij} - h(x'_{ij}\beta + \tilde{\alpha}_i)\}x_{ijk} = 0, \quad 1 \leq k \leq p, \qquad (3.10)$$

where $\tilde{\alpha}_i = \phi_i(y_{i\cdot}, \beta)$ with $y_{i\cdot} = \sum_{j=1}^{n_i} y_{ij}$. Denote this solution by $\hat{\beta}$.

Suppose that $\hat{\beta}$ is consistent; that is, $\hat{\beta} \xrightarrow{P} \beta$. Hereafter in this section β denotes the true parameter vector. Let $\xi_{i,k}$ denote the inside summation in (3.10); that is, $\xi_{i,k} = \sum_{j=1}^{n_i}\{y_{ij} - h(x'_{ij}\beta + \tilde{\alpha}_i)\}x_{ijk}$. Then, by (3.9),

$$\frac{1}{m}\sum_{i=1}^{m}\xi_{i,k} = \frac{1}{m}\sum_{i=1}^{m}\sum_{j=1}^{n_i}\{\psi_{ij}(\hat{\beta}) - \psi_{ij}(\beta)\},$$

where $\psi_{ij}(\beta) = h(x'_{ij}\beta + \tilde{\alpha})$. Now, by the Taylor expansion, we have

$$\psi_{ij}(\hat{\beta}) - \psi_{ij}(\beta) = \sum_{k=1}^{p}\left.\frac{\psi_{ij}}{\partial\beta_k}\right|_{\tilde{\beta}}(\tilde{\beta}_k - \beta_k),$$

where $\tilde{\beta}$ lies between β and $\hat{\beta}$. It can be derived from (3.9) that

$$\frac{\partial\tilde{\alpha}_i}{\partial\beta_k} = -\frac{\sum_{j=1}^{n_i} h'(x'_{ij}\beta + \tilde{\alpha}_i)x_{ijk}}{\sigma^{-2} + \sum_{j=1}^{n_i} h'(x'_{ij}\beta + \tilde{\alpha}_i)}. \qquad (3.11)$$

Thus, it can be shown that

$$\left|\psi_{ij}(\hat{\beta}) - \psi_{ij}(\beta)\right| \leq \frac{p}{4}\left(1 + \frac{\sigma^2}{4}n_i\right)\left(\max_{i,j,k}|x_{ijk}|\right)\max_{1\leq k\leq p}|\hat{\beta}_k - \beta_k|. \qquad (3.12)$$

It follows that $m^{-1}\sum_{i=1}^{m}\xi_{i,k} \xrightarrow{P} 0$, as $m \to \infty$.

On the other hand, $\xi_{1,k}, \ldots, \xi_{m,k}$ are independent random variables; so it follows by the law of large numbers (LLN) that $m^{-1}\sum_{i=1}^{m}\{\xi_{i,k} - \mathrm{E}(\xi_{i,k})\} \xrightarrow{P} 0$ as $m \to \infty$. If we combine the results, the conclusion is that

$$\frac{1}{m}\sum_{i=1}^{m}\mathrm{E}(\xi_{i,k}) \longrightarrow 0, \quad 1 \leq k \leq p. \qquad (3.13)$$

However, (3.13) is not true, in general. For example, [128] considered a special case in which the limit in (3.13) is a continuous non-zero function.

The contradiction shows that $\hat{\beta}$ cannot be consistent in general. The inconsistency is caused by the bias induced in the approximation to the likelihood that does not vanish as the sample size increases. It might be thought that, perhaps, a higher order Laplace approximation would solve the inconsistency problem. In fact, Lin and Breslow [16] proposed a bias correction to PQL based on second-order Laplace approximation. The second-order approximation reduces the bias of the PQL estimator; however, it does not make the bias vanish asymptotically. In other words, the bias-corrected PQL estimator is still inconsistent. In fact, no matter to what order the Laplace approximation is carried, the bias-corrected PQL estimator is never consistent. Of course, as the Laplace approximation is carried to even higher order, the bias may be reduced to such a level that is acceptable from a practical point of view. On the other hand, a main advantage of PQL is that it is computationally easy to operate. As the Laplace approximation is carried to higher order, the computational difficulty increases; thus, the computational advantage of PQL is diminishing. Note that, if the computation required for an approximate method is comparable to that for the exact (maximum likelihood) method (see Section 3.3), the approximate method may not be worth pursuing.

It may be wondered in what situation the Laplace approximation will lead to consistent estimators. This is equivalent to asking in what situation the Laplace approximation becomes asymptotically accurate. A sufficient condition is that, as the sample size increases, the integrand of the integral that one wishes to approximate becomes more and more concentrate near its maximum. The latter would be the case if, for example, the number of observations associated with each random effect goes to infinity, but this is not very interesting, so far as the random effects models are concerned. In by far most practical situations where a random effects model is used, there is insufficient information for estimating the individual random effects. In other words, if the random effects can be estimated individually, with accuracy, they may be better off treated as fixed effects.

3.2 Generalized estimating equations

We have seen some asymptotic deficiency of the PQL in terms of consistency. Thus, naturally, the focus now is to develop procedures that lead to, at least, consistent estimators. For the most part, there are two main approaches in achieving this goal. The first is to stay with PQL's intention in developing computationally attractive methods, but making sure that the resulting estimator is consistent. This approach may be broadly considered as estimating equations, which we discuss in this section. The second is to continue to develop computational procedure for the MLE, with the assumption that the latter is a consistent estimator. However, the latter assumption is not that obvious, especially for GLMMs involving crossed random effects. This approach will be considered in Section 3.3.

3.2.1 Estimating equations and GEE

An estimating equation is defined through an estimating function, whose general framework was set up by V. P. Godambe some 30 years before the GLMMs [129]. In [130], the author viewed the approach as an extension of the Gauss–Markov theorem. An estimating function is a function, possibly vector valued, that depends both on $y = (y_i)_{1 \leq i \leq N}$, a vector of observations, and ψ, a vector of parameters. Denoted by

$g(y, \psi)$, the estimating function is required to satisfy

$$E_\psi\{g(y, \psi)\} \quad = \quad 0 \qquad (3.14)$$

for every ψ. For simplicity, let us first consider the case that $y_1, \ldots, y_N$ are independent with $E(y_i) = \psi$, a scalar. Let $\mathscr{G}$ denote the class of estimating functions

$$g(y, \psi) \quad = \quad \sum_{i=1}^{N} a_i(\psi)(y_i - \psi),$$

where $a_i(\psi)$ are differentiable functions such that

$$\liminf_{n \to \infty} \frac{1}{N} \sum_{i=1}^{N} a_i(\psi) > 0 \qquad (3.15)$$

for every ψ. Then, an extension of the Gauss–Markov theorem states the following [130]: If $\text{var}(y_i) = \sigma^2$, $1 \le i \le N$, the $g^* = \sum_{i=1}^{N}(y_i - \psi)$ is an optimal estimating function within $\mathscr{G}$ and the equation $g^* = 0$ provides $\bar{y}$, the sample mean, as an estimator of ψ. Here, and throughout this section, the optimality is in the sense that the solution to the estimating equation,

$$g(y, \psi) \quad = \quad 0, \qquad (3.16)$$

has minimum asymptotic variance. In case that ψ is multivariate, the asymptotic covariance matrix will be considered. Let $\xi = (\xi_j)_{1 \le j \le k}$ be a random vector. Then, the covariance matrix of ξ, denoted by $\text{Var}(\xi)$, is the $k \times k$ matrix whose (s,t) element is $\text{cov}(\xi_s, \xi_t)$, $1 \le s,t \le k$. Two symmetric matrices, A, B, satisfy $A \ge B$ is $A - B$ is nonnegative definite. An estimator, $\hat{\psi}$, has the minimum covariance matrix within a class of estimators, $\mathscr{G}$, if $\text{Var}(\tilde{\psi}) \ge \text{Var}(\hat{\psi})$ holds for every $\tilde{\psi} \in \mathscr{G}$. Similarly, $\hat{\psi}$ is said to have the minimum asymptotic covariance matrix if $V(\tilde{\psi}) \ge V(\hat{\psi})$ holds for every $\tilde{\psi} \in \mathscr{G}$, where $V(\tilde{\psi}), V(\hat{\psi})$ denote the asymptotic covariance matrices of $\tilde{\psi}, \hat{\psi}$ (defined in a suitable way), respectively. To see why the above extension of the Gauss-Markov theorem holds, consider a first-order Taylor expansion of the equation

$$0 \quad = \quad \sum_{i=1}^{N} a_i(\hat{\psi})(y_i - \hat{\psi})$$

$$\approx \quad \sum_{i=1}^{N} a_i(\psi)(y_i - \psi) - \left\{ \sum_{i=1}^{N} a_i(\psi) - \sum_{i=1}^{N} \dot{a}_i(\psi)(y_i - \psi) \right\} (\hat{\psi} - \psi)$$

$$\approx \quad \sum_{i=1}^{N} a_i(\psi)(y_i - \psi) - \left\{ \sum_{i=1}^{N} a_i(\psi) \right\} (\hat{\psi} - \psi),$$

where ψ denotes the true parameter, and $a_i(\psi) = \partial a_i / \partial \psi$ evaluated at ψ. In each step of the above approximations, the dropped term is of lower order than the remaining terms, taking into account of (3.15). It follows that

$$\hat{\psi} - \psi \quad \approx \quad \frac{\sum_{i=1}^{N} a_i(\psi)(y_i - \psi)}{\sum_{i=1}^{N} a_i(\psi)}.$$

By taking the variance on both sides of the above approximation, we get

$$\text{var}(\hat{\psi}) \;\approx\; \sigma^2 \frac{\sum_{i=1}^{N} a_i^2(\psi)}{\{\sum_{i=1}^{N} a_i(\psi)\}^2} \;\geq\; \frac{\sigma^2}{N},$$

using the Cauchy-Schwarz inequality, and the equality holds if and only if all of the $a_i(\psi)$'s are equal (e.g., [1], p. 130), in which case one has the estimating equation that is equivalent to $\sum_{i=1}^{N}(y_i - \psi) = 0$, leadig to the sample mean.

Note that, for (3.16) to be an estimating equation, the corresponding estimating function, $g(y,)$, should be as close to zero as possible, if ψ is the true parameter. In view of (3.14), this means that one needs to minimize $\text{var}(g)$. On the other hand, in order to distinguish the true θ from a false one, it makes sense to maximize $\partial g/\partial \psi$, or the absolute value of its expected value. When both are put on the same scale, the two criteria for optimality can be combined by considering

$$\frac{\text{var}(g)}{\{\text{E}(\partial g/\partial \theta)\}^2} \;=\; \text{var}(g_s), \tag{3.17}$$

where $g_s = g/\text{E}(\partial g/\partial \theta)$ is a standardized version of g. Thus, the optimality in the above is also in the sense that

$$\text{var}(g_s^*) \;\leq\; \text{var}(g_s) \quad \text{for any } g \in \mathscr{G}.$$

It should be noted that (3.14) is the key to the consistency of the solution to (3.16), while the optimality part is simply aimed at reducing the variability of the estimator.

Now consider a multivariate version of the estimating function. Let y be a vector of responses that is associated with a vector x of explanatory variables. Here we allow x to be random as well. Suppose that the (conditional) mean of y given x is associated with ψ, a vector of unknown parameters. For notational simplicity, write $\mu = \text{E}_\psi(y|x) = \mu(x, \psi)$, and $V = \text{Var}(y|x)$. Here, again, Var denotes the covariance matrix, and Var or E without subscript ψ means to be taken at the true ψ. Let $\dot{\mu}$ denote the matrix of partial derivatives; that is, $\dot{\mu} = \partial \mu/\partial \psi'$. Consider the following class of vector-valued estimating functions $\mathscr{H} = \{G = B(y - \mu)\}$, where $B = B(x, \psi)$, such that $\text{E}(\dot{G})$ is nonsingular. We have the following theorem.

Theorem 3.1. Suppose that V is known, and that $\text{E}(\dot{\mu}'V^{-1}\dot{\mu})$ is nonsingular. Then, the optimal estimating function within $\mathscr{H}$ is given by $G^* = \dot{\mu}'V^{-1}(y - \mu)$, that is, with $B = B^* = \dot{\mu}'V^{-1}$.

A proof is given in [2], Section 4.5.1. The optimality in Theorem 3.1 can also be derived using a similar argument as above for the simple case. Equivalently, the latter argument leads to minimization of the following generalized information criterion

$$\mathscr{I}(G) \;=\; \{\text{E}(\dot{G})\}'\{\text{E}(GG')\}^{-1}\{\text{E}(\dot{G})\}, \tag{3.18}$$

where $\dot{G} = \partial G/\partial \theta'$. It is easy to see that (3.18) is, indeed, the Fisher information matrix when G is the score function corresponding to a likelihood—that is, $G = \partial l/\partial \theta$, where l is the log-likelihood function—and this provides another view of Godambe's criterion of optimality. Also, (3.18) is equal to the reciprocal of (3.17) in the univariate case, so that maximizing (3.18) is equivalent to minimizing (3.17).

In order to develop estimating equations for GLMM, let us first consider a special case called *longitudinal GLMM*. This means that the GLMM involves clustered random effects such that the responses are correlated within each cluster, but independent between different clusters. For example, in medical studies repeated measures are often collected from different patients over time. If the random effects are patient-specific, and independent for different patients, a longitudinal GLMM may be appropriate, especially if the measurements are discrete or categorical. In fact, this is the main reason why these models are called longitudinal.

For a longitudinal GLMM, the optimal estimating function according to Theorem 3.1 can be expressed as

$$ G^* = \sum_{i=1}^{m} \dot{\mu}_i' V_i^{-1} (y_i - \mu_i), $$

where $y_i = (y_{ij})_{1 \le j \le n_i}$, $\mu_i = E(y_i)$, and $V_i = \text{Var}(y_i)$. Here, the covariates x_i are considered fixed rather than random. The corresponding estimating equation is known as generalized estimating equation, or GEE [9], given by

$$ \sum_{i=1}^{m} \dot{\mu}_i' V_i^{-1} (y_i - \mu_i) = 0. \qquad (3.19) $$

In (3.19), it is assumed that V_i, $1 \le i \le m$ are known because; otherwise, the equation cannot be solved. However, the true V_is are unknown in practice. Note that, under a GLMM, the V_is may depend on a vector of variance components θ in addition to β; that is, $V_i = V_i(\beta, \theta)$, $1 \le i \le m$. If a parametric model, such as GLMM, is not assumed, the V_is may be completely unknown. Liang and Zeger [9] came up with a brilliant idea called "working" covariance matrix. They showed that one can substitute the V_is in (3.19) by any known covariance matrices and, under mild conditions, the resulting estimator, that is, the solution to the GEE, is still a consistent estimator of β. For example, one may use the identity matrices as the working covariance matrices that correspond to a model assuming independent errors with equal variance. The latter model may be wrong, but the impact is only on the efficiency of the resulting estimator. In fact, with any fixed covariance matrices V_i, the GEE (3.19) is unbiased in the sense that the expected value of the left side, evaluated at the same parameters as those involved in the μ_is, is equal to zero. As noted, this is the key for the consistency of the resulting estimator.

Alternatively, one may replace the V_is in (3.19) by their consistent estimators, say, $\hat{V}_i$s. For example, under a GLMM, suppose that θ is replaced by a $\sqrt{m}$-consistent estimator, say, $\hat{\theta}$. The latter means that $\sqrt{m}(\hat{\theta} - \theta)$ is bounded in probability ([1], Section 3.4). Then, under mild conditions, the resulting estimator is asymptotically as efficient as the GEE estimator, that is, solution to (3.19) with the true V_is. This means that $\sqrt{m}(\hat{\beta} - \beta)$ has the same asymptotic covariance matrix as $\sqrt{m}(\tilde{\beta} - \beta)$, where $\hat{\beta}$ is the solution to (3.19) with θ replaced by $\hat{\theta}$, and $\tilde{\beta}$ is the solution to (3.19) with the true V_is (e.g., [9]). Of course, $\tilde{\beta}$ is not an estimator unless the V_is are known. This nice feature of $\hat{\beta}$ is sometimes called oracle property [117]. However, to find a $\sqrt{m}$-consistent estimator one typically needs to assume a parametric model for the

V_is, which increases the risk of model misspecifications. Even under a parametric co-variance model, the $\sqrt{m}$-consistent estimator may not be easy to compute, especially if the model is complicated.

Another strategy to improve the efficiency of GEE was proposed by Qu, Lindsay, and Li [131], known as quadratic inference function (QIF) . This requires a special form for the V_i in (3.19). First, it assumes that $n_i = \ldots = n_m$, that is, the number of observations is the same, denoted by k, for all of the subjects. Secondly, $V_i = A_i^{1/2} R A_i^{1/2}$, where A_i is a diagonal matrix whose diagonal elements are the variances of $y_{ij}, 1 \leq j \leq k$, and R is a working correlation matrix. In other words, only the variances depend on i (and j), but the correlation matrix is the same for all of the subjects. Note that, under a generalized linear model (GLM) [6], the variances are functions of the means, μ_i, which depends only on β; sometimes, an additional dispersion parameter is involved, which does not influence the solution to (3.19). Thus, the focus is to find a good working correlation matrix, R, so that the solution to the GEE is more efficient.

Qu et al. [131] suggests to model R by a class of matrices expressed as

$$R^{-1} = \sum_{h=1}^{H} a_h M_h, \qquad (3.20)$$

where $M_1, \ldots, M_H$ are known base matrices, and $a_1, \ldots, a_H$ are unknown constants. The authors discussed several examples to show how to choose the base matrices. In order to determine the constants $a_1, \ldots, a_N$, the authors considered the following extended score: $g(\beta) = m^{-1} \sum_{i=1}^{m} g_i(\beta)$, where

$$g_i(\beta) = [\dot{\mu}_i' A_i^{1/2} M_h A_i^{1/2} (y_i - \mu_i)]_{1 \leq h \leq H}. \qquad (3.21)$$

The authors then applied an optimality theory developed by Hansen [53] for generalized method of moments, which suggests to consider $\hat{\beta} = \operatorname{argmin}_\beta (g' C^{-1} g)$, where C is a known matrix, and that the optimal C is $C = \operatorname{Var}(g)$. Because $E\{g_i(\beta)\} = 0$ when β is the true parameter vector, an "observed covariance matrix,"

$$\tilde{C} = \frac{1}{m^2} \sum_{i=1}^{m} g_i(\beta) g_i(\beta)', $$

is used to substitute the optimal C. The QIF is then defined as

$$Q(\beta) = g' \tilde{C}^{-1} g. \qquad (3.22)$$

It can be shown that, asymptotically, the minimizer of the QIF is a solution to

$$\{E(\dot{g})\}' \{\operatorname{Var}(g)\}^{-1} g \approx g' \tilde{C}^{-1} g = 0. \qquad (3.23)$$

It follows, by Theorem 3.1, that Equation (3.23) is optimal. On the other hand, it is easy to see that (the left side of) (3.23) is in the form of (3.19), with $V_i = A_i^{1/2} R A_i^{1/2}$, and R^{-1} given by (3.21). Thus, in view of (3.23), the optimal constants, $a_1, \ldots, a_H$, are asymptotically determined as $(a_1, \ldots, a_H)' = \dot{g}' \tilde{C}^{-1}$.

In Section 3.4.3, we extend the QIF idea to joint estimation of fixed and random effects. In the next subsection, we propose another alternative that offers a more robust and computationally attractive solution.

As noted, GEE has been used in the analysis of longitudinal data, in which mean responses, which are associated with β, are often of primary interest. It should be noted that the GEE method is based on marginal models, rather than conditional models. This is because only the marginal means and covariance matrices are involved in GEE; the random effects, for example, are involved in the conditional model, not marginal one. Thus, for example, one cannot estimate the random effects using the GEE (3.19). On the other hand, GEE is applicable to more general models than GLMM. For example, in longitudinal studies there often exists serial correlation among the repeated measures from the same subject. Such a serial correlation may not be taken into account by a GLMM. Note that, under the GLMM assumption, the repeated measures are conditionally independent given the random effects, which means that no additional correlation exists once the values of the random effects are specified. However, this may not be reasonable in some cases.

Example 3.1. Consider the salamander mating example discussed in Section 1.3. McCullagh and Nelder [6] proposed a GLMM for analyzing the data, which involved female and male random effects. The dataset and model have since been extensively studied. However, in most cases it was assumed that a different set of animals (20 for each sex) was used in each mating experiment, although, in reality, the same set of animals was repeatedly used in two of the three mating experiments ([6], Section 14.5). Furthermore, most of the GLMMs used in this context (with the exception of, perhaps, [19]) assumed that no further correlation among the data exists given the random effects. However, the responses in this case should be considered longitudinal, because repeated measures were collected from the same subjects (once in the summer, and once in the autumn of 1986). Therefore, serial correlation may exist among the repeated responses even given the random effects (i.e., the salamanders).

The GEE method is computationally more attractive than maximum likelihood. More important, GEE does not require a full specification of the distribution of the data. In fact, consistency of the GEE estimator only requires correct specification of the mean functions; that is, μ_i, $1 \leq i \leq m$. However, for the estimator to maintain the (asymptotic) optimality (in the sense of Theorem 3.1), the covariance matrices V_i, $1 \leq i \leq m$ also need to be correctly specified, but such assumptions are still much weaker than the full specification of the distribution of y. A key to obtaining an efficient GEE estimator is to consistently estimate the covariance matrices V_is. A standard approach to the latter is parametric models for the variance–covariance structure of the data in order to obtain a $\sqrt{m}$-consistent estimator of θ (see earlier discussion). Such a method is sensitive to model misspecifications, and may be difficult to operate computationally under a GLMM. Below we present a different approach.

3.2.2 *Iterative estimating equations*

We describe the method under a semiparametric regression model and then discuss its application to longitudinal GLMMs. Consider a follow-up study conducted over a set of prespecified visit times $t_1,\ldots,t_b$. Suppose that the responses are collected from subject i at the visit times t_j, $j \in J_i \subset J = \{1,\ldots,b\}$. Let $y_i = (y_{ij})_{j \in J_i}$. Here we allow the visit times to be dependent on the subject. This enables us to include some cases with missing responses. Let $X_{ij} = (X_{ijl})_{1 \le l \le p}$ represent a vector of explanatory variables associated with y_{ij} so that $X_{ij1} = 1$. Write $X_i = (X_{ij})_{j \in J_i} = (X_{ijl})_{j \in J_i, 1 \le l \le p}$. Note that X_i may include both time-dependent and independent covariates so that, without loss of generality, it may be expressed as $X_i = (X_{i1}, X_{i2})$, where X_{i1} does not depend on j (i.e., time) whereas X_{i2} does. We assume that (X_i, Y_i), $i = 1,\ldots,m$ are independent. Furthermore, it is assumed that

$$E(Y_{ij}|X_i) = g_j(X_i, \beta), \qquad (3.24)$$

where β is a $p \times 1$ vector of unknown regression coefficients and $g_j(\cdot, \cdot)$ are fixed functions. We use the notation $\mu_{ij} = E(Y_{ij}|X_i)$ and $\mu_i = (\mu_{ij})_{j \in J_i}$. Note that $\mu_i = E(Y_i|X_i)$. In addition, denote the (conditional) covariance matrix of Y_i given X_i as

$$V_i = \text{Var}(Y_i|X_i), \qquad (3.25)$$

whose (j,k)th element is $v_{ijk} = \text{cov}(Y_{ij}, Y_{ik}|X_i) = E\{(Y_{ij} - \mu_{ij})(Y_{ik} - \mu_{ik})|X_i\}$, $j,k \in J_i$. Note that the dimension of V_i may depend on i. Let $D = \{(j,k) : j,k \in J_i \text{ for some } 1 \le i \le n\}$. Our main interest is to estimate β, the vector of regression coefficients. According to the earlier discussion, if the V_is are known, β may be estimated by the GEE (3.19). On the other hand, if β is known, the covariance matrices V_i can be estimated by the method of moments (MoM), as follows.

Note that for any $(j,k) \in D$, some of the v_{ijk}s may be the same, either by the nature of the data or by the assumptions. Let L_{jk} denote the number of different v_{ijk}s. Suppose that $v_{ijk} = v(j,k,l)$, $i \in I(j,k,l)$, where $I(j,k,l)$ is a subset of $\{1,\ldots,m\}$, $1 \le l \le L_{jk}$. For any $(j,k) \in D$, $1 \le l \le L_{jk}$, define

$$\hat{v}(j,k,l) = \frac{1}{n(j,k,l)} \sum_{i \in I(j,k,l)} \{Y_{ij} - g_j(X_i, \beta)\}\{Y_{ik} - g_k(X_i, \beta)\}, \qquad (3.26)$$

where $n(j,k,l) = |I(j,k,l)|$, the cardinality. Then, define $\hat{V}_i = (\hat{v}_{ijk})_{j,k \in J_i}$, where $\hat{v}_{ijk} = \hat{v}(j,k,l)$, if $i \in I(j,k,l)$.

The main points of the above may be summarized as follows. If the V_is were known, one could estimate β by the GEE; on the other hand, if β were known, one could estimate the V_is by the MM. It is clear that there is a cycle here, which motivates the following iterative procedure. Starting with an initial estimator of β, use (3.26), with β replaced by the initial estimator, to obtain the estimators of the V_is; then use (3.19) to update the estimator of β, and repeat the process. We call such a procedure iterative estimating equations, or IEE . If the procedure converges, the limiting estimator is called the IEE estimator, or IEEE. We consider an example.

Example 3.2. A special case of the semiparametric regression model is the linear model with $E(y_i) = X_i\beta$ and $\text{Var}(y_i) = V_0$, where X_i is a matrix of known covariates, and $V_0 = (v_{qr})_{1 \le q,r \le k}$ is an unknown covariance matrix. Let $y = (y_i)_{1 \le i \le m}$ and assume that y_i, $1 \le i \le m$ are independent. Then, $V = \text{Var}(y) = \text{diag}(V_0,\ldots,V_0)$. For this special case, the IEE can be formulated as follows. If β were known, a simple consistent estimator of V would be

$$\hat{V} = \text{diag}(\hat{V}_0, \dots, \hat{V}_0), \quad \text{where} \quad \hat{V}_0 = \frac{1}{m} \sum_{i=1}^{m} (y_i - X_i\beta)(y_i - X_i\beta)' . \tag{3.27}$$

On the other hand, if V were known, the BLUE of β is given by (2.46). When both β and V are unknown, when iterates between these two steps, starting with $V_0 = I_k$. This procedure is called iterative WLS, or I-WLS [132].

To apply IEE to a longitudinal GLMM, let us denote the responses by y_{ij}, $i = 1, \dots, m$, $j = 1, \dots, n_i$, and let $y_i = (y_{ij})_{1 \le j \le n_i}$. We assume that each y_i is associated with a vector of random effects, α_i, that has dimension d such that

$$g(\mu_{ij}) \quad = \quad x_{ij}'\beta + z_{ij}'\alpha_i, \tag{3.28}$$

where $\mu_{ij} = E(y_{ij}|\alpha_i)$, g is the link function, and x_{ij}, z_{ij} are known vectors. Furthermore, we assume that the responses from different clusters $y_1, \dots, y_m$ are independent. Finally, suppose that

$$\alpha_i \sim f(u|\theta), \tag{3.29}$$

where $f(\cdot|\theta)$ is a d-variate pdf known up to a vector of dispersion parameters θ such that $E_\theta(\alpha_i) = 0$. Let $\psi = (\beta', \theta')'$. Then, we have

$$\begin{aligned}
E(y_{ij}) \quad &= \quad E\{E(y_{ij}|\alpha_i)\} \\
&= \quad E\{h(x_{ij}'\beta + z_{ij}'\alpha_i)\} \\
&= \quad \int h(x_{ij}'\beta + z_{ij}'u)f(u|\theta)du,
\end{aligned}$$

where $h = g^{-1}$. Let $W_i = (X_i\ Z_i)$, where $X_i = (x_{ij}')_{1 \le j \le n_i}$, $Z_i = (z_{ij}')_{1 \le j \le n_i}$. For any vectors $a \in R^p$, $b \in R^d$, define

$$\mu_1(a, b, \psi) \quad = \quad \int h(a'\beta + b'u)f(u|\theta)du.$$

Furthermore, for any $n_i \times p$ matrix A and $n_i \times d$ matrix B, let $C = (A\ B)$, and $g_j(C, \psi) = \mu_1(a_j, b_j, \psi)$, where a_j' and b_j' are the jth rows of A and B, respectively. Then, it is easy to see that

$$E(y_{ij}) \quad = \quad g_j(W_i, \psi). \tag{3.30}$$

It is clear that (3.30) is simply (3.24) with X_i replaced by W_i, and β replaced by ψ. Note that, because W_i is a fixed matrix of covariates, we have $E(y_i|W_{ij}) = E(y_{ij})$. In other words, the longitudinal GLMM satisfies the semiparametric regression model introduced above, hence IEE applies. Again, we consider an example.

Example 3.3. Consider a random-intercept model with binary responses. Let y_{ij} be the response for subject i collected at time t_j. We assume that given a subject-specific random effect α_i, binary responses y_{ij}, $j = 1, \ldots, k$ are conditionally independent with conditional probability $p_{ij} = P(y_{ij} = 1|\alpha_i)$, which satisfies $\text{logit}(p_{ij}) = \beta_0 + \beta_1 t_j + \alpha_i$, where β_0, β_1 are unknown coefficients. Furthermore, we assume that $\alpha_i \sim N(0, \sigma^2)$, where $\sigma > 0$ and is unknown. Let $y_i = (y_{ij})_{1 \leq j \leq k}$. It is assumed that $y_1, \ldots, y_m$ are independent, where m is the number of subjects. It is easy to show that, under the assumed model, one has

$$
\begin{aligned}
E(y_{ij}) &= \int_{-\infty}^{\infty} h(\beta_0 + \beta_1 t_j + \sigma u) f(u) du \\
&\equiv \mu(t_j, \psi),
\end{aligned}
$$

where $h(x) = e^x / (1 + e^x)$, $f(u) = (1/\sqrt{2\pi}) e^{-u^2/2}$, and $\psi = (\beta_0, \beta_1, \sigma)'$. Write $\mu_j = \mu(t_j, \psi)$, and $\mu = (\mu_j)_{1 \leq j \leq k}$. We have

$$
\begin{aligned}
\frac{\partial \mu_j}{\partial \beta_0} &= \int_{-\infty}^{\infty} h'(\beta_0 + \beta_1 t_j + \sigma u) f(u) du, \\
\frac{\partial \mu_j}{\partial \beta_1} &= t_j \int_{-\infty}^{\infty} h'(\beta_0 + \beta_1 t_j + \sigma u) f(u) du, \\
\frac{\partial \mu_j}{\partial \sigma} &= \int_{-\infty}^{\infty} h'(\beta_0 + \beta_1 t_j + \sigma u) u f(u) du.
\end{aligned}
$$

Also, it is easy to see that the y_is have the same (joint) distribution, hence $V_i = \text{Var}(y_i) = V_0$, an unspecified $k \times k$ covariance matrix, $1 \leq i \leq m$. Thus, the GEE equation for estimating ψ is given by

$$
\sum_{i=1}^{m} \dot\mu' V_0^{-1} (y_i - \mu) = 0, \tag{3.31}
$$

provided that V_0 is known. On the other hand, if ψ is known, V_0 can be estimated by the method of moments as follows,

$$
\hat{V}_0 = \frac{1}{m} \sum_{i=1}^{m} (y_i - \mu)(y_i - \mu)'. \tag{3.32}
$$

The IEE procedure then iterates between the two steps when both V_0 and ψ are unknown, starting with $V_0 = I$, the k-dimensional identity matrix.

Note that the mean function, μ_j, in (3.31) and (3.32), is a one-dimensional integral, which can be approximated by a simple Monte Carlo method, namely,

$$
\mu_j \approx \frac{1}{L} \sum_{l=1}^{L} h(\beta_0 + \beta_1 t_j + \sigma \xi_l), \tag{3.33}
$$

where the ξ_l's are independent $N(0,1)$ random variables generated by a computer. Similar approximations can be obtained for $\dot\mu_j$. This idea will be extended later.

Two questions, both of theoretical and practical interest, are: (i) Does the IEE algorithm converge (numerically)? and (ii) if the IEE converges, how does the limit

of the convergence behave asymptotically, as an estimator? The short answers to
these questions are: (i) Yes, not only the IEE converges, it converges linearly (see
below); (ii) Yes, not only the limit of the convergence is a consistent estimator, it is
asymptotically as efficient as the solution to the GEE (3.19), as if the true V_i's are
known. We discuss these results below, and refer further details to [132].

We adapt a term from numerical analysis. An iterative algorithm that results in a
sequence $x^{(k)}$, $k = 1, 2, \ldots$ converges linearly to a limit x^*, if there is $0 < \rho < 1$ such
that $\sup_{k \geq 1} \{|x^{(k)} - x^*|/\rho^k\} < \infty$ (e.g., [133]).

Let $L_1 = \max_{1 \leq i \leq m} \max_{j \in J_i} s_{ij}$ and $s_{ij} = \sup_{|\tilde{\beta} - \beta| \leq \varepsilon_1} |(\partial/\partial\beta)g_j(X_i, \tilde{\beta})|$, β the
true parameter vector, ε_1 is any positive constant, and

$$(\partial/\partial\beta)f(\tilde{\beta}) = (\partial f/\partial\beta)|_{\beta=\tilde{\beta}}.$$

Similarly, let $L_2 = \max_{1 \leq i \leq m} \max_{j \in J_i} w_{ij}$, where

$$w_{ij} = \sup_{|\tilde{\beta} - \beta| \leq \varepsilon_1} \|(\partial^2/\partial\beta\partial\beta')g_j(X_i, \tilde{\beta})\|.$$

Also, let $\mathscr{V} = \{v : \lambda_{\min}(V_i) \geq \lambda_0, \lambda_{\max}(V_i) \leq M_0, 1 \leq i \leq m\}$, where $\lambda_{\min}$ and $\lambda_{\max}$
represent the smallest and largest eigenvalues, respectively, and δ_0 and M_0 are given
positive constants. Note that $\mathscr{V}$ is a nonrandom set. An array of nonnegative definite
matrices $\{A_{m,i}\}$ is bounded from above if $\|A_{m,i}\| \leq c$ for some constant c; the array is
bounded from below if $A_{m,i}^{-1}$ exists and $\|A_{m,i}^{-1}\| \leq c$ for some constant c. We also refer
the notion of O_P, including that for random vectors and matrices, to [1] (Section 3.4).
Let p and R be the dimensions of β and v, respectively. We assume the following.

A1. For any $(j, k) \in D$, the number of different v_{ijk}s is bounded, that is, for each
$(j, k) \in D$, there is a set of numbers $\mathscr{V}_{jk} = \{v(j, k, l), 1 \leq l \leq L_{jk}\}$, where L_{jk} is
bounded, such that $v_{ijk} \in \mathscr{V}_{jk}$ for any $1 \leq i \leq n$ with $j, k \in J_i$.

A2. The functions $g_j(X_i, \beta)$ are twice continuously differentiable with respect to
β; $E(|Y_i|^4)$, $1 \leq i \leq m$ are bounded; and $L_1, L_2, \max_{1 \leq i \leq n}(\|V_i\| \vee \|V_i^{-1}\|)$ are $O_P(1)$.

A3 (Consistency of GEE estimator). For any given V_i, $1 \leq i \leq m$ bounded from
above and below, the GEE equation (3.19) has a unique solution $\hat{\beta}$ that is consistent.

A4 (Differentiability of GEE solution). For any v, the solution to (3.19), $\beta(v)$, is
continuously differentiable with respect to v, and $\sup_{v \in \mathscr{V}} \|\partial\beta/\partial v\| = O_P(1)$.

A5. $n(j, k, l) \to \infty$ for any $1 \leq l \leq L_{jk}$, $(j, k) \in D$, as $m \to \infty$.

Our first result is regarding the numerical convergence of IEE. Let $\hat{\beta}^{(k)}, \hat{v}^{(k)}$ denote the updates of β and v, respectively, at the kth iteration.

Theorem 3.2. Under assumptions A1–A5, we have

$$P(\text{IEE converges}) \quad \to \quad 1$$

as $m \to \infty$. Furthermore, we have

$$P\left[\sup_{k \geq 1}\{|\hat{\beta}^{(k)} - \hat{\beta}^*|/(p\eta)^{k/2}\} < \infty\right] \quad \to \quad 1,$$

$$P\left[\sup_{k \geq 1}\{|\hat{v}^{(k)} - \hat{v}^*|/(R\eta)^{k/2}\} < \infty\right] \quad \to \quad 1$$

as $n \to \infty$ for any $0 < \eta < (p \vee R)^{-1}$, where $(\hat{\beta}^*, \hat{v}^*)$ is the (limiting) IEEE.

Note 1. It is clear that the restriction $\eta < (p \vee R)^{-1}$ is unnecessary (because, for example, $(p\eta_1)^{-k/2} < (p\eta_2)^{-k/2}$ for any $\eta_1 \geq (p \vee R)^{-1} > \eta_2$), but linear convergence would only make sense when $\rho < 1$ (see the definition above).

Note 2. The proof of Theorem 3.2 (see [132]), in fact, demonstrated that for any $\delta > 0$, there are constants $M_{1,\delta}$, $M_{2,\delta}$, and integer m_δ such that, for all $m \geq m_\delta$,

$$\mathrm{P}\left[\sup_{k \geq 1} \left\{ \frac{|\hat{\beta}^{(k)} - \hat{\beta}^*|}{(p\eta)^{k/2}} \right\} \leq M_{1,\delta} \right] > 1 - \delta,$$

$$\mathrm{P}\left[\sup_{k \geq 1} \left\{ \frac{|\hat{v}^{(k)} - \hat{v}^*|}{(R\eta)^{k/2}} \right\} \leq M_{2,\delta} \right] > 1 - \delta.$$

Next, we consider asymptotic behavior of the limiting IEEE. For simplicity, the latter is simply called IEEE. The first result is about consistency.

Theorem 3.3. Under the assumptions of Theorem 3.2, the IEEE is consistent.

To establish the asymptotic efficiency of IEEE, we need to strengthen assumptions A2 and A5 a little. Define $L_{2,0} = \max_{1 \leq i \leq m} \max_{j \in J_i} \|\partial^2 \mu_{ij}/\partial\beta\partial\beta'\|$, and $L_3 = \max_{1 \leq i \leq m} \max_{j \in J_i} d_{ij}$, where

$$d_{ij} = \max_{1 \leq a,b,c \leq p} \sup_{|\tilde{\beta}-\beta| \leq \varepsilon_1} \left| \frac{\partial^3}{\partial\beta_a\partial\beta_b\partial\beta_c} g_j(X_i, \tilde{\beta}) \right|.$$

A2′. Same as A2 except that $g_j(X_i, \beta)$ are three-times continuously differentiable with respect to β, and that $L_2 = O_\mathrm{P}(1)$ is replaced by $L_{2,0} \vee L_3 = O_\mathrm{P}(1)$.

A5′. There is a positive integer γ such that $m/\{n(j,k,l)\}^\gamma \to 0$ for any $1 \leq l \leq L_{jk}$, $(j,k) \in D$, as $m \to \infty$.

We also need the following additional assumption.

A6. $m^{-1} \sum_{i=1}^{m} \dot{\mu}_i' V_i^{-1} \dot{\mu}_i$ is bounded away from zero in probability.

Let $\tilde{\beta}$ be the solution to (3.19), where the V_is are the true covariance matrices. Note that $\tilde{\beta}$ is efficient, or optimal, in the sense discussed earlier (see Section 3.2.1); however, $\tilde{\beta}$ is not computable, unless the true V_is are known.

Theorem 3.4. Under assumptions *A1*, *A2′*, *A3*, *A4*, *A5′* and *A6*, we have

$$\sqrt{m}(\hat{\beta}^* - \tilde{\beta}) \longrightarrow 0 \text{ in probability.}$$

Thus, asymptotically, $\hat{\beta}^*$ is as efficient as $\tilde{\beta}$.

Note. The proof of Theorem 3.4 also reveals the following asymptotic expansion,

$$\hat{\beta}^* - \beta = \left(\sum_{i=1}^{m} \dot{\mu}_i' V_i^{-1} \dot{\mu}_i \right)^{-1} \sum_{i=1}^{m} \dot{\mu}_i' V_i^{-1}(Y_i - \mu_i) + \frac{o_\mathrm{P}(1)}{\sqrt{m}}, \qquad (3.34)$$

where $o_\mathrm{P}(1)$ represents a term that converges to zero (vector) in probability (e.g., [1], Section 3.4). By Theorem 3.4, (3.34) also holds with $\hat{\beta}^*$ replaced by $\tilde{\beta}$, even though the latter is typically not computable.

Example 3.4 (real-data example). We use a real-data example to illustrate the convergence of IEE, or I-WLS (see Example 3.2) in this occasion. [132] analyzed a data set from [134] regarding hip replacements of thirty patients (also see [2], Section 1.7.2). Each patient was measured four times, once before the operation and three times after, for hematocrit, TPP, vitamin E, vitamin A, urinary zinc, plasma zinc, hydroxyprolene (in milligrams), hydroxyprolene (index), ascorbic acid, carotine, calcium, and plasma phosphate (12 variables). An important feature of the data is that there is considerable amount of missing observations. In fact, most of the patients have at least one missing observation for all 12 measured variables. As a result, the observational times are very different for different patients.

Two of the variables are considered: hematocrit and calcium. The first variable was considered by [134], who used the data to assess age, sex, and time differences. The authors assumed an equicorrelated model and obtained Gaussian estimates of regression coefficients and variance components (i.e., MLE under normality). Here we take a robust approach without assuming a parametric covariance structure. The covariates consist of the same variables as suggested by [134]. The variables include an intercept, sex, occasion dummy variables (three), sex by occasion interaction dummy variables (three), age, and age by sex interaction. For the hematocrit data, the I-WLS algorithm converged in seven (7) iterations. Here the criterion for the convergence is that the Euclidean distance between consecutive updates of the parameters is less than 10^{-5}. For the calcium data, the I-WLS in thirteen (13) iterations. The results were presented in Tables 1.5 and 1.6 of [2].

3.2.3 Method of simulated moments

Apparently, the first attempt in obtaining consistent estimators in GLMM was made by [17], who proposed a method of simulated moments (MSM) for GLMM. The MSM had been known to econometricians since the late 1980s. See, for example, [135] and [136]. The method applies to cases where moments of random variables cannot be expressed as analytic functions of parameters; therefore, direct computation of the MoM estimators is not possible. In MSM, the moments are approximated by Monte Carlo methods, and this is the only difference between MSM and MoM. To develop a MSM for GLMMs, let us first consider a simple example.

Example 3.5. Let y_{ij} be a binary outcome with logit$\{P(y_{ij} = 1 | \alpha)\} = \mu + \alpha_i$, $1 \leq i \leq m$, $1 \leq j \leq k$, where $\alpha_1, \ldots, \alpha_m$ are independent random effects with $\alpha_i \sim N(0, \sigma^2)$, $\alpha = (\alpha_i)_{1 \leq i \leq m}$, and μ, σ are unknown parameters with $\sigma \geq 0$. It is more convenient to use the following expression: $\alpha_i = \sigma u_i$, $1 \leq i \leq m$, where $u_1, \ldots, u_m$ are independent $N(0, 1)$ random variables. It is easy to show that a set of sufficient statistics (e.g., [11], Section 1.6) for μ and σ are $y_{i\cdot} = \sum_{j=1}^{k} y_{ij}$, $1 \leq i \leq m$. Thus, we consider the following MM estimating equations based on the sufficient statistics,

$$\frac{1}{m} \sum_{i=1}^{m} y_{i\cdot} = E(y_{1\cdot}),$$

$$\frac{1}{m} \sum_{i=1}^{m} y_{i\cdot}^2 = E(y_{1\cdot}^2).$$

Table 3.1 *Simulated Mean and Standard Error*

m	k	Estimator of μ		Estimator of σ^2	
		Mean	SE	Mean	SE
20	2	0.31	0.52	2.90	3.42
20	6	0.24	0.30	1.12	0.84
80	2	0.18	0.22	1.08	0.83
80	6	0.18	0.14	1.03	0.34

It is easy to show that $E(y_{1.}) = kE\{h_\theta(\zeta)\}$ and $E\{y_{1.}^2\} = kE\{h_\theta(\zeta)\} + k(k-1)E\{h_\theta^2(\zeta)\}$, where $h_\theta(x) = \exp(\mu + \sigma x)/\{1 + \exp(\mu + \sigma x)\}$ and $\zeta \sim N(0,1)$. It is more convenient to consider the following equivalent equations:

$$\frac{y_{..}}{mk} = E\{h_\theta(\zeta)\}, \tag{3.35}$$

$$\frac{1}{mk(k-1)} \sum_{i=1}^{m} (y_{i.}^2 - y_{i.}) = E\{h_\theta^2(\zeta)\}, \tag{3.36}$$

where $y_{..} = \sum_{i=1}^{m} y_{i.}$. Let $u_1, \ldots, u_L$ be a sequence of $N(0,1)$ random variables generated by a computer. Then, the right sides of (3.35) and (3.36) may be approximated by $L^{-1} \sum_{l=1}^{L} h_\theta(u_l)$ and $L^{-1} \sum_{l=1}^{L} h_\theta^2(u_l)$, respectively. The equations then get solved to obtain the MSM estimators of μ and σ.

To see how the MSM estimators perform, a small simulation study was carried out in [17] with $m = 20$ or 80 and $k = 2$ or 6. The true parameters are $\mu = 0.2$ and $\sigma^2 = 1.0$. The results in Table 3.1 were based on 1000 simulations, where the estimator of σ^2 is the square of the estimator of σ. The results show consistency of the MSM estimator, that is, as m increases, the performance of the estimators gets better in both Mean and SE.

To describe the general procedure for MSM, we assume that the conditional density of y_i given the vector of random effects α has the following form,

$$f(y_i|\alpha) = \exp[(w_i/\phi)\{y_i\xi_i - b(\xi_i)\} + c_i(y_i, \phi)], \tag{3.37}$$

where ϕ is a dispersion parameter, and w_is are known weights. Typically, $w_i = 1$ for ungrouped data; $w_i = n_i$ for grouped data if the response is an average, where n_i is the group size; and $w_i = 1/n_i$ if the response is a sum. Here $b(\cdot)$ and $c_i(\cdot, \cdot)$ are the same as in (3.1). As for ξ_i, we assume a canonical link, that is, (3.3) with $\eta_i = \xi_i$. Furthermore, we consider the ANOVA GLMM[137], that is, GLMM with $\alpha = (\alpha_1', \ldots, \alpha_q')'$, where α_r is a random vector whose components are independent and distributed as $N(0, \sigma_r^2)$, $1 \leq r \leq q$; and $Z = (Z_1, \ldots, Z_q)$ so that $Z\alpha = Z_1\alpha_1 + \cdots + Z_q\alpha_q$. Furthermore, it is assumed that $Z_r, 1 \leq r \leq q$ are standard design matrices in that each consists of 0s and 1s; there is exactly one 1 in each row, and at least one 1 in each column. Let z_{ir}' denote the ith row of Z_r. The following expression of α is sometimes more convenient,

$$\alpha = Du, \tag{3.38}$$

where D is blockdiagonal with the diagonal blocks $\sigma_r I_{m_r}$, $1 \leq r \leq q$, and $u \sim N(0, I_m)$ with $m = m_1 + \cdots + m_q$. First assume that ϕ is known. Let $\theta = (\beta', \sigma_1, \ldots, \sigma_q)'$. Consider an unrestricted parameter space $\theta \in \Theta = R^{p+q}$. This allows computational convenience for using MSM because, otherwise, there will be constraints on the parameter space. Of course, this raises the issue of identifiability, because both $(\beta', \sigma_1, \ldots, \sigma_q)'$ and $(\beta', -\sigma_1, \ldots, -\sigma_q)$ correspond to the same model. Nevertheless, it suffices to make sure that β and $\sigma^2 = (\sigma_1^2, \ldots, \sigma_q^2)'$ are identifiable. In fact, as we shall see, under suitable conditions, the MSM estimators of β and σ^2 are consistent; thus, the conditions also ensure identifiability of β and σ^2.

Under these set-up, [17] (also see [2], pp. 192–193) derived the following MM equations for estimating the GLMM parameters:

$$\sum_{i=1}^{N} w_i x_{ij} y_i = \sum_{i=1}^{N} w_i x_{ij} E_\theta(y_i), \qquad 1 \leq j \leq p, \qquad (3.39)$$

$$\sum_{(s,t) \in \mathscr{I}_r} w_s w_t y_s y_t = \sum_{(s,t) \in \mathscr{I}_r} w_s w_t E_\theta(y_s y_t), \qquad 1 \leq r \leq q, \qquad (3.40)$$

where $\mathscr{I}_r = \{(s,t) : 1 \leq s \neq t \leq N, z'_{sr} z_{tr} = 1\} = \{(s,t) : 1 \leq s \neq t \leq N, z_{sr} = z_{tr}\}$. The first and second moments on the right sides of (3.39) and (3.40) are then replaced by the simulated moments, leading to the MSM equations [e.g., [2], (4.48)].

To state a result on the consistency of MSM in estimating the parameters, $\varphi = (\beta', \sigma_1^2, \ldots, \sigma_q^2)'$, under the ANOVA GLMM, let N be the (data) sample size and L the Monte Carlo sample size. We first state a lemma that establishes convergence of the simulated moments to the corresponding moments after suitable normalizations. Let Q be the set of row vectors v whose components are positive integers ordered decreasingly. Let Q_d be the subset of vectors in Q, whose sum of the components is equal to d. For example, $Q_2 = \{2, (1,1)\}$, $Q_4 = \{4, (3,1), (2,2), (2,1,1), (1,1,1,1)\}$. For $v \in Q$ and $v = (v_1, \ldots, v_s)$, define $b^{(v)}(\cdot) = b^{(v_1)}(\cdot) \cdots b^{(v_s)}(\cdot)$, where $b^{(k)}(\cdot)$ represents the kth derivative. For $1 \leq r \leq q$, $1 \leq u \leq N$, let $I_{r,u} = \{1 \leq v \leq N : (u,v) \in \mathscr{I}_r\}$, $J_r = \{(u,v,u',v') : (u,v), (u',v') \in \mathscr{I}_r, (z_u, z_v)'(z_{u'}, z_{v'}) \neq 0\}$. Let $S = \cup_{r=1}^{q} \mathscr{I}_r = \{(u,v) : 1 \leq u \neq v \leq N, z'_u z_v \neq 0\}$.

Lemma 3.1. Suppose that (i) $b(\cdot)$ is four times differentiable such that

$$\limsup_{N \to \infty} \max_{1 \leq i \leq N} \max_{v \in Q_d} E|b^{(v)}(\xi_i)| < \infty, \qquad d = 2, 4;$$

(ii) the sequences $\{a_{Nj}\}$, $1 \leq j \leq p$ and $\{b_{Nr}\}$, $1 \leq r \leq q$ are chosen such that the following converge to zero when divided by a_{Nj}^2,

$$\sum_{i=1}^{N} w_i x_{ij}^2, \qquad \sum_{(u,v) \in S} w_u w_v |x_{uj} x_{vj}|, \qquad (3.41)$$

$1 \leq j \leq p$, and the following converge to zero when divided by b_{Nr}^2,

$$\sum_{(u,v) \in \mathscr{I}_r} w_u w_v, \qquad \sum_{u=1}^{N} w_u \left(\sum_{v \in I_{r,u}} w_v\right)^2, \qquad \sum_{(u,v,u',v') \in J_r} w_u w_v w_{u'} w_{v'}, \qquad (3.42)$$

$1 \leq r \leq q$. Then, the following converges to zero in L^2 when divided by a_{Nj},

$$\sum_{i=1}^{N} w_i x_{ij} \{ y_i - E_\theta(y_i) \},$$

$1 \leq j \leq p$, and the following converges to zero in L^2 when divided by b_{Nr},

$$\sum_{(u,v) \in \mathscr{I}_r} w_u w_v \{ y_u y_v - E_\theta(y_u y_v) \},$$

$1 \leq r \leq q$.

The proof is given in [17]. We now define the normalized moments, simulated moments, and sample moments. Let

$$M_{N,j}(\theta) = \frac{1}{a_{Nj}} \sum_{i=1}^{N} w_i x_{ij} E\{ b'(\xi_i) \},$$

$$\tilde{M}_{N,j} = \frac{1}{a_{Nj}L} \sum_{i=1}^{N} w_i x_{ij} \sum_{l=1}^{L} b'(\xi_{il}),$$

$$\hat{M}_{N,j} = \frac{1}{a_{Nj}} \sum_{i=1}^{N} w_i x_{ij} y_i,$$

$1 \leq j \leq p$, where $\xi_{il} = x_i'\beta + z_i'Du^{(l)}$, and $u^{(1)}, \ldots, u^{(L)}$ are generated independently from the m-dimensional standard normal distribution. Here the subscript N refers to normalization. Similarly, we define

$$M_{N,p+r} = \frac{1}{b_{Nr}} \sum_{(u,v) \in \mathscr{I}_r} w_u w_v E\{ b'(\xi_u) b'(\xi_v) \},$$

$$\tilde{M}_{N,p+r} = \frac{1}{b_{Nr}L} \sum_{(u,v) \in I_r} w_u w_v \sum_{l=1}^{L} b'(\xi_{ul}) b'(\xi_{vl}),$$

$$\hat{M}_{N,p+r} = \frac{1}{b_{Nr}} \sum_{(u,v) \in I_r} w_u w_v y_u y_v,$$

$1 \leq r \leq q$. Let A_{Nj}, $1 \leq j \leq p$ and B_{Nr}, $1 \leq r \leq q$ be sequences of positive numbers such that $A_{Nj} \to \infty$ and $B_{Nr} \to \infty$ as $N \to \infty$. Let $\hat{\theta}$ be any $\theta \in \Theta_N = \{ \theta : |\beta_j| \leq A_{Nj}, 1 \leq j \leq p; |\sigma_r| \leq B_{Nr}, 1 \leq r \leq q \}$ satisfying

$$|\tilde{M}_N(\theta) - \hat{M}_N| \leq \delta_N, \tag{3.43}$$

where $\tilde{M}_N(\theta)$ is the $(p+q)$-dimensional vector whose jth component is $\tilde{M}_{N,j}(\theta)$, $1 \leq j \leq p+q$, $\hat{M}_N$ is defined similarly, and $\delta_N \to 0$ as $N \to \infty$. For any vector $v = (v_r)_{1 \leq r \leq s}$, define $\|v\| = \max_{1 \leq r \leq s} |v_r|$.

Theorem 3.5. Suppose that the conditions of Lemma 3.1 are satisfied.

a. Let ε_N be the maximum of the terms in (3.41) divided by a_{Nj}^2 and the terms in (3.42) divided by b_{Nr}^2 over $1 \leq j \leq p$ and $1 \leq r \leq q$. If $\varepsilon_N/\delta_N^2 \to 0$, $\hat{\theta}$ exists with probability tending to one as $N \to \infty$.

b. If, furthermore, the first derivatives of $E_\theta(y_i)$ and $E(y_u y_v)$ $(u \neq v)$ with respect to components of θ can be taken under the expectation sign; the quantities

$$\sup_{\|\theta\|\leq B} E\{b'(\xi_i)\}^4, \ E\left\{\sup_{\|\theta\|\leq B} |b''(\xi_i)|\right\}, \ E\left\{\sup_{\|\theta\|\leq B} |b''(\xi_u)b''(\xi_v)|\right\},$$

$1 \leq i \leq N$, $(u,v) \in S$ are bounded for any $B > 0$; and

$$\liminf_{n\to\infty} \inf_{\|\tilde\varphi-\varphi\|>\varepsilon} |M_N(\tilde\theta) - M_N(\theta)| > 0 \tag{3.44}$$

for any $\varepsilon > 0$, there exists a sequence $\{d_N\}$ such that, as $N, L \to \infty$ with $L \geq d_N$, $\hat\varphi$ is a consistent estimator of φ.

Note that condition (3.44) ensures the identifiability of the true θ. Similar conditions can be found in, for example, [135] and [136]. Suppose that the function $M_N(\cdot)$ is continuous and injective. Then, we have

$$\inf_{\|\tilde\varphi-\varphi\|>\varepsilon} |M_N(\tilde\theta) - M_N(\theta)| > 0.$$

If the lower bound stays away from zero as $n \to \infty$, then (3.44) is satisfied. We consider an example.

Example 3.5 (Continued). Suppose that $\sigma^2 > 0$, $m \to \infty$ and k remains fixed and $k > 1$. Then, it can be shown that all of the conditions of Theorem 3.5 are satisfied. In particular, we verify condition (3.44) here. Note that in this case $M_N(\cdot)$ does not depend on N. Write $M_1(\theta) = E\{h_\theta(\zeta)\}$, $M_2(\theta) = E\{h_\theta^2(\zeta)\}$. It is easy to show that $\sup_\mu |M_1(\theta) - M_2(\theta)| \to 0$ as $\sigma \to \infty$ and $\sup_\mu |M_1^2(\theta) - M_2(\theta)| \to 0$ as $\sigma \to 0$. Therefore, there exist $0 < a < b$ and $A > 0$ such that $\inf_{\tilde\theta \notin [-A,A]\times[a,b]} |M(\tilde\theta) - M(\theta)| > 0$, where $M(\theta) = (M_1(\theta), M_2(\theta))'$. By continuity, it suffices to show that $M(\cdot)$ in injective. Let $0 < c < 1$ and consider the equation

$$M_1(\theta) = c. \tag{3.45}$$

For any $\sigma > 0$, there is a unique $\mu = \mu_c(\sigma)$ that satisfies (3.45). The function $\mu_c(\cdot)$ is continuously differentiable. Write $\mu_c = \mu_c(\sigma)$, $\mu_c' = \mu_c'(\sigma)$. By (3.45), one has

$$E\left[\frac{\exp(\mu_c + \sigma\zeta)}{\{1+\exp(\mu_c+\sigma\zeta)\}^2}(\mu_c' + \zeta)\right] = 0. \tag{3.46}$$

Now consider $M_2(\theta)$ along the curve determined by (3.45); that is, $M_c(\sigma) = M_2(\mu_c, \sigma)$. We use the following covariance inequality [e.g., [1], (5.64)]. For continuous functions f. g and h with f and g strictly increasing and $h > 0$, we have

$$\int f(x)g(x)h(x)dx \int h(x)dx > \int f(x)h(x)dx \int g(x)h(x)dx,$$

provided that the integrals are finite. By (3.46) and the covariance inequality, we have

$$M_c'(\sigma) = 2\mathrm{E}\left[\frac{\{\exp(\mu_c + \sigma\zeta)\}^2}{\{1 + \exp(\mu_c + \sigma\zeta)\}^3}(\mu_c' + \zeta)\right] > 0.$$

The injectivity of $M(\cdot)$ then follows.

The constant d_N in Theorem 3.5 can be determined by the proof of the theorem.

In practice, it is often desirable to obtain the standard errors of the estimator. We discuss how to do this for the MSM estimator, using an asymptotic argument. Define $\hat{\psi} = (\hat{\beta}', |\hat{\sigma}_1|, \ldots, |\hat{\sigma}_q|)'$, where $\hat{\theta} = (\hat{\beta}', \hat{\sigma}_1, \ldots, \hat{\sigma}_q)'$ is the MSM estimator of θ. Write the MSM equations as $\hat{M} = \tilde{M}(\hat{\theta})$, where $\tilde{M}$ is the vector of simulated moments. Similarly, let $M(\theta)$ denote the vector of moments. We assume, without loss of generality, that $\sigma_r \geq 0$, $1 \leq r \leq q$ for the true θ. Because for large N and L, the simulated moments approximate the corresponding moments, and $\hat{\psi}$ is a consistent estimator of θ, we have, by the Taylor expansion,

$$
\begin{aligned}
\hat{M} &= \tilde{M}(\hat{\theta}) \\
&\approx M(\hat{\theta}) \\
&= M(\hat{\psi}) \approx M(\theta) + \dot{M}(\theta)(\hat{\psi} - \theta) \\
&\approx M(\theta) + \dot{M}(\theta)J^{-1}(\theta)(\hat{\varphi} - \varphi),
\end{aligned}
$$

where $\dot{M}(\cdot)$ is the matrix of first derivatives, $\varphi = (\beta', \sigma_1^2, \ldots, \sigma_q^2)'$, $\hat{\varphi}$ is the corresponding MSM estimator of φ, and $J(\theta) = \mathrm{diag}(1, \ldots, 1, 2\sigma_1, \ldots, 2\sigma_q)$. Thus, an approximate covariance matrix of $\hat{\varphi}$ is given by

$$\mathrm{Var}(\hat{\varphi}) \approx J(\theta)\{\dot{M}(\theta)^{-1}\}\mathrm{Var}(\hat{M})\{\dot{M}(\theta)^{-1}\}'J(\theta). \tag{3.47}$$

In practice, $J(\theta)$ can be estimated by $J(\hat{\psi})$, and $\dot{M}(\theta)$ can be estimated by first replacing θ by $\hat{\psi}$ and then approximating the moments by simulated moments, as we did earlier. As for $\mathrm{Var}(\hat{M})$, although one could derive its parametric form, the latter is likely to involve ϕ, the dispersion parameter which is sometimes unknown. Alternatively, the covariance matrix of $\hat{M}$ can be estimated using a parametric bootstrap method as follows. First generate data from the GLMM, treating $\hat{\varphi}$ as the true θ. The generated data are a bootstrap sample, denoted by $y_{i,k}^*$, $1 \leq i \leq n$, $1 \leq k \leq K$. Then, compute the vector of sample moments based on the bootstrap sample, say, $\hat{M}_k^*$, $1 \leq k \leq K$. A bootstrap estimate of $\mathrm{Var}(\hat{M})$ is then given by

$$\widehat{\mathrm{Var}}(\hat{M}) = \frac{1}{K-1}\sum_{k=1}^{K}\left(\hat{M}_k^* - \overline{\hat{M}^*}\right)\left(\hat{M}_k^* - \overline{\hat{M}^*}\right)',$$

where $\overline{\hat{M}^*} = K^{-1}\sum_{k=1}^{K}\hat{M}_k^*$.

We use another example to illustrate the MSM, including the s.e. computation, and compare the performance of MSM with other methods.

Example 3.6. The example was considered by [138] and [139] in their simulation studies. Suppose that, given the random effects $u_1, \ldots, u_{15}$, which are independent and distributed as $N(0, 1)$, responses y_{ij}, $i = 1, \ldots, 15$, $j = 1, 2$ are conditionally independent such that $y_{ij}|u \sim \mathrm{binomial}(6, \pi_{ij})$, where $u = (u_i)_{1 \leq i \leq 15}$,

Table 3.2 *Comparison of Estimators*

True Parameter	Average of Estimator			SE of Estimator		
	MSM	AREML	IBC	MSM	AREML	IBC
$\beta_0 = 0.2$	0.20	0.25	0.19	0.32 (0.31)	0.31	0.26
$\beta_1 = 0.1$	0.10	0.10	0.10	0.04 (0.04)	0.04	0.04
$\sigma^2 = 1.0$	0.93	0.91	0.99	0.63 (0.65)	0.54	0.60
$\beta_0 = 0.2$	0.21	0.11	0.20	0.42 (0.37)	0.35	0.36
$\beta_1 = 0.1$	0.10	0.10	0.10	0.05 (0.05)	0.05	0.05
$\sigma^2 = 2.0$	1.83	1.68	1.90	1.19 (1.04)	0.80	0.96

$\text{logit}(\pi_{ij}) = \beta_0 + \beta_1 x_{ij} + \sigma u_i$ with $x_{i1} = 2i - 16$ and $x_{i2} = 2i - 15$. The MSM equations for estimating β_0, β_1, and σ take the following form

$$\sum_{i=1}^{15}(y_{i1} + y_{i2}) = \frac{6}{L}\sum_{l=1}^{L}\sum_{i=1}^{15}(\pi_{i1l} + \pi_{i2l}),$$

$$\sum_{i=1}^{15}(x_{i1}y_{i1} + x_{i2}y_{i2}) = \frac{6}{L}\sum_{l=1}^{L}\sum_{i=1}^{15}(x_{i1}\pi_{i1l} + x_{i2}\pi_{i2l}),$$

$$\sum_{i=1}^{15}y_{i1}y_{i2} = \frac{36}{L}\sum_{l=1}^{L}\sum_{i=1}^{15}\pi_{i1l}\pi_{i2l},$$

where $\pi_{ijl} = h(\beta_0 + \beta_1 x_{ij} + \sigma u_{il})$ with $h(x) = e^x/(1 + e^x)$, and u_{il}, $1 \le i \le 15$, $1 \le l \le L$ are random variables generated independently from an $N(0,1)$ distribution.

A simulation study was carried out for the model considered here with $L = 100$. Two sets of true parameters were considered: (i) $\sigma^2 = 1.0$ and (ii) $\sigma^2 = 2.0$; in both cases $\beta_0 = 0.2$, $\beta_1 = 1.0$. The results based on 1000 simulations are summarized in Table 3.2, which is copied from [17], and compared with the approximate REML (AREML) estimator of [138] and the iterative bias correction (IBC) estimator of [139]. The numbers in the parentheses are averages of the estimated s.e.(SE)s. The AREML method is similar to the PQL of [14], discussed earlier. For the most part, AREML is based on a link between BLUP and REML (e.g., [140]) and a quadratic approximation to the conditional log-density of the responses given the random effects. The IBC method iteratively corrects the bias of PQL, which results in an asymptotically unbiased estimator. However, IBC may be computationally intensive.

It appears that MSM is doing quite well in terms of the bias, especially compared with AREML. On the other hand, the standard errors of MSM estimators seem larger than those of AREML and IBC estimators. Finally, the estimated SEs are very close to the simulated ones, an indication of good performance of the above method of standard error estimation.

3.2.4 Robust estimation in GLMM

The latest simulation results show that, although the MSM estimators are consistent, they may be inefficient in the sense that the variances of the estimators are relatively large. In this subsection, we consider an improvement of the MSM.

Let us first consider an extension of GLMM. Recall that it is assumed in a GLM [6] that the distribution of the response is a member of a known exponential family. Thus, for a linear model to fit within the GLM, one needs to assume that the distribution of the response is normal. However, the definition of a linear model does not have to require normality, and many of the methods, such as the least squares, developed in linear models do not require the normality assumption. Thus, in a way, GLM has not fully extended the linear model.

In view of this, we consider a broader class of models than the GLMM, in which the form of the conditional distribution, such as the exponential family, is not required. The method can be described under an even broader framework. Suppose that, given a vector $\alpha = (\alpha_k)_{1 \leq k \leq m}$ of random effects, responses $y_1, \ldots, y_N$ are conditionally independent such that

$$E(y_i|\alpha) = h(\xi_i), \tag{3.48}$$
$$\text{var}(y_i|\alpha) = a_i(\phi)v(\eta_i), \tag{3.49}$$

where $h(\cdot)$, $v(\cdot)$, and $a_i(\cdot)$ are known functions, ϕ is a dispersion parameter,

$$\xi_i = x_i'\beta + z_i'\alpha, \tag{3.50}$$

where β is a vector of unknown fixed effects, and $x_i = (x_{ij})_{1 \leq j \leq p}$, $z_i = (z_{ik})_{1 \leq k \leq m}$ are known vectors. Finally, assume that

$$\alpha \sim F_\theta, \tag{3.51}$$

where F_θ is a multivariate distribution known up to a vector $\theta = (\theta_r)_{1 \leq r \leq q}$ of disperson parameters, or variance components. Note that we do not require that the conditional density of y_i given α is a member of the exponential family; instead, only up to second conditional moments are specified, by (3.48) and (3.49). In fact, as will be seen, to obtain consistent estimators, only (3.48) is needed.

We now consider estimation under the extended GLMM. Let ψ be a vector of parameters under an assumed model. Suppose that there is a vector of base statistics, say, S, which typically is of higher dimension than ψ. We assume that the following conditions are satisfied:

(i) The mean of S is a known function of ψ.

(ii) The covariance matrix of S is a known function of ψ, or at least is consistently estimable.

(iii) Certain smoothness and regularity conditions hold.

Condition (iii) of the above requirement is a bit vague at this point, but it will be specified later when we discuss asymptotic theory. Let the dimension of ψ and S be r and N, respectively. If only (i) is assumed, an estimator of ψ may be obtained by solving the following equation,

$$BS = Bu(\psi), \tag{3.52}$$

where B is a $r \times N$ matrix and $u(\psi) = E_\psi(S)$. This is called the first-step estimator, in which the choice of B is quite arbitrary. It can be shown (see below) that, under suitable conditions, the first-step estimator is consistent, although it may not be efficient. To improve the efficiency, we further require (ii). Denote the first-step estimator by $\tilde{\psi}$, and consider a Taylor expansion around the true ψ, which leads to

$$\tilde{\psi} - \psi \approx (BU)^{-1} Q(\psi),$$

where $U = \partial u / \partial \psi'$ and $Q(\psi) = B\{S - u(\psi)\}$. Note that $Q(\tilde{\psi}) = 0$. Denote the covariance matrix of S by V. Then, we have

$$\mathrm{Var}(\tilde{\psi}) \approx \{(BU)^{-1}\} BVB' \{(BU)^{-1}\}'. \tag{3.53}$$

By Theorem 3.1, the optimal B, in the sense of minimizing the right side of (3.53), is $U'V^{-1}$. Unfortunately, this optimal B depends on ψ, which is exactly what we wish to estimate. Our approach is to replace ψ in the optimal B by $\tilde{\psi}$, the first-step estimator. This leads to the second-step estimator, denoted by $\hat{\psi}$, obtained by solving

$$\tilde{B} S = \tilde{B} u(\psi), \tag{3.54}$$

where $\tilde{B} = U'V^{-1}|_{\psi = \tilde{\psi}}$. It can be shown that, under suitable conditions, the second-step estimator is consistent and asymptotically efficient in the sense that its asymptotic covariance matrix is the same as that of the solution to the optimal estimating equation, that is, (3.52) with $B = U'V^{-1}$.

Note. It might appear that one could do better by allowing B in (3.52) to depend on θ, that is, $B = B(\theta)$. However, Theorem 3.1 shows that the asymptotic covariance matrix of the estimator corresponding to the optimal $B(\psi)$ is the same as that corresponding to the optimal B (which is a constant matrix). Thus, the complication does not result in a real gain, at least asymptotically.

To apply the method to the extended GLMMs, we need to first select the base statistics. By similar arguments as in the previous subsection, a natural choice seems to be the following,

$$S_j = \sum_{i=1}^{N} w_i x_{ij} y_i, \qquad 1 \le j \le p,$$

$$S_{p+j} = \sum_{s \ne t} w_s w_t z_{sk} z_{tk} y_s y_t, \qquad 1 \le k \le m. \tag{3.55}$$

In fact, if $Z = (z_{ik})_{1 \le i \le n, 1 \le k \le m} = (Z_1 \cdots Z_q)$, where each Z_r is an $N \times m_r$ standard design matrix in that it consists of zeros and ones; there is exactly one 1 in each row, and at least one 1 in each column. Then, by choosing $B = \mathrm{diag}(I_p, 1'_{m_1}, \dots, 1'_{m_q})$, one obtains the MM eqnations of the previous subsection. Thus, the latter is a special case of the first-step estimation. However, the following examples show that the second-step estimators can be considerably more efficient than the first-step ones.

Example 3.7 (mixed logistic model). Consider an extension of Example 3.4 by allowing the number of binary responses to be different for different subjects, that is,

Table 3.3 *Simulation Results: Mixed Logistic Model*

Method of	Estimator of μ			Estimator of σ			Overall
Estimation	Mean	Bias	SD	Mean	Bias	SD	MSE
1st-step	.21	.01	.16	.98	−.02	.34	.15
2nd-step	.19	−.01	.16	.98	−.02	.24	.08

replacing k by k_i. The rest of the assumptions remain the same. It is easy to see that (3.55) reduce to $y_{..}$ and $y_{i.}^2 - y_{i.}$, $1 \le i \le m$, where $y_{i.} = \sum_{j=1}^{n_i} y_{ij}$ and $y_{..} = \sum_{i=1}^{m} y_{i.}$.

If $k_i = k, 1 \le i \le m$, that is, we are in the situation of Example 3.4, also known as balanced data, the first-step estimating equations can be shown to be equivalent to the second-step ones. In other words, in the case of balanced data, there is no gain by doing the second-step, and the first-step estimators are already (asymptotically) optimal. However, when the data are unbalanced, the first and second-step estimators are no longer equivalent, and there is a real gain by doing the second-step. To see this, a simulation was carried out [19], in which $m = 100$, $n_i = 2$, $1 \le i \le 50$, and $n_i = 6$, $51 \le i \le 100$. The true parameters were chosen as $\mu = 0.2$ and $\sigma = 1.0$. The results based on 1000 simulations are summarized in Table 3.3, where SD represents the simulated standard deviation, and the overall MSE is the MSE of the estimator of μ plus that of the estimator of σ. Overall, there is about a 43% reduction of the MSE of the second-step estimators over the first-step ones.

Because the first- and second-step estimators are developed under the assumption of the extended GLMM, the methods apply to some situations beyond (the classical) GLMM. The following is an example.

Example 3.8 (Beta-binomial). If $Y_1, \ldots, Y_l$ are correlated Bernoulli random variables, the distribution of $Y = Y_1 + \cdots + Y_l$ is not binomial, and therefore does not belong to the exponential family. Here we consider a special case. Let p be a random variable with a beta$(\pi, 1 - \pi)$ distribution, where $0 < \pi < 1$. Suppose that, given p, $Y_1, \ldots, Y_l$ are independent Bernoulli(p) random variables, so that $Y|p \sim$ binomial(l, p). It can be shown that the marginal distribution of Y is given by

$$P(Y = k) = \frac{\Gamma(k + \pi)\Gamma(l - k + 1 - \pi)}{k!(l - k)!\Gamma(\pi)\Gamma(1 - \pi)}, \qquad 1 \le k \le l. \qquad (3.56)$$

This distribution is called beta-binomial(l, π). It follows that $E(Y) = l\pi$ and $Var(Y) = \phi l\pi(1 - \pi)$, where $\phi = (l + 1)/2$. It is seen that the mean function under beta-binomial(l, π) is the same as that of binomial(l, π), but the variance function is different. In other words, there is an overdispersion.

Now, suppose that, given the random effects $\alpha_i, 1 \le i \le m$, which are independent and distributed as $N(0, \sigma^2)$, responses y_{ij}, $1 \le i \le m$, $1 \le j \le n_i$ are independent and distributed as beta-binomial(l, π_i), where $\pi_i = h(\mu + \alpha_i)$ with $h(x) = e^x/(1 + e^x)$. Note that this is not a GLMM under the classical definition, because the conditional distribution of y_{ij} is not a member of the exponential family. However, the model falls within the extended definition, because

$$E(y_{ij}|\alpha) = l\pi_i, \qquad (3.57)$$

Table 3.4 *Simulation Results: Beta-Binomial*

Method of Estimation	Estimation of μ			Estimation of σ			Overall MSE
	Mean	Bias	SD	Mean	Bias	SD	
1st-step	.25	.05	.25	1.13	.13	.37	.22
2nd-step	.25	.05	.26	1.09	.09	.25	.14

$$\text{var}(y_{ij}|\alpha) = \phi l \pi_i (1 - \pi_i). \tag{3.58}$$

If only (3.57) is assumed, one may obtain the first-step estimator of (μ, σ), for example, by choosing $B = \text{diag}(1, 1'_m)$. If, in addition, (3.58) is assumed, one may obtain the second-step estimator. To see how much difference there is between the two, a simulation study was carried out with $m = 40$ ([19]). Again, an unbalanced situation was considered: $n_i = 4$, $1 \le i \le 20$ and $n_i = 8$, $21 \le i \le 40$. We took $l = 2$, and the true parameters $\mu = 0.2$ and $\sigma = 1.0$. The results based on 1000 simulations are summarized in Table 3.4. Overall, we see about 36% improvement in MSE of the second-step estimators over the first-step ones.

The improvements of the second-step estimators over the first-step ones in the precedent examples are not incidental. We now discuss an asymptotic theory related to this improvement. First, we specify condition (iii) of the requirements for the base statistics. The results established here are actually more general than GLMM or extended GLMM.

Let the responses be $y_1, \ldots, y_N$, whose distribution depends on a parameter vector, ψ. Let Ψ be the parameter space. First, note that B, S, and $u(\psi)$ in (3.52) may depend on N, the sample size; hence in the subsection we use the notation B_N, S_N, and $u_N(\psi)$. Also, the solution to (3.52) is unchanged when B_N is replaced by $C_N^{-1} B_N$, where $C_N = \text{diag}(c_{N,1}, \ldots, c_{N,r})$, $c_{N,j}$ is a sequence of positive constants, $1 \le j \le r$, and r is the dimension of ψ. Write $M_N = C_N^{-1} B_N S_N$, and $M_N(\psi) = C_N^{-1} B_N u_N(\psi)$. Then the first-step estimator, $\tilde{\psi} = \tilde{\psi}_N$ is the solution to the equation

$$M_N(\psi) = M_N. \tag{3.59}$$

Consider $M_N(\cdot)$ as a map from Ψ to a subset of R^r. Let ψ denote the true ψ everywhere except when defining a function of ψ, and $M_N(\Psi)$ be the image of Ψ under $M_N(\cdot)$. For $x \in R^r$ and $A \subset R^r$, define $d(x, A) = \inf_{y \in A} |x - y|$. Obviously, $M_N(\psi) \in M_N(\Psi)$. Furthermore, if $M_N(\psi)$ is in the interior of $M_N(\Psi)$, we have $d(M_N(\psi), M_N^c(\Psi)) > 0$. In fact, the latter essentially ensures the existence of the solution to (3.59), as shown by the following theorem. The proof is given in [19].

Theorem 3.6. Suppose that, as $N \to \infty$,

$$M_N - M_N(\psi) \longrightarrow 0 \tag{3.60}$$

in probability, and

$$\liminf d\{M_N(\psi), M_N^c(\Psi)\} > 0. \tag{3.61}$$

Then, with probability tending to one, the solution to (3.59) exists and is in Ψ. If, in addition, there is a sequence $\Psi_N \subset \Psi$ such that

$$\liminf_{\psi_* \notin \Psi_N} \inf |M_N(\psi_*) - M_N(\psi)| > 0, \qquad (3.62)$$

$$\liminf_{\psi_* \in \Psi_N, \psi_* \neq \psi} \inf \frac{|M_N(\psi_*) - M_N(\psi)|}{|\psi_* - \psi|} > 0, \qquad (3.63)$$

then, any solution $\tilde{\psi}_N$ to (3.59) is consistent.

The lemmas below give sufficient conditions for (3.60)–(3.63). The proofs are fairly straightforward. Let V_N denote the covariance matrix of S_N.

Lemma 3.2. (3.60) holds provided that, as $N \to \infty$,

$$\text{tr}(C_N^{-1} B_N V_N B_N' C_N^{-1}) \longrightarrow 0.$$

Lemma 3.3. Suppose that there is a vector-valued function $M_0(\psi)$ such that $M_N(\psi) \to M_0(\psi)$ as $N \to \infty$. Furthermore, suppose that there exist $\varepsilon > 0$ and $N_\varepsilon \geq 1$ such that $y \in M_N(\Psi)$ whenever $|y - M_0(\psi)| < \varepsilon$ and $N \geq N_\varepsilon$. Then (3.61) holds. In particular, if $M_N(\psi)$ does not depend on N, that is, $M_N(\psi) = M(\psi)$, say, then (3.61) holds provided that $M(\psi)$ is in the interior of $M(\Psi)$, the image of $M(\cdot)$.

Lemma 3.4. Suppose that there are continuous functions $f_j(\cdot), g_j(\cdot), 1 \leq j \leq r$, such that $f_j\{M_N(\psi)\} \to 0$ if $\psi \in \Psi$ and $\psi_j \to -\infty$, $g_j\{M_N(\psi)\} \to 0$ if $\psi \in \Psi$ and $\psi_j \to \infty$, $1 \leq j \leq r$, uniformly in N. If, as $N \to \infty$,

$$\limsup |M_N(\psi)| < \infty,$$
$$\liminf \min[|f_j\{M_N(\psi)\}|, |g_j\{M_N(\psi)\}|] > 0, \ 1 \leq j \leq r,$$

then there is a compact subset $\Psi_0 \subset \Psi$ such that (3.62) holds with $\Psi_N = \Psi_0$.

Write $U_N = \partial u_N / \partial \psi'$. Let $H_{N,j}(\psi) = \partial^2 u_{N,j} / \partial \psi \partial \psi'$, where $u_{N,j}$ is the jth component of $u_N(\psi)$, and $H_{N,j,\varepsilon} = \sup_{|\psi_* - \psi| \leq \varepsilon} \|H_{N,j}(\psi_*)\|$, $1 \leq j \leq L_N$, where L_N is the dimension of u_N.

Lemma 3.5. Suppose that $M_N(\cdot)$ is twice continuously differentiable, and that

$$\liminf \lambda_{\min}(U_N' B_N' C_N^{-2} B_N U_N) > 0,$$

and there is $\varepsilon > 0$ such that

$$\limsup \frac{\max_{1 \leq i \leq r} c_{N,i}^{-2}(\sum_{j=1}^{L_N} |b_{N,ij}| H_{N,j,\varepsilon})^2}{\lambda_{\min}(U_N' B_N' C_N^{-2} B_N U_N)} < \infty,$$

where $b_{N,ij}$ is the (i, j) element of B_N. Furthermore, suppose, for any compact subset $\Psi_1 \subset \Psi$ such that $d(\psi, \Psi_1) > 0$, we have

$$\liminf_{\psi_* \in \Psi_1} \inf |M_N(\psi_*) - M_N(\psi)| > 0.$$

Then (3.63) holds for $\Psi_N = \Psi_0$, where Ψ_0 is any compact subset of Ψ that includes ψ as an interior point.

Once again, we consider a specific example.

Example 3.5 (continued). As noted (see Example 3.7), in this case, the first and second-step estimators of $\theta = (\mu, \sigma)'$ are the same, and they both correspond to $B_N = \text{diag}(1, 1'_m)$. It can be shown that, by choosing $C_N = \text{diag}\{mk, mk(k-1)\}$, all of the conditions of Lemmas 3.2–3.5 are satisfied.

We now consider the asymptotic normality of the first-step estimator. We say that an estimator $\tilde{\psi}_N$ is asymptotically normal with mean ψ and asymptotic covariance matrix $(\Gamma'_N \Gamma_N)^{-1}$ if $\Gamma_N(\tilde{\psi}_N - \psi) \longrightarrow N(0, I_r)$ in distribution, where $r = \dim(\psi)$. Here it is understood that Γ_N is $r \times r$ and non-singular. Let

$$\lambda_{N,1} = \lambda_{\min}(C_N^{-1} B_N V_N B'_N C_N^{-1}),$$

$$\lambda_{N,2} = \lambda_{\min}\{U'_N B'_N (B_N V_N B'_N)^{-1} B_N U_N\}.$$

Theorem 3.7. Suppose that (i) the components of $u_N(\psi)$ are twice continuously differentiable; (ii) $\tilde{\psi}_N$ satisfies (3.59) with probability tending to one and is consistent; (iii) there exists $\varepsilon > 0$ such that

$$\frac{|\tilde{\psi}_N - \psi|}{(\lambda_{N,1} \lambda_{N,2})^{1/2}} \max_{1 \le i \le r} c_{N,i}^{-1} \left(\sum_{j=1}^{L_N} |b_{N,ij}| H_{N,j,\varepsilon} \right) \longrightarrow 0$$

in probability; and (iv)

$$\{C_N^{-1} B_N V_N B'_N C_N^{-1}\}^{-1/2} [M_N - M_N(\psi)] \longrightarrow N(0, I_r)$$

in distribution. Then $\tilde{\psi}$ is asymptotically normal with mean ψ and asymptotic covariance matrix

$$(B_N U_N)^{-1} B_N V_N B'_N (U'_N B'_N)^{-1}. \tag{3.64}$$

Sufficient conditions for existence, consistency, and asymptotic normality of the second-step estimators can be obtained by replacing the conditions of Theorems 3.6 and 3.7 by the corresponding conditions with a probability statement. For example, let ξ_N be a sequence of nonnegative random variables. We say that $\liminf \xi_N > 0$ with probability tending to one if for any $\varepsilon > 0$ there is $\delta > 0$ such that $P(\xi_N > \delta) \ge 1 - \varepsilon$ for all sufficiently large N. Note that this is equivalent to $\xi_N^{-1} = O_P(1)$. Then, (3.62) is replaced by (3.62) with probability tending to one. Note that the asymptotic covariance matrix of the second-step estimator is given by (3.64) with $B_N = U'_N V_N^{-1}$, which is $(U'_N V_N U_N)^{-1}$. This is the same as the asymptotic covariance matrix of the solution to (3.54), or (3.59), with the optimal B (B_N). In other words, the second-step estimator is asymptotically optimal. See [19] for details.

3.3 Maximum likelihood estimation

As noted at the beginning of this chapter, the likelihood function under a GLMM typically does not have an analytic expression. This feature of GLMM not only presents a computational challenge, as discussed earlier, but also makes it difficult to study asymptotic behavior of the MLE under a GLMM, well, at least for certain types of GLMMs. To see for what types of GLMM this is most challenging, let us recall the salamander data, discussed in Section 1.3. As pointed out, asymptotic analysis for GLMM is most difficult if the model involves crossed random effects. On the other hand, it is fairly straightforward to develop asymptotic theory for GLMMs that only involve clustered random effects, such as in Example 1.2. A reference for the latter case can be found, for example, in [141]. Thus, in this section, we mainly focus on the more difficult case, that is, GLMMs with crossed random effects.

Before we develop the asymptotic theory, we would like to provide a brief review on the history of computational developments, especially for GLMMs with crossed random effects, which, naturally, are computationally much more challenging than GLMMs with clustered random effects. The review will also include historical efforts in establishing consistency of the MLE for GLMMs with crossed random effects.

3.3.1 *Historical notes*

If the likelihood function only involves low-dimensional integrals (typically one or two-dimensional), those integrals can be evaluated by numerical integration techniques. See, for example, [142], [143]. However, numerical integration is generally intractable if the dimension of integrals involved is greater than two. Alternatively, the integrals may be evaluated by Monte Carlo methods. It should be pointed out that, for problems involving irreducibly high-dimensional integrals, naive Monte Carlo usually does not work. For example, the high-dimensional integral in (1.10) cannot be evaluated by a naive Monte Carlo method. To see this, assume that $m = n = 40$, as in the salamander data, and that $S = \{(i, j) : 1 \leq i \leq m, 1 \leq j \leq n\}$. Then, the integrand in (1.10) involves a product of 1600 terms with each term less than one (and greater than zero). Such a term is numerically zero. Thus, if one simulates i.i.d. random variables of such terms from the normal distribution, and use the LNN to approximate the integral, the average of such terms will not yield anything but zero without a huge simulation sample size. For such reasons, more advanced Monte Carlo techniques have been developed for computing the MLE in GLMM.

One idea is to use the E-M algorithm [144]. The latter is known as a general procedure for computing the MLE, which amounts to iterate between an E-step and a M-step. In the mixed effects model context, the E-step consists of taking conditional expectation of the "complete-data" log-likelihood, corresponding to the joint distribution of y and α, given y; the conditional expectation is then maximized in the M-step with respect to the parameters to get an update of the latter. This works fairly easily under the LMM and, in fact, results in closed-form expressions for the parameter updates (e.g., [2], p. 166). However, under a non-LMM GLMM, there is a complication in the E-step.

The difficulty is that the conditional expectation involves the same kind of integrals that we encounter in the GLMM likelihood that cannot be computed analytically. McCulloch [145] proposed to use Monte Carlo method, namely, the Gibbs sampler (e.g., [146]), to approximate the E-step and his idea was followed by similar approaches, known as Monte Carlo E-M (MCEM). In particular, [147] used the Metropolis–Hastings algorithm to extend the MCEM algorithm of [145]. Booth and Hobert [18] proposed to use i.i.d. sampling, instead of Markov-chain Monte Carlo (e.g., [146]), in their MCEM algorithm. More specifically, Booth and Hobert [18] compared two sampling strategies, *importance sampling* and *rejection sampling* (e.g., [146]), and proposed an automated method that, at each iteration of the MCEM, determines the appropriate Monte Carlo sample size. See, for example, [2] (pp. 164–173) for more detail.

Another, more recent, computational advance is called *data cloning* (DC; [27], [28]). It uses Bayesian computational approach for frequentist purposes. Let π denote the prior density function of ψ. Then, one has the posterior,

$$\pi(\psi|y) = \frac{p_\psi(y))\pi(\psi)}{p(y)}, \tag{3.65}$$

where $p(y)$ is the integral of the numerator with respect to ψ, which does not depend on ψ. There are computational tools using the Markov chain Monte Carlo for posterior simulation that generate random variables from the posterior without having to compute the numerator or denominator of (3.65) (e.g., [148]; [149]). Thus, we can assume that one can generate random variables from the posterior. If the observations y were repeated independently from K different individuals such that all of these individuals result in exactly the same data, y, denoted by $y^{(K)} = (y, \ldots, y)$, then the posterior based on $y^{(K)}$ is given by

$$\pi_K\{\psi|y^{(K)}\} = \frac{\{p_\psi(y)\}^K \pi(\psi)}{\int \{p_\psi(y)\}^K \pi(\psi)d\psi}. \tag{3.66}$$

Lele et al. [27] and [28] showed that, as K increases, the right side of (3.66) converges to a multivariate normal distribution whose mean vector is equal to the MLE, $\hat{\psi}$, and whose covariance matrix is approximately equal to $K^{-1}I_f^{-1}(\hat{\psi})$. For the most part, this result is due to the fact that, as K increases, the numerator in (3.66), which is the integrand of the denominator, becomes more and more concentrate near its maximum. As noted in Section 3.1 (last paragraph), this is when Laplace approximation becomes asymptotically accurate.

Thus, for large K, one can approximate the MLE by the sample mean vector of, say, $\psi^{(1)}, \ldots, \psi^{(B)}$ generated from the posterior distribution (3.66). Denoted the sample mean by $\bar{\psi}^{(\cdot)}$, and call it the DC MLE. Furthermore, the estimated Fisher information matrix, $I_f^{-1}(\hat{\psi})$ can be approximated by K times the sample covariance matrix of $\psi^{(1)}, \ldots, \psi^{(B)}$. As an application of the DC method, [150] used the DC method to obtain the MLE for the salamander-mating data.

Note that the DC MLE is an approximate, rather than exact, MLE, in the sense that, as $K \to \infty$, the difference between $\bar{\psi}^{(\cdot)}$ and the exact MLE vanishes. It should also be noted that, although there have been significant computation advances in the computation of the MLE in GLMM, the computation advances had not led to a major theoretical breakthrough for the MLE in GLMM in terms of the asymptotic properties, such as consistency.

The problem regarding consistency of the MLE in GLMMs with crossed random effects began to draw attention in early 1997. Over the years, the problem had been discussed with numerous researchers as an effort to get help in solving the problem. In addition, the problem was presented as open problems in [2] (p. 173) and [1] (p. 541), such as the one stated near the end of Section 1.3. In almost all of the discussions, the researchers showed strong interest in the problem, and willingness to tackle it. Yet, the problem remained unsolved. In [1] (p. 550), the author provided evidence on why he thinks the answer has to be positive for the open problem presented near the end of Section 1.3, as follows.

Let $k = m \wedge n$. Consider a subset of the data, $y_{ii}, i = 1, \ldots, k$. Note that the subset is a sequence of i.i.d. random variables. It follows, by the standard arguments, that the MLE of μ based on the subset, denoted by $\tilde{\mu}$, is consistent. Let $\hat{\mu}$ denote the MLE of μ based on the full data, $y_{ij}, i = 1, \ldots, m, j = 1, \ldots, n$. The point is that even the MLE based on a subset of the data, $\tilde{\mu}$, is consistent; and if one has more data (information), one should do better. Therefore, $\hat{\mu}$ has to be consistent as well.

3.3.2   The subset argument

A major breakthrough came in 2012, when Jiang [20] came up with an idea called *subset argument*. The idea was actually hinted by the "evidence" provided at the end of the previous subsection that supports a positive answer to the open problem. Basically, the idea suggests that, perhaps, one could use the fact that the MLE based on the subset data is consistent to argue that the MLE based on the full data is also consistent. The question is how to execute the idea.

Recall that, in the original proof of Wald ([39]; also see [40]), the focus was on the likelihood ratio $p_{\psi}(y)/p_{\psi_0}(y)$, and showing that the ratio converges to zero outside any (small) neighborhood of ψ_0, the true parameter vector. Can we execute the subset idea in terms of the likelihood ratio? This is where the breakthrough came by considering the relationship between the likelihood ratio under the full data and that under the subset data. It is expressed in the following *subset inequality* , which is the key to the proof.

Let us focus, for now, on the simple case of the open problem presented near the end of Section 1.3. Let $y_{[1]}$ denote the (row) vector of $y_{ii}, i = 1, \ldots, m \wedge n$, and $y_{[2]}$ the (row) vector of the rest of the $y_{ij}, i = 1, \ldots, m, j = 1, \ldots, n$. Let $p_{\mu}(y_{[1]}, y_{[2]})$ denote the pmf of $(y_{[1]}, y_{[2]})$, $p_{\mu}(y_{[1]})$ the pmf of $y_{[1]}$,

$$p_{\mu}(y_{[2]}|y_{[1]}) = \frac{p_{\mu}(y_{[1]}, y_{[2]})}{p_{\mu}(y_{[1]})} \tag{3.67}$$

the conditional pmf of $y_{[2]}$ given $y_{[1]}$, and P_{μ} the probability distribution, respectively, when μ is the true parameter. For any $\varepsilon > 0$, we have

$$P_{\mu}\{p_{\mu}(y_{[1]}, y_{[2]}) \leq p_{\mu+\varepsilon}(y_{[1]}, y_{[2]})|y_{[1]}\} = P_{\mu}\left\{ \frac{p_{\mu+\varepsilon}(y_{[1]}, y_{[2]})}{p_{\mu}(y_{[1]}, y_{[2]})} \geq 1 \middle| y_{[1]} \right\}$$

$$\leq \quad E\left\{\frac{p_{\mu+\varepsilon}(y_{[1]}, y_{[2]})}{p_{\mu}(y_{[1]}, y_{[2]})}\bigg| y_{[1]}\right\}$$

$$= \quad \sum_{y_{[2]}} \frac{p_{\mu+\varepsilon}(y_{[1]}, y_{[2]})}{p_{\mu}(y_{[1]}, y_{[2]})} p_{\mu}(y_{[2]}|y_{[1]})$$

$$= \quad \sum_{y_{[2]}} \frac{p_{\mu+\varepsilon}(y_{[1]}, y_{[2]})}{p_{\mu}(y_{[1]})}$$

$$= \quad \frac{p_{\mu+\varepsilon}(y_{[1]})}{p_{\mu}(y_{[1]})}, \tag{3.68}$$

using (3.67). Note that there is a likelihood ratio (LR) on each side of inequality (3.68); however, there is a huge difference between these two LRs. The first one is the LR based on the full data, which is a piece of "mess", in view of (1.10); the second is a LR based on independent (actually, i.i.d.) observations (see below), which is "nice and clean". The series of equalities and inequality in (3.68) is what we call the subset argument, while the following arguments are fairly standard, which we present simply for the sake of completeness.

On the other hand, by the standard asymptotic arguments (e.g., [1], p. 9), it can be shown that the likelihood ratio $p_{\mu+\varepsilon}(y_{[1]})/p_{\mu}(y_{[1]})$ converges to zero in probability, as $m \wedge n \to \infty$. Here we use the fact that the components of $y_{[1]}$, that is, $y_{ii}, 1 \leq i \leq m \wedge n$ are independent Bernoulli random variables. It follows that, for any $\eta > 0$, there is $N_\eta \geq 1$ such that, with probability $\geq 1 - \eta$, we have

$$\begin{aligned}
\zeta_N &= P_\mu\{p_\mu(y_{[1]}, y_{[2]}) \\
&\leq p_{\mu+\varepsilon}(y_{[1]}, y_{[2]})|y_{[1]}\} \\
&\leq \gamma^{m \wedge n}
\end{aligned}$$

for some $0 < \gamma < 1$, if $m \wedge n \geq N_\eta$. The argument shows that $\zeta_N = O_P(\gamma^{m \wedge n})$, hence converges to 0 in probability. It follows, by the dominated convergence theorem (e.g., [1], p. 32), that

$$\begin{aligned}
E_\mu(\zeta_N) &= P_\mu\{p_\mu(y_{[1]}, y_{[2]}) \\
&\leq p_{\mu+\varepsilon}(y_{[1]}, y_{[2]})\} \to 0.
\end{aligned}$$

Similarly, it can be shown that

$$P_\mu\{p_\mu(y_{[1]}, y_{[2]}) \leq p_{\mu-\varepsilon}(y_{[1]}, y_{[2]})\} \to 0.$$

The rest of the proof follows by the standard arguments (e.g., [1], pp. 9–10), which leads to the following.

Theorem 3.8. There is, with probability tending to one, a root to the likelihood equation, $\hat{\mu}$, such that $\hat{\mu} \xrightarrow{P} \mu$.

Note that the idea of the proof has not followed the traditional path of attempting to develop a (computational) procedure to approximate the MLE, such as those discussed in the previous subsection. This might explain why the computational advances over the past two decades had not led to a major theoretical breakthrough. In fact, the subset argument also plays a key role in the subsequent results presented in the next subsection, which extend Theorem 3.8, in different ways. In principle, the subset argument may be viewed as a general method that allows one to argue consistency of the MLE in any situation of dependent data, not necessarily under a GLMM, provided that one can identify suitable subset(s) of the data whose asymptotic properties are easier to handle, such as collections of independent random vectors. The connection between the full data and subset data is made by the subset inequality, which, in a more general form, is a consequence of the martingale property of the likelihood-ratio (e.g., [1], pp. 244–246). We state this extended subset inequality as follows. Suppose that Y_1 is a subvector of a random vector Y. Let $p_\psi(\cdot)$ and $p_{1,\psi}(\cdot)$ denote the pdf's of Y and Y_1, respectively, with respect to a σ-finite measure ν, under the parameter vector ψ. For simplicity, suppose that p_{ψ_0}, p_{1,ψ_0} are positive a.e. ν, and $\lambda(\cdot)$ is a positive, measurable function. Then, for any ψ, we have

$$P_{\psi_0}\{p_{\psi_0}(Y) \leq \lambda(Y_1)p_\psi(Y)|Y_1\} \;\leq\; \lambda(Y_1)\frac{p_{1,\psi}(Y_1)}{p_{1,\psi_0}(Y_1)} \quad \text{a.e. } \nu, \qquad (3.69)$$

where P_{ψ_0} denotes the probability distribution corresponding to p_{ψ_0}.

3.3.3 Two types of consistency; extensions

The result of Theorem 3.8 is usually referred to as Cramér-type consistency ([38]), which states that a root to the likelihood equation is consistent. However, it does not always imply that the MLE, which by definition is the (global) maximizer of the likelihood function, is consistent. See earlier discussions in Section 2.1. A stronger result is called Wald-type consistency [39]; also see [40], which states that the MLE is consistent. Also note that the limiting process in Theorem 3.8 is $m, n \to \infty$, or, equivalently, $m \wedge n \to \infty$. With a slightly more restrictive limiting process, Wald-consistency can actually be established, as follows.

Theorem 3.9 (Wald consistency). If $(m \wedge n)^{-1} \log(m \vee n) \to 0$ as $m, n \to \infty$, then the MLE of μ is consistent.

Again, the subset argument plays a key role in the proof. Basically, the argument allows one to argue that, beyond a small neighborhood of the true parameter, the likelihood ratio is uniformly smaller than that at the true parameter. Thus, combined with the result of Theorem 3.9, which takes care of the small neighborhood, one gets the wald consistency. See [20] for details.

We now consider a few more applications of the subset argument to GLMM under a general setting. The detailed proofs can be found in [20]. Note that it is typically possible to find a subset of the data that are independent, in some way, under a GLMM. For example, under the ANOVA GLMM (see Section 3.2.3), a subset of independent data can always be found. The "trick" is to select a subset, or more than one subsets if necessary, with the following desirable properties:

(I) the subset(s) can be divided into independent clusters with the number(s) of clusters increasing with the sample size; and

(II) the combination of the subset(s) jointly identify all the unknown parameters. More specifically, let $y_i^{(a)}, i = 1, \ldots, N_a$ be the ath subset of the data, $1 \leq a \leq b$, where b is a fixed positive integer. Suppose that, for each a, there is a partition, $\{1, \ldots, N_a\} = \cup_{j=1}^{m_a} S_{a,j}$. Let $y_{a,j} = [y_i^{(a)}]_{i \in S_{a,j}}$, and $p_\psi(y_{a,j})$ be pdf of $y_{a,j}$, with respect to the measure ν (or the product measure induced by μ if $y_{a,j}$ is multivariate), when ψ is the true parameter vector. Let Ψ denote the parameter space, and ψ_0 the true parameter vector. Then, (I) and (II) can be formally stated as follows:

A1. $y_{a,j}, 1 \leq j \leq m_a$ are independent with $m_a \to \infty$ as $N \to \infty, 1 \leq a \leq b$, where N is the total sample size;

A2. for every $\psi \in \Psi \setminus \{\psi_0\}$, we have

$$\min_{1 \leq a \leq b} \limsup_{N \to \infty} \frac{1}{m_a} \sum_{j=1}^{m_a} \mathrm{E}_{\psi_0} \left[\log \left\{ \frac{p_\psi(y_{a,j})}{p_{\psi_0}(y_{a,j})} \right\} \right] < 0.$$

Note that A2 controls the average Kullback-Leibler information ([151]); thus, the inequality always holds if $<$ is replaced by $\leq$.

First, let us consider a simpler case by assuming that Ψ is finite. Although the assumption may seem restrictive, it is not totally unrealistic. For example, any computer system only allows a finite number of digits. This means that the parameter space that is practically stored in a computer system is finite. Using the subset argument, it is fairly straightforward to prove the following.

Theorem 3.10. Under assumptions A1 and A2, if, in addition,

A3. for every $\psi \in \Psi \setminus \{\psi_0\}$, we have

$$\frac{1}{m_a^2} \sum_{j=1}^{m_a} \mathrm{var}_{\psi_0} \left[\log \left\{ \frac{p_\psi(y_{a,j})}{p_{\psi_0}(y_{a,j})} \right\} \right] \longrightarrow 0, \quad 1 \leq a \leq b,$$

then $P_{\psi_0}(\hat{\psi} = \psi_0) \to 1$, as $N \to \infty$, where $\hat{\psi}$ is the MLE of ψ.

We now consider the case that Ψ is a convex subspace of R^d, the d-dimensional Euclidean space, in the sense that $\psi_1, \psi_2 \in \Psi$ implies $(1-t)\psi_1 + t\psi_2 \in \Psi$ for every $t \in (0,1)$. We need to strenthen assumptions A2 and A3 to the following:

B2. $\psi_0 \in \Psi^\circ$, the interior of Ψ, and there is $0 < M < \infty$ (same as in B3 below) such that, for every $\varepsilon > 0$, we have

$$\limsup_{N \to \infty} \sup_{\psi \in \Psi, \varepsilon \leq |\psi - \psi_0| \leq M} \min_{1 \leq a \leq b} \frac{1}{m_a} \sum_{j=1}^{m_a} \mathrm{E}_{\psi_0} \left[\log \left\{ \frac{p_\psi(y_{a,j})}{p_{\psi_0}(y_{a,j})} \right\} \right] < 0. \quad (3.70)$$

B3. There are positive constant sequences $s_N, s_{a,N}, 1 \leq a \leq b$ such that

$$\sup_{\psi \in \Psi, |\psi - \psi_0| \leq M} \max_{1 \leq c \leq d} \left| \frac{\partial}{\partial \psi_c} \log\{p_\psi(y)\} \right| = O_P(s_N) \quad (3.71)$$

with $\log(s_N) / \min_{1 \leq a \leq b} m_a \to 0$, where $p_\psi(y)$ is the pdf of of $y = (y_i)_{1 \leq i \leq N}$ given that $\psi = (\psi_c)_{1 \leq c \leq d}$ is the true parameter vector;

$$\sup_{\psi \in \Psi, |\psi - \psi_0| \leq M} \frac{1}{m_a} \sum_{j=1}^{m_a} \max_{1 \leq c \leq d} \left| \frac{\partial}{\partial \psi_c} \log\{p_\psi(y_{a,j})\} \right| = O_P(s_{a,N}) \quad (3.72)$$

with $\log(s_{a,N})/m_a \to 0$; and (for the same $s_{a,N}$)

$$\sup_{\psi \in \Psi, |\psi - \psi_0| \leq M} \frac{s_{a,N}^{d-1}}{m_a^2} \sum_{j=1}^{m_a} \mathrm{var}_{\psi_0}\left[\log\left\{\frac{p_\psi(y_{a,j})}{p_{\psi_0}(y_{a,j})}\right\}\right] \longrightarrow 0, \quad 1 \leq a \leq b. \quad (3.73)$$

Theorem 3.11. *Under assumptions A1, B2 and B3, there is, with probability $\to 1$, a root to the likelihood equation, $\hat{\psi}$, such that $\hat{\psi} \xrightarrow{P} \psi_0$, as $N \to \infty$.*

Again, Theorem 3.11 is a Cramér-consistency result. A Wald-consistency result can be established under additional assumptions that control the behavior of the likelihood function in a neighborhood of infinity. For example, the following result may be viewed as an extension of Theorem 3.9. Once again, the subset argument plays a critical role in the proof. For simplicity, we focus on the case of discrete responses, which is typical for GLMMs. In addition, we assume the following. For any $0 \leq v < w$, define $S_d[v, w) = \{x \in R^d : v \leq |x| < w\}$ and write, in short, $S_d(k) = S_d[k, k+1)$ for $k = 1, 2, \ldots$.

C1. There are sequences of constants, $b_k, c_N \geq 1$, and random variables, ζ_N, where c_N, ζ_N do not depend on k, such that $\zeta_N = O_P(1)$ and

$$\sup_{\psi \in \Psi \cap S_d[k-1,k+2)} \max_{1 \leq c \leq d} \left|\frac{\partial}{\partial \psi_c} \log\{p_\psi(y)\}\right| \leq b_k c_N \zeta_N, \quad k = 1, 2, \ldots$$

C2: There is a subset of independent data vectors, $y_{(j)}, 1 \leq j \leq m_N$ (not necessarily among those in *A1*) so that:
(i) $E_{\psi_0}|\log\{p_{j,\psi_0}(y_{(j)})\}|$ is bounded, $p_{j,\psi}(\cdot)$ being the pmf of $y_{(j)}$ under ψ;
(ii) there is a sequence of positive constants, γ_k, with $\lim_{k\to\infty} \gamma_k = \infty$, and a subset $\mathcal{T}_N$ of possible values of $y_{(j)}$, such that for every $k \geq 1$ and $\psi \in \Psi \cap S_d(k)$, there is $t \in \mathcal{T}_N$ satisfying $\max_{1 \leq j \leq m_N} \log\{p_{j,\psi}(t)\} \leq -\gamma_k$;
(iii) $\inf_{t \in \mathcal{T}_N} m_N^{-1} \sum_{j=1}^{m_N} p_{j,\psi_0}(t) \geq \rho$ for some constant $\rho > 0$; and
(iv) $|\mathcal{T}_N|/m_N = o(1)$, and $c_N^d \sum_{k=K}^{\infty} k^{d_1} b_k^d e^{-\delta m_N \gamma_k} = o(1)$ for some $K \geq 1$ and $\delta < \rho$, where $d_1 = d 1_{(d>1)}$.

It is easy to verify that the new assumptions *C1, C2* are satisfied in the case of Theorem 3.9. Another example is considered in the next subsection.

Theorem 3.12 (Wald consistency). *Suppose that A1 holds; B2, B3 hold for any fixed $M > 0$ (instead of some $M > 0$), and with the $s_{a,N}^{d-1}$ in (3.73) replaced by $s_{a,N}^d$. In addition, suppose that C1, C2 hold. Then, the MLE of ψ_0 is consistent.*

3.3.4 Example

Let us consider a special case of the mixed logistic model with crossed random effects considered in Section 1.3, with $x'_{ijk}\beta = \mu$ and σ^2 and τ^2 unknown. We change the notation slightly, namely, $y_{i,j,k}$ instead of y_{ijk}. Suppose that $S = S_1 \cup S_2$ such that $c_{ij} = r, (i,j) \in S_r, r = 1, 2$ (as in the case of the salamander data). We use two subsets to jointly identify all the unknown parameters. The first subset is similar to that used in the proofs of Theorems 3.8, namely, $y_{i,i} = (y_{i,i,k})_{k=1,2}, (i,i) \in S_2$. Let m_1 be the total number of such (i,i)'s, and assume that $m_1 \to \infty$, as $m,n \to \infty$. Then, the subset satisfies $A1$. Let $\psi = (\mu, \sigma^2, \tau^2)'$. It can be shown that the sequence $y_{i,i}, (i,i) \in S_2$ is a sequence of i.i.d. random vectors with the probability distribution given by

$$p_\psi(y_{i,i}) = E\left[\frac{\exp\{y_{i,i,\cdot}(\mu+\xi)\}}{\{1+\exp(\mu+\xi)\}^2}\right], \tag{3.74}$$

where $\xi \sim N(0, \omega^2)$, with $\omega^2 = \sigma^2 + \tau^2$, and $y_{i,i,\cdot} = y_{i,i,1} + y_{i,i,2}$. By the strict concavity of the logarithm, we have

$$E_{\psi_0}\left[\log\left\{\frac{p_\psi(y_{i,i})}{p_{\psi_0}(y_{i,i})}\right\}\right] < 0 \tag{3.75}$$

unless $p_\psi(y_{i,i})/p_{\psi_0}(y_{i,i})$ is a.s. P_{ψ_0} a constant, which must be one because both p_ψ and p_{ψ_0} are probability distributions. It is easy to show that the probability distribution of (3.74) is completely determined by the function $M(\vartheta) = [M_r(\vartheta)]_{r=1,2}$, where $M_r(\vartheta) = E\{h^r_\vartheta(\zeta)\}$ with $\vartheta = (\mu, \omega)'$, $h_\vartheta(\zeta) = \exp(\mu + \omega\zeta)/\{1 + \exp(\mu + \omega\zeta)\}$, and $\zeta \sim N(0,1)$. In other words, $p_\psi(y_{i,i}) = p_{\psi_0}(y_{i,i})$ for all values of $y_{i,i}$ if and only if $M(\vartheta) = M(\vartheta_0)$. [17] showed that the function $M(\cdot)$ is injective (also see [2], p. 221). Thus, (3.75) holds unless $\mu = \mu_0$ and $\omega^2 = \omega_0^2$.

It remains to deal with a ψ that satisfies $\mu = \mu_0$, $\omega^2 = \omega_0^2$, but $\psi \neq \psi_0$. For such a ψ, we use the second subset, defined as $y_i = (y_{i,2i-1,1}, y_{i,2i,1})'$ such that $(i, 2i-1) \in S$ and $(i, 2i) \in S$. Let m_2 be the total number of all such i's, and assume that $m_2 \to \infty$ as $m, n \to \infty$. It is easy to see that $A1$ is, again, satisfied for the new subset. Note that any ψ satisfying $\mu = \mu_0$ and $\omega^2 = \omega_0^2$ is completely determined by the parameter $\gamma = \sigma^2/\omega^2$. Furthermore, the new subset is a sequence of i.i.d. random vectors with the probability distribution, under such a ψ, given by

$$p_\gamma(y_i) = E\left[\frac{\exp\{y_{i,2i-1,1}(\mu_0+X)\}}{1+\exp(\mu_0+X)} \cdot \frac{\exp\{y_{i,2i,1}(\mu_0+Y)\}}{1+\exp(\mu_0+Y)}\right], \tag{3.76}$$

where (X, Y) has the bivariate normal distribution with $\text{var}(X) = \text{var}(Y) = \omega_0^2$ and $\text{cor}(X, Y) = \gamma$. Similar to (3.75), we have

$$E_{\gamma_0}\left[\log\left\{\frac{p_\gamma(y_i)}{p_{\gamma_0}(y_i)}\right\}\right] < 0 \tag{3.77}$$

unless $p_\gamma(y_i) = p_{\gamma_0}(y_i)$ for all values of y_i. Consider (3.76) with $y_i = (1,1)$ and let P_γ denote the probability distribution of (X, Y) with the correlation coefficient γ. By Fubini's theorem, it can be shown that

$$p_\gamma(1,1) = \int_0^\infty \int_0^\infty P_\gamma\{X \geq \text{logit}(s) - \mu_0, Y \geq \text{logit}(t) - \mu_0\} ds dt. \tag{3.78}$$

Furthermore, by Slepian's inequality (e.g., [1], pp. 157–158), the integrand on the right side of (3.78) is strictly increasing with γ, hence so is the integral. Thus, if $\gamma \neq \gamma_0$, at least we have $p_\gamma(1,1) \neq p_{\gamma_0}(1,1)$, hence (3.77) holds.

In summary, for any $\psi \in \Psi$, $\psi \neq \psi_0$, we must have either (3.75) or (3.77) hold. Therefore, by continuity, assumption $B2$ holds, provided that true variances, σ_0^2, τ_0^2

are positive. Note that, in the current case, the expectations involved in *B2* do not depend on either j or N, the total samlpe size.

To verify *B3*, it can be shown that $|(\partial/\partial\mu)\log\{p_\psi(y)\}| \leq N$. Furthermore, we have $|(\partial/\partial\sigma^2)\log\{p_\psi(y)\}| \vee |(\partial/\partial\tau^2)\log\{p_\psi(y)\}| \leq (A+C+1)N$ in a neighborhood of θ_0, $\mathcal{N}(\theta_0)$. Therefore, (3.71) holds with $s_N = N$.

As for (3.72), it is easy to show that the partial derivatives involved are uniformly bounded for $\psi \in \mathcal{N}(\psi_0)$. Thus, (3.72) holds for any $s_{a,N}$ such that $s_{a,N} \to \infty$, $a = 1,2$. Furthermore, the left side of (3.73) is bounded by $c_a s_{a,N}^2/m_a$ for some constant $c_a > 0$, $a = 1,2$ (note that $d = 3$ in this case). Thus, for example, we may choose $s_{a,N} = \sqrt{m_a/\{1+\log(m_a)\}}$, $a = 1,2$, to ensure that $\log(s_{a,N})/m_a \to 0$, $a = 1,2$, and (3.73) holds.

In conclusion, all of the assumptions of Theorem 3.11 hold provided that $\sigma_0^2 > 0$, $\tau_0^2 > 0$, and $(m_1 \wedge m_2)^{-1}\log(N) \to 0$.

Similarly, the conditions of Theorem 3.12 can be verified. Essentially, what is new is to check assumptions *C1* and *C2*. See the Supplementary Material of [20].

3.4 Conditional inference

In Section 2.4, we briefly discussed asymptotic properties of the EBLUPs (see the paragraph following Example 2.13). This section is intended to extend these results to GLMMs, in a certain sense. More specifically, it is about asymptotics in estimating jointly the fixed and random effects in GLMMs.

3.4.1 *Maximum posterior*

The BLUPs and EBLUPs make complete sense in estimating the random effects under the LMM; however, they are not necessarily meaningful in GLMM. The reason is that, in the latter case, the model is non-linear; thus, it is no longer natural to focus on linear mixed effects, that is, linear combination of fixed and random effects. We illustrate with a real-data example.

Example 3.9 (non-linear mixed effecs). Jiang [152] considered data from the Behavioral Risk Factor Surveillance System (BRFSS). The latter is a Center of Disease Control and Prevention coordinated, state-based random-digit-dialing telephone survey. More specifically, the authors were interested in a data set regarding the use of mammography among women aged 40 or older, from 1993 to 1995, and for areas from three Federal Regional Offices: Boston (Maine, Vermont, Massachusetts, Connecticut, Rhode Island, and New Hampshire), New York (New York and New Jersey), and Philadelphia (Pennsylvania, Delaware, Washington D.C., Maryland, Virginia, and West Virginia). More specifically, there was interest in estimating the proportions of women in different health service areas (HSAs) who had mammography during the period, and the proportion of interest was expressed, in the logit scale, as

$$\text{logit}(p) = \beta_0 + \beta_1 * \text{age} + \beta_2 \text{age}^2 + \beta_3 * \text{race} + \beta_4 * \text{edu\%}$$
$$+ \text{HSA effect}, \tag{3.79}$$

under the mixed logistic model proposed by [152]. Here, in (3.79), the β's were the fixed effects while HSA effect was considered an area-specific random effect, with the area being the HSA. There were a total of 118 HSAs involved in the study.

It is clear that the right side of (3.79) is a linear mixed effect. However, the quantity of main interest, p, is a non-linear mixed effect.

Let y be a vector of observations, γ a vector of unobserved "random variables", and ψ a vector of parameters. Here, γ may be a vector of random effects, or a vector of fixed and random effects; correspondingly, ψ may be a vector of fixed effects and variance components, or a vector of variance components only. Let $p(y, \gamma|\psi)$ denote the joint pdf of y and γ, with respect to a σ-finite measure, ν, given that ψ is the true parameter vector. Note that, in case that γ involves the fixed effects, one may need to define what is meant by the distribution of γ, perhaps, under a Bayesian framework, but this is not important, at least at this point. What is important is the following well-known relationship:

$$p(y, \gamma|\psi) \quad = \quad p(y|\psi)p(\gamma|y, \psi), \tag{3.80}$$

where $p(y|\psi)$ and $p(\gamma|y, \psi)$ denote the marginal pdf of y and conditional pdf of γ given y, respectively, given that ψ is the true parameter vector. Using a Bayesian term, $p(\gamma|y, \psi)$ is called a posterior. Henderson's original idea ([153] of BLUP, was to find $\hat{\gamma} = \hat{\gamma}(y, \psi)$ that maximizes the left side of (3.80) (e.g., [2], p. 76). From the right side of the same equation, this is equivalent to finding $\hat{\gamma}$ that maximizes $p(\gamma|y, \psi)$, the posterior. In other words, the BLUP may be regarded as a maximum posterior (MPE) estimator of γ, and this concept is not restricted to linear models.

To be specific, let $\gamma = (\beta', \alpha')'$, where β and α are vectors of fixed and random effects, respectively. The MPE of β and α are typically obtained by solving a system of equations that equal the derivatives of $l(y, \gamma|\psi) = \log\{p(y, \gamma|\psi)\}$ to zero, that is,

$$\frac{\partial l}{\partial \beta} \quad = \quad 0, \tag{3.81}$$

$$\frac{\partial l}{\partial \alpha} \quad = \quad 0, \tag{3.82}$$

In practice, there are often a large number of random effects involved in a GLMM. For example, the number of HSA-specific random effects involved in Example 3.9 is 118. This means that one has to simultaneously solve a large number of nonlinear equations. It is well known that standard methods of solving nonlinear systems, such as Newton–Raphson (N-R), may be inefficient and extremely slow when the dimension of the solution is high. In fact, even in the linear case directly solving the BLUP equations may involve inversion of a large matrix, which can be computationally burdensome. There is another disadvantages of N-R, that is, its convergence is sensitive to the initial value. In fact, when the dimension of the solution is high, it can be very difficult to find an initial values that will result in convergence. Jiang [154] proposed a nonlinear Gauss–Seidel algorithm (NLGSA), and proved global convergence of NLGSA, that is, that the algorithm converges with any initial value. We use an example to illustrate.

Example 3.10. Consider a special case of the mixed logistic model associated with the likelihood function (1.10). Namely, given the random effects $u_1,\ldots,u_m$ and $v_1,\ldots,v_n$, binary responses y_{ij}, $i=1,\ldots,m$, $j=1,\ldots,n$ are conditionally independent such that, with $p_{ij}=P(y_{ij}=1|u,v)$, one has

$$\mathrm{logit}(p_{ij}) = \mu+u_i+v_j, \tag{3.83}$$

where μ is an unknown parameter, $u=(u_i)_{1\le i\le m}$, and $v=(v_j)_{1\le j\le n}$. Assume that the random effects are independent with $u_i\sim N(0,\sigma^2)$ and $v_j\sim N(0,\tau^2)$. To compute the MPE, one may solve the equation (3.82), with $\alpha=(u',v')'$, as functions of μ; the solution $\tilde\alpha=\tilde\alpha(\mu)$ is brought to (3.81), with $\beta=\mu$, to solve for μ, etc. Thus, the focus is to solve the following system of nonlinear equations given μ:

$$\frac{u_i}{\sigma^2}+\sum_{j=1}^{n}\frac{\exp(\mu+u_i+v_j)}{1+\exp(\mu+u_i+v_j)} = y_{i\cdot}, \quad 1\le i\le m, \tag{3.84}$$

$$\frac{v_j}{\tau^2}+\sum_{i=1}^{m}\frac{\exp(\mu+u_i+v_j)}{1+\exp(\mu+u_i+v_j)} = y_{\cdot j}, \quad 1\le j\le n, \tag{3.85}$$

where $y_{i\cdot}=\sum_{j=1}^{n}y_{ij}$ and $y_{\cdot j}=\sum_{i=1}^{m}y_{ij}$. Note that given the v_js, each equation in (3.84) is univariate, which can be easily solved. A similar observation is made regarding (3.85). This motivates the following algorithm. Starting with initial values $v_j^{(0)}$, $1\le j\le n$, solve (3.84) with $v_j^{(0)}$ in place of v_j, $1\le j\le n$ to get $u_i^{(1)}$, $1\le i\le m$; then (3.85) with $u_i^{(1)}$ in place of u_i, $1\le i\le m$ to get $v_j^{(1)}$, $1\le j\le n$; and so on. It is clear that each step of the algorithm involves solving a univariate equation that has a unique solution, and is very easy to operate; the algorithm is guaranteed convergent, and the convergence is not affected by the initial values [154].

The convergence of NLGSA implies existence of a solution to (3.81) and (3.82); on the other hand, it is known that the solution to those equations, if exists, is unique and is equal to the MPE ([152], Lemma 2.1). Thus, in conclusion, the MPE exists, which is the unique solution to (3.81) and (3.82), which can be solved via the NLGSA, which converges globally.

Jiang [152] also studied asymptotic behavior of MPE under the ANOVA GLMM with the standard design [see Section 3.2.4, below (3.55)]. The results are similar to those presented in the next subsection under the notion of *conditional inference*.

Note that MPE is in the same spirit of PQL in terms of estimating the fixed effects, by estimating the latter jointly with the random effects. However, the main focus of PQL is estimating the fixed parameters, which also include the variance components. Another similar method is maximum hierarchical likelihood (MHL), proposed by [155]. The method may be regarded as an extension of PQL in that it allows the random effects to have certain nonnormal distributions. More specifically, the random effects are assumed to have a conjugate distribution. Consider, for simplicity, the case where the responses may be expressed as y_{ij}, $i=1,\ldots,m$, $j=1,\ldots,n_i$, and u_i is a random effect associated with the ith cluster. Consider the canonical link function such that $\xi_{ij}=\xi(\mu_{ij})=\xi(g^{-1}(x_{ij}'\beta))+\alpha_i$ with $\alpha_i=\xi(u_i)$.

The hierarchical likelihood, or h-likelihood, is defined as the logarithm of the joint density function of y and α; that is,

$$h = l(\xi,\phi;y|\alpha) + l(v;\alpha), \tag{3.86}$$

where $l(\xi,\phi;y|v) = \sum_{ij} l_{ij}$ with l_{ij} being (3.1) with the index i replaced by ij, and v is an additional parameter. As for the second term on the right side of (3.86), under the conjugate distribution, it is assumed that the kernel of $l(v;v)$ can be expressed as,

$$\sum_i \{a_1(v)\alpha_i - a_2(v)b(\alpha_i)\}, \tag{3.87}$$

where $a_1(\cdot)$ and $a_2(\cdot)$ are some functions. Note that the function $b(\cdot)$ is the same as that in (3.1). [155] noted that, although expression (3.87) takes the form of the Bayesian conjugate prior (e.g., [156], pp. 370), it is only for α; that is, no priors are specified for β, ϕ, or v. By maximizing the h-likelihood, one obtains the equations $\partial h/\partial\beta = 0$, $\partial h/\partial\alpha = 0$, which are similat to (3.81) and (3.82).

It is clear that, when normality, instead of conjugate distribution, is assumed for the random effects, MHL is the same as the PQL, or joint estimating the fixed and random effects. As noted earlier, the joint estimating idea was originated from Henderson [153]. An advantage of the conjugate distribution is that the MHL estimators of the random effects are simple on the u-scale. To see this, note that under the assumed model, and using properties of the exponential family (e.g., [2], Section C.5), one has $(\partial b/\partial\xi = \mu$, so that $\partial b/\partial\alpha = u$. Thus, by differentiating the h-likelihood with respect to α_i and letting the derivative equal zero, one obtains

$$u_i = \frac{y_{i\cdot} - \mu_{i\cdot} + \phi a_1(v)}{\phi a_2(v)}, \tag{3.88}$$

where $y_{i\cdot} = \sum_j y_{ij}$ and $\mu_{i\cdot} = \sum_j \mu_{ij}$. Note that (3.38) is not a closed-form expression for u_i, because $\mu_{i\cdot}$ also involves u_i.

Lee and Nelder showed that, in some cases, the MHL estimator for β is the same as the (marginal) MLE for β. One such case is, of course, the normal–normal case, or Gaussian mixed models, discussed in Chapter 2. Another example is the Poisson-gamma model (see [2], Example 3.6 and Exercise 3.7).

In terms of asymptotic properties, Lee and Nelder showed that the MHL estimators of the fixed effects are asymptotically equivalent to the (marginal) MLE of the fixed effects. However, the asymptotics is in the sense that $n_i \to \infty$, $1 \le i \le m$ at the same rate whereas t, the number of clusters, remains constant. Such a condition is often unrealistic in a mixed model situation. For example, in small area estimation (see the next chapter), n_i corresponds to the sample size for the ith small area, which may be quite small, whereas the number of small areas, m, can be quite large. In fact, as Lee and Nelder pointed out, the MHL equations are the first-order approximation to the ML equations. What they did not mention is that such an approximation becomes accurate only when the cluster sizes n_i become large. As for the MHL estimators of the random effects, Lee and Nelder showed that they are asymptotically best unbiased predictors, under the same asymptotic assumption. Although the latter may not be realistic, it might seem reasonable if the objective were to consistently estimate the individual random effects. The only part of this assumption that could be weakened is that the number of clusters, m, should be allowed to increase, too. See the next subsection.

3.4.2   Conditional inference

In the case of GLMM, if we assume that $\alpha \sim N(0, G)$, where the covariance $G = G(\theta)$ depends on a vector θ of variance components, the objective function, $l = \log\{p(y, \gamma | \psi)\}$, for the MPE can be expressed as

$$l \; = \; c(y; \theta, \phi) + \sum_{i=1}^{N} \frac{y_i \xi_i - b(\xi_i)}{a_i(\phi)} - \frac{1}{2} \alpha' G^{-1} \alpha, \qquad (3.89)$$

where $c(y; \theta, \phi)$ is a function of y, θ and ψ; thus, $\psi = (\theta', \phi)'$ in this case. One complication is that, usually, the variance components in θ are unknown (ϕ may or may not be known); however, they are involved in (3.89). So, a question is how one can maximize (3.89) without knowing θ. In this regard, a somewhat striking result proved by [152] helps. It was shown that, under suitable conditions, the MPE has similar consistency properties as Theorem 3.13 below regardless at what θ (3.89) is evaluated. So, for example, one can use whatever reasonable guess of θ in (3.89), and the consistency property of the resulting MPE is not affected. In a way, this desirable asymptotic property of MPE is similar to the consistency of GEE estimator under working covariance matrices (see Section 3.2.1).

On the other hand, once one obtains the MPE, a consistent estimator of θ may be easily obtained. For example, suppose that one component of θ is the variance of the random effects, $\alpha_i, 1 \le i \le m$, say σ^2. Then, the latter can be estimated by $\hat{\sigma}^2 = m^{-1} \sum_{i=1}^{m} \hat{\alpha}_i^2$, where $\hat{\alpha}_i$ is the MPE of α_i. Under suitable conditions, $\hat{\sigma}^2$ is a consistent estimator of σ^2 (see [152], Section 5). Furthermore, one may replace θ in (3.89) by its estimator obtained based on the MPE, and iterate between the two steps. This is simialr to the IEEE procedure (see Section 3.2.2).

[157] goes one-step further to propose an extension of (3.89) to avoid dealing with θ altogether. For simplicity, assume that ϕ is known. First, one can drop the c function in (3.89) for estimating γ. Secondly, we can extend the summation in (3.89) by allowing the function $b(\cdot)$ to depend on i, and write the summation as

$$\sum_{i=1}^{N} w_i \{ y_i \xi_i - b_i(\xi_i) \}, \qquad (3.90)$$

where the weights w_i are known. Finally, the last term in (3.89) may be viewed as a penalty, say, on larger value of α. In fact, there are more reasons to "penalize" than just large values. One of those is that, conditionally, the individual fixed and random effects in a GLMM may not be identifiable. To see this, consider Example 3.10. Note that one can write (3.83) as

$$\text{logit}(p_{ij}) \; = \; (\mu + c + d) + (u_i - c) + (v_j - d)$$

for any c and d. Of course, such a problem occurs in linear models as well, in which case there are two remedies: (i) reparameterization; and (ii) constraints. Let us focus on remedy (ii). A set of linear constraints on α may be expressed as $P\alpha = 0$ for some matrix P. By Lagrange's method of multipliers, maximizing (3.90) subject to $P\alpha = 0$ is equivalent to maximizing

$$\sum_{i=1}^{N} w_i\{y_i\xi_i - b_i(\xi_i)\} - \lambda|P\alpha|^2 \tag{3.91}$$

without constraints, where λ is the Lagrange multiplier. On the other hand, for fixed λ the last term in (3.91) may be regarded as a penalizer. The only thing that needs to be specified is the matrix P.

For any matrix M and vector space V, let $\mathscr{B}(V) = \{B : B$ is a matrix whose columns constitute a base for $V\}$; $\mathscr{N}(M) = $ the null-space of $M = \{v : Mv = 0\}$; $P_M = M(M'M)^-M'$, where A^- denotes the generalized inverse of A (e.g., [2], §B.4); and $P_{M^\perp} = I - P_M$. Let $A \in \mathscr{B}\{\mathscr{N}(P_{X^\perp}Z)\}$, where the ith row of X and Z are x_i' and z_i', respectively [see (3.2)]. We consider an example.

Example 3.11. Suppose that the linear predictor in the GLMM can be expressed as $\eta_{ij} = x_{ij}'\beta + \alpha_i$, $1 \le i \le m, 1 \le j \le n_i$, where the first component of x_{ij} is 1 corresponding to the intercept, and the matrix $X = (X_i)_{1 \le i \le m}$, with $X_i = (x_{ij}')_{1 \le j \le n_i}$, is full rank. Then, one may choose $A = 1_m$; hence, $P_A\alpha = \bar{\alpha}1_m$, where $\bar{\alpha} = m^{-1}\sum_{i=1}^{m}\alpha_i$. Thus, $P_A\alpha = 0$ is equivalent to $\alpha. = \sum_{i=1}^{m}\alpha_i = 0$.

In general, we define a penalized generalized WLS (PGWLS) estimator of $\gamma = (\beta', \alpha')'$ as the maximizer of

$$l_P(\gamma) = \sum_{i=1}^{N} w_i\{y_i\xi_i - b_i(\xi_i)\} - \frac{\lambda}{2}|P_A\alpha|^2, \tag{3.92}$$

where λ is a positive constant. The term PGWLS refers to the fact that, in the case of linear models without penalty, the procedure is known as WLS [see (2.62)].

We now further explain the choice of the penalty in (3.92), and present some asymptotic theory about PGWLS. In some ways, asymptotic theory regarding random effects is different from that about fixed parameters. First, the individual random effects are typically not identifiable [see the discussion below (3.90)]. Therefore, any asymptotic theory must take care, in particular, of the identifiability issue. Second, the number of random effects, m, should be allowed to increase with the sample size, N. For example, asymptotic properties of estimators of fixed parameters when the number of parameters increases with N have been studied by Portnoy in a series of papers (e.g., [158]).

To explore the asymptotic behavior of PGWLS estimators, we need to assume that m increases at a slower rate than N; that is, $m/N \to 0$. We also need to further explain how the matrix P_A in (3.92) is chosen. For simplicity, assume canonical link, which means that

$$\xi_i = \eta_i \tag{3.93}$$

holds for all i (e.g., [2], p. 121), where η_i is the linear prediction in (3.2). Note that the first term in (3.92), that is,

$$l_C(\gamma) = \sum_{i=1}^{N} w_i\{y_i\xi_i - b_i(\xi_i)\},$$

depends on $\gamma = (\beta', \alpha')'$ only through $\eta = X\beta + Z\alpha$. However, γ cannot be identified by η, so there may be many vectors γ corresponding to the same η. The idea is therefore to consider a restricted space $S = \{\gamma : P_A\alpha = 0\}$, such that within the subspace γ is uniquely determined by η. Define a map $T : \gamma \mapsto \tilde{\gamma} = (\tilde{\beta}, \tilde{\alpha})'$ as follows:

$$\tilde{\alpha} = P_{A\perp}\alpha,$$
$$\tilde{\beta} = \beta + (X'X)^{-1}X'ZP_A\alpha.$$

Obviously, T does not depend on the choice of A. Because

$$X\tilde{\beta} = X\beta + Z\alpha - P_{X\perp}ZP_A\alpha$$
$$= X\beta + Z\alpha,$$

we have $l_C(\gamma) = l_C(\tilde{\gamma})$. Let $G_A = \begin{pmatrix} X & Z \\ 0 & A' \end{pmatrix}$. The proofs of the following results can be found in [157].

Lemma 3.6. We have $\mathrm{rank}(G_A) = p + m$, where $p = \dim(\beta)$.

Corollary 3.1. Suppose that $b_i''(\cdot) > 0$, $1 \leq i \leq N$. Then, there can be only one maximizer of l_P.

Let B be a matrix, v a vector, and V a vector space. Define $\lambda_{\min}(B)|_V = \inf_{v \in V\setminus\{0\}}(v'Bv/v'v)$. Also, let $H = (X\ Z)'(X\ Z)$.

Lemma 3.7. For any positive numbers b_j, $1 \leq j \leq p$ and a_k, $1 \leq k \leq m$, we have

$$\lambda_{\min}(W^{-1}HW^{-1})|_{WS} \geq \frac{\lambda_{\min}(G_A'G_A)}{(\max_{1\leq j\leq p}b_j^2) \vee (\max_{1\leq k\leq m}a_k^2)} > 0,$$

where $W = \mathrm{diag}(b_1, \ldots, b_p, a_1, \ldots, a_m)$ and $WS = \{W\gamma : \gamma \in S\}$.

Let X_j (Z_k) denote the jth (kth) column of X (Z). We have the following result.

Theorem 3.13. Let $b_i(\cdot)$ be two-times continuously differentiable,

$$\max_{1\leq i\leq N}[w_i^2 E\{\mathrm{var}(y_i|\alpha)\}] \text{ be bounded,}$$

and

$$\frac{|P_A\alpha|^2}{N}\left\{\left(\max_{1\leq j\leq p}|X_j|^2\right)\|(X'X)^{-1}X'Z\|^2 + \left(\max_{1\leq k\leq m}|Z_k|^2\right)\right\}$$

converge to zero in probability. Let c_N, d_N be any sequences such that

$$\limsup(\max_{1\leq j\leq p}|\beta_j|/c_N) < 1,$$
$$P(\max_{1\leq k\leq m}|\alpha_k|/d_N < 1) \to 1.$$

Also, let $M_i \geq c_N \sum_{j=1}^{p} |x_{ij}| + d_N \sum_{k=1}^{m} |z_{ik}|$, $1 \leq i \leq N$, and $\hat{\gamma}$ be the maximizer of l_P over $\Gamma(M) = \{\gamma : |\eta_i| \leq M_i, 1 \leq i \leq N\}$. Then, we have

$$\frac{1}{N} \left\{ \sum_{j=1}^{p} |X_j|^2 (\hat{\beta}_j - \beta_j)^2 + \sum_{k=1}^{m} |Z_k|^2 (\hat{\alpha}_k - \alpha_k)^2 \right\} \longrightarrow 0$$

in probability, provided that

$$\frac{p+m}{N} = o(\omega^2), \tag{3.94}$$

where, with $W = \text{diag}(|X_1|, \ldots, |X_p|, |Z_1|, \ldots, |Z_m|)$, we have

$$\omega = \lambda_{\min}(W^{-1}HW^{-1})|_{WS} \min_{1 \leq i \leq n} \{w_i \inf_{|u| \leq M_i} |b_i''(u)|\}.$$

To better understand the result of Theorem 3.13, let us consider some special cases. The first two results, namely, (i) and (ii) of the following corollary, follow directly from the theorem.

Corollary 3.2. Suppose that the conditions of Theorem 3.13 and (3.94) hold.

(i) If p is fixed and $\liminf \lambda_{\min}(X'X)/N > 0$, $\hat{\beta}$ is consistent, where $\hat{\beta}$ is the β part of $\hat{\gamma}$.

(ii) Under the ANOVA GLMM where each Z_u ($1 \leq u \leq q$) is a standard design matrix [see above (3.38)], we have

$$\left(\sum_{v=1}^{m_u} n_{uv} \right)^{-1} \sum_{v=1}^{m_u} n_{uv} (\hat{\alpha}_{uv} - \alpha_{uv})^2 \xrightarrow{P} 0, \tag{3.95}$$

where $\alpha_u = (\alpha_{uv})_{1 \leq v \leq m_u}$, $\hat{\alpha}_u = (\hat{\alpha}_{uv})_{1 \leq v \leq m_u}$ is the α_u part of $\hat{\gamma}$, and n_{uv} is the number of appearances of α_{uv} in the model, $1 \leq u \leq q$.

The last result shows that, under suitable conditions, the PGWLS estimators of the fixed effects are consistent; the PGWLS of the random effects are consistent in some overall sense (but not necessarily individually). For example, in Example 3.5, where $q = 1$, $\alpha_1 = \alpha = (\alpha_i)_{1 \leq i \leq m}$ (the notation is a bit confusing here, as the index i corresponds to the component of α, not the vector α_1 as in the ANOVA GLMM), $\hat{\alpha}_1 = \hat{\alpha} = (\hat{\alpha}_i)_{1 \leq i \leq m}$, $m_1 = m$, and $n_{1v} = k, 1 \leq v \leq m$. Thus, (3.95) reduces to

$$\frac{1}{m} \sum_{i=1}^{m} (\hat{\alpha}_i - \alpha_i)^2 \xrightarrow{P} 0. \tag{3.96}$$

To discuss the next special case, consider the longitudinal GLMM , in which the responses are clustered groups with each group associated with a single (possibly vector-valued) random effect. Suppose that the random effects $\alpha_1, \ldots, \alpha_m$ satisfy $E(\alpha_i) = 0$. The responses are y_{ij}, $1 \leq i \leq m$, $1 \leq j \leq n_i$, such that, given the random that, given the random effects, y_{ij}s are (conditionally) independent with $E(y_{ij}|\alpha) = b_{ij}'(\eta_{ij})$, where $b_{ij}(\cdot)$ is differentiable. Furthermore, we have

$$\eta_{ij} = \mu + x_{ij}'\beta + z_i'\alpha_i,$$

where μ is an unknown intercept, $\beta = (\beta_j)_{1\leq j\leq s}$ (s is fixed) is an unknown vector of regression coefficients, and x_{ij} and z_i are known vectors. Such models are useful, for example, in the context of small-area estimation (see Chapter 3), in which α_i represents a random effect associated with the ith small area. Here we are interested in the estimation of μ, β as well as $v_i = z_i'\alpha$, the area-specific random effects. Thus, without loss of generality, we shall focus on the latter case, with

$$\eta_{ij} = \mu + x_{ij}'\beta + v_i,$$

where $v_1,\ldots,v_m$ are independent random effects with $E(v_i) = 0$. It is clear that the model is a special case of GLMM. Following the earlier notation, it can be shown that, in this case, $A = 1_m \in \mathcal{B}(\mathcal{N}(P_{X^\perp}Z))$, $S = \{\gamma : v_. = 0\}$, where $v_. = \sum_{i=1}^m v_i$. Thus, (3.92) as a more explicit expression:

$$l_P(\gamma) = \sum_{i=1}^m \sum_{j=1}^{n_i} w_{ij}\{y_{ij}\eta_{ij} - b_{ij}(\eta_{ij})\} - \frac{\lambda}{2}m\bar{v}^2,$$

where $\bar{v} = v_./m$. Write

$$\delta_N = \min_{i,j}\left\{w_{ij}\inf_{|u|\leq M_{ij}} b_{ij}''(u)\right\},$$

$$\lambda_N = \lambda_{\min}\left\{\sum_{i=1}^m\sum_{j=1}^{n_i}(x_{ij}-\bar{x}_i)(x_{ij}-\bar{x}_i)'\right\},$$

where $\bar{x}_i = n_i^{-1}\sum_{j=1}^{n_i} x_{ij}$. For this special type of longitudinal GLMM, we have the following more explicit result.

Theorem 3.14. Suppose that $b_{ij}''(\cdot)$ is continuous. Furthermore, suppose that $w_{ij}^2 E\{\text{var}(y_{ij}|v)\}$, $|x_{ij}|$ are bounded, $\liminf(\lambda_N/N) > 0$, and $\bar{v} \to 0$ in probability. Let c_N, d_N be such that $\limsup\{(|\mu|\vee|\beta|)/c_N\} < 1$ and $P(\max_i|v_i|/d_N < 1) \to 1$. Let M_{ij} satisfy $M_{ij} \geq c_N(1+|x_{ij}|) + d_N$. Finally, let $\hat{\gamma} = (\hat{\mu},\hat{\beta}',\hat{v}')'$ be the maximizer of l_P over $\Gamma(M) = \{\gamma : |\eta_{ij}| \leq M_{ij}, \forall i, j\}$. Then, we have $\hat{\beta} \to \beta$ in probability, and

$$\frac{1}{N}\sum_{i=1}^m n_i(\hat{a}_i - a_i)^2 \longrightarrow 0$$

in probability, where $a_i = \mu + v_i$ and $\hat{a}_i = \hat{\mu} + \hat{v}_i$, provided that $m/N = o(\delta_N^2)$. If the latter condition is strengthened to $(\min_{1\leq i\leq m} n_i)^{-1} = o(\delta_N^2)$, then we have, in addition, $\hat{\mu} \to \mu$, $N^{-1}\sum_{i=1}^m n_i(\hat{v}_i - v_i)^2 \to 0$, and $m^{-1}\sum_{i=1}^m(\hat{v}_i - v_i)^2 \to 0$ in probability.

3.4.3 Quadratic inference function

The QIF was introduced earlier in Section 3.2.1 for estimating fixed parameters. The method was extended by Wang, Tsai and Qu [159] to jointly estimating fixed and random effects. Recall a key to QIF is the score, which was defined by (3.21) for the fixed parameter case. To jointly estimate the fixed and random effects in GLMM, the new QIF starts with an objective function similar to (3.92); however, the quasi-deviance in the PQL, defined above (3.4), is used instead of the summand in (3.92). The objective function now takes the form

$$l_Q(\gamma) \quad = \quad -\frac{1}{2\phi}\sum_{i=1}^{m}d_i(y_i,\mu_i) - \frac{\lambda}{2}|P_A\alpha|^2, \tag{3.97}$$

where ϕ is a dispersion parameter, $y_i = (y_{ij})_{1\leq j\leq T}$, $\mu_i = (\mu_{ij})_{1\leq j\leq T}$ with $\mu_{ij} = E(y_{ij}|\alpha_i)$, $d_i(y_i,\mu_i)$ is a qausi-deviance that satisfies that, for all $1 \leq i \leq m$,

$$\frac{\partial d_i}{\partial \beta} \quad = \quad \frac{\partial \mu_i'}{\partial \beta}W_i^{-1}(y_i - \mu_i), \tag{3.98}$$

$$\frac{\partial d_i}{\partial \alpha_i} \quad = \quad \frac{\partial \mu_i'}{\partial \alpha_i}W_i^{-1}(y_i - \mu_i), \tag{3.99}$$

where $W_i = \mathrm{Var}(y_i|\alpha_i)$ and α_i is the ith component of α. Note that the PQL quasi-deviance [14] implies that the covariance matrix W_i is diagonal; however, this is not necessarily the case here, which allows to incorporate correlations among the y_{ij}'s given α_i. Also note that the penalty in (3.97) is the same as (3.92), with the same matrix A defined above the latter equation.

Another (important) difference from the previous subsection is that the GLMM assumption is relaxed, in two ways. First, the observations are not assumed to be conditionally independent given the random effects. This is practical because, for example, in longitudinal studies, there may be serial correlation over time, for each subject. This means that the correlations among the longitudinal observations from the same subject cannot be fully explained but the subject-specific random effect. In fact, in the analysis of longitudinal data (e.g., [7]), serial correlations are often modeled in addition to the random effects. As an example of LMM, note that, in Example 2.4, the covariance matrix of ε_i, R_i, is not necessarily diagonal. Second, the distribution of the random effects, α, is not assumed to be multivariate normal. In fact, the normality assumption does not play a big role for PGWLS either (see the previous subsection), and this has not changed in the development of QIF.

But before the QIF can be fully developed, there is one business that needs to be taken care of first. As can be seen, the set-up so far assumes that the number of observations in each cluster is the same for different clusters. In other words, the dimension of y_i is the same for different i. This may not be practical because, for example, in longitudinal studies, the observational times for different subjects may be different. To deal with this issue, [159] applied the following transformation for the case of longitudinal data. First, let $1,\ldots,T$ be the indexes corresponding to all possible observational times (i.e., those at which there is at least one observation, for some subject). Let n_i be the number of observations for the ith cluster, which may be different for different i's. Define a $T \times n_i$ transformation, Λ_i, by removing the columns of the $T \times T$ identity matrix, I_T, that correspond to the "missing" observations, for subject i. Then, define $y_i^* = \Lambda_i y_i$, $\mu_i^* = \Lambda_i \mu_i$, thus $\dot{\mu}_i^* = \Lambda_i \dot{\mu}_i$. Now recall the expression, $V_i = A_i^{1/2}RA_i^{1/2}$, used in developing the QIF in Section 3.2.1. Define

$A_i^* = \Lambda_i A_i \Lambda_i'$, and $(A_i^*)^{-1} = \Lambda_i A_i^{-1} \Lambda_i'$. Note that the latter is not really the inverse matrix of A_i^*, but it satisfies $(A_i^*)^{-1} A_i^* = \Lambda_i \Lambda_i'$. Recall that A_i is a diagonal matrix whose diagonal elements correspond to the variances of the observations in the ith cluster. It follows that A_i^* is also a diagonal matrix, whose diagonal elements are zero for the missing observations; the nonzero diagonal elements of A_i^* are the same as those of A_i. With the above transformation, the cluster sizes become equal; therefore, one can focus on the case of equal cluster size, T.

By taking the partial derivatives of (3.97), with respect to β and each component of $\alpha = (\alpha_i)_{1 \le i \le m}$, using (3.98), (3.99), one gets

$$\sum_{i=1}^{m} \frac{\partial \mu_i'}{\partial \beta} W_i^{-1}(y_i - \mu_i) = 0, \tag{3.100}$$

$$\frac{\partial \mu_i'}{\partial \alpha_i} W_i^{-1}(y_i - \mu_i) - \lambda \left(\frac{\partial P_A \alpha}{\partial \alpha_i} \right) P_A \alpha = 0, \ 1 \le i \le m. \tag{3.101}$$

Note that (3.100) involves a summation but not in (3.101) due to the fact that μ_i depends on β and α_i only. Then, proceeding as in Section 3.2.1, the focus is the correlation matrix R, which is modeled as (3.20). The coefficients $a_1, \ldots, a_H$ are determined by minimizing two quadratic functions. The first, corresponding to β, is defined by

$$g_f' C_f^{-1} g_f, \tag{3.102}$$

where $C_f = m^{-2} \sum_{i=1}^{m} g_{f,i} g_{f,i}'$ with

$$g_{f,i} = \left[\frac{\partial \mu_i'}{\partial \beta} A_i^{-1/2} M_h A_i^{-1/2} (y_i - \mu_i) \right]_{1 \le h \le H}.$$

The subscript f refers to fixed effects. The second quadratic function is defined as

$$g_r' g_r, \tag{3.103}$$

where $g_r' = [g_{r,1}', \lambda \alpha_1', \ldots, g_{r,m}', \lambda \alpha_m', \lambda (P_A \alpha)']$ with

$$g_{r,i} = \frac{\partial \mu_i'}{\partial \alpha_i} A_i^{-1/2} M_1 A_i^{-1/2} (y_i - \mu_i).$$

The subscript r refers to random effects. Note that only the matrix M_1 is involved in g_r. In [159], the authors argued that, in most cases, correlation for the random-effects modeling is not as critical as that for the fixed-effects modeling, hence only the first of the base matrices, $M_1, \ldots, M_H$, is used in (3.103). Also, here, the random effects α_i are allowed to be vector-valued, say, $\dim(\alpha_i) = b$, hence the transposes are used in the definition of g_r.

To solve (3.102) and (3.103), [159] proposes an iterative procedure: 1. start with an initial vector, $\hat{\beta}$, which is obtained by fitting the GLM assuming independent correlation structure, and set $\hat{\alpha} = 0$; 2. minimize (3.102), with β replaced by $\hat{\beta}$, to obtain the random-effects estimator, $\hat{\alpha}$; 3. minimize (3.103), with α replaced by $\hat{\alpha}$, to obtain the updated fixed-effects estimator, $\hat{\beta}$; 4. iterate between steps 2 and 3 until the convergence criterion

$$|\hat{\beta} - \beta| + |\hat{\alpha} - \alpha| \quad < \quad \varepsilon$$

is reached, where ε is a small positive number and is typically chosen as 10^{-6}. However, convergence of the iterative algorithm is not proved.

Wang et al. [159] studied large-sample properties of the QIF estimators. Let $\hat{\beta}$ and $\hat{\alpha}$ denote the QIF estimators of β and α, respectively, that is, $\hat{\gamma} = (\hat{\beta}', \hat{\alpha}')'$ jointly minimizes (3.102) and (3.103). Let $\hat{\beta}_{[1]}$ denote the minimizer of (3.102) with $\alpha = \hat{\alpha}$, and $\hat{\beta}_{[0]}$ the minimizer of (3.102) with α being the true random effects vector. Note that $\hat{\beta}_{[0]}$ is not computable, but is used in describing the asymptotic behavior of $\hat{\beta}_{[1]}$. Namely, the authors first showed that, under suitable conditions, one has

$$\sqrt{m}(\hat{\beta}_{[0]} - \beta) \quad \xrightarrow{d} \quad N(0, \Omega_0), \tag{3.104}$$

as $m \to \infty$, where β is the true parameter vector, and Ω_0 is the limit of the matrix of second derivatives, with respect to β and evaluated at $\hat{\beta}_0$, of the QIF (3.102), in which α is the true random effects vector. The existence of the limit Ω_0 is implied by the conditions imposed. Then, Theorem 1 of [159] states that, under suitable conditions, $\hat{\beta}_{[1]}$ is a consistent estimator of β; furthermore, one has

$$\sqrt{m}(\hat{\beta}_{[1]} - \beta) \quad \xrightarrow{d} \quad N(0, \Omega_1), \tag{3.105}$$

as $m \to \infty$, where β is the true parameter vector, and Ω_1 is a certain covariance matrix which is not necessarily equal to Ω_0. However, if $\hat{\alpha}$ is a consistent estimator of α, then one has $\Omega_1 = \Omega_0$.

In a way, Ω_0 is considered an optimal asymptotic covariance matrix for estimating β, because it is based on knowing the true α. Thus, if $\hat{\alpha}$ is a consistent estimator of α, $\hat{\beta}_{[1]}$ would be considered efficient, because its asymptotic covariance matrix is the same as that of $\hat{\beta}_{[0]}$, that is, Ω_0. However, $\hat{\alpha}$ is a consistent estimator of α only under some rather restricted conditions. In the supplementary material of [159], the authors discussed conditions under which $\hat{\alpha}$ is consistent. Essentially, the conditions assume that the number of observations corresponding to each individual random effect goes to infinity. Note that, in most cases where mixed effects models are used, there is not sufficient data information for estimating each individual random effect (otherwise, the random effects might, in fact, be treated as fixed effects). See similar discussion in Section 2.4.1 regarding the consistency of EBLUP.

3.5   Maximization by parts

Song et al. ([160]) proposed an interesting idea for computing MLE in some complicated situations, called *maximization by parts* (MBP). More specifically, the standard Newton–Raphson procedure requires calculations of the first and second derivatives. If the likelihood function is complicated, the derivation and calculation of its derivatives, especially the second derivatives, can be both analytically and computationally challenging. The idea of MBP is easy to illustrate. Suppose that the log-likelihood function is decomposed as

$$l(\psi)l(\psi) = l_w(\psi) + l_e(\psi) \tag{3.106}$$

so that, for any given vector, a, the equation $\cot l_w(\psi) = a$ is (relatively) easy to solve. Then, the likelihood equation

$$\dot{l}(\psi) = 0 \tag{3.107}$$

can be written as

$$\dot{l}_w(\psi) = -\dot{l}_e(\psi). \tag{3.108}$$

Here the ψs on both sides of (3.108) are supposed to be the same, but they do not have to be so in an iterative equation, and this is the key. The initial estimator, $\hat{\psi}_{[0]}$, is a solution to $\dot{l}_w(\psi) = 0$. Then, use the equation

$$\dot{l}_w(\hat{\psi}_{[1]}) = -\dot{l}_e\{\hat{\psi}_{[0]}\} \tag{3.109}$$

to update the estimator, and so on. It is easy to see that, if the sequence $\hat{\psi}_{[l]}, l = 0, 1, \ldots$ converges, the limit, say, $\hat{\psi}$, satisfies (3.107). Jiang [161] observed that the left side of (3.107), evaluated at the sequence $\hat{\psi}_{[l]}$, has absolute values

$$|\dot{l}_e(\hat{\psi}_{[1]}) - \dot{l}_e(\hat{\psi}_{[0]})|, \ |\dot{l}_e(\hat{\psi}_{[2]}) - \dot{l}_e(\hat{\psi}_{[1]})|, \ldots. \tag{3.110}$$

Suppose that the function $l_e(\cdot)$ is at least locally uniformly continuous. Now consider the distances between consecutive points in $\hat{\psi}_{[l]}$, that is,

$$|\hat{\psi}_{[1]} - \hat{\psi}_{[0]}|, \ |\hat{\psi}_{[2]} - \hat{\psi}_{[1]}|, \ldots \tag{3.111}$$

If (3.111) shows sign of vanishing, which will be the case if the sequence $\hat{\psi}_{[l]}$ is indeed convergent, (3.110) is expected to do the same. This means that the left side of (3.107) vanishes in absolute value along the sequence $\hat{\psi}_{[l]}$. Note that, because the MLE satisfies (3.107), the absolute value of the left side of the equation, evaluated at an estimator, may be used as a measure of "closeness" of the estimator to the MLE; and the efficiency of the estimator is expected to increase as it gets closer to the MLE. However, convergence of the MBP algorithm was not proved.

Nevertheless, [160] investigated asymptotics associated with MBP. A key condition for the asymptotics is the information dominance, that is,

$$\|\mathscr{I}_w^{-1}\mathscr{I}_e\| < 1, \tag{3.112}$$

where $\mathscr{I}_w = -n^{-1}\mathrm{E}\{\ddot{l}_w(\psi_0)\}$, ψ_0 being the true parameter vector, $\mathscr{I}_e = -n^{-1}\mathrm{E}(\{\ddot{l}_e(\psi_0)\}$, and $\|\cdot\|$ denotes the spectral norm of matrix. Song et al. [160] interprets (3.112) as that l_w contains more information about ψ_0 than l_e does, and notes a connection between (3.112) and contraction mapping in the fixed point theory (e.g., [162]). Then, Theorem 1 of [160] states that, if $l_w(\cdot)$ and $l_e(\cdot)$ are both twice continuously differentiable in a neighborhood of the true ψ, then consistency of $\hat{\psi}_{[0]}$, the initial estimator, implies consistency of $\hat{\psi}_{[l]}$ for every fixed $l = 1, 2, \ldots$. Theorem 2 of [160] states that, under certain regularity conditions, $\hat{\psi}_{[l]}$ is asymptotically normal with mean equal to the true ψ and covariance matrix $n^{-1}\Sigma_l$ with $\Sigma_l = A_l'\Omega A_l$,

$$
A_l = \left[\begin{array}{c} (I - \tau^l)\mathscr{I}^{-1} \\ (I - \tau^{l-1})\mathscr{I}^{-1} \end{array} \right],
$$

I being the identity matrix of the same dimension as ψ, $\tau = -\mathscr{I}_w^{-1}\mathscr{I}_e$, $\mathscr{I}^{-1}$ being the inverse of the Fisher information matrix, and

$$
\Omega = \lim_{n \to \infty} \frac{1}{n} \left[\begin{array}{cc} \mathrm{E}(\dot{l}_w\dot{l}_w') & \mathrm{E}(\dot{l}_w\dot{l}_e') \\ \mathrm{E}(\dot{l}_e\dot{l}_w') & \mathrm{E}(\dot{l}_e\dot{l}_e') \end{array} \right].
$$

Moreover, the theorem states that, under the condition of information dominance, one has, as $l \to \infty$, that $\Sigma_l \to \mathscr{I}^{-1}$. In the way, the theorems are mostly about asymptotic behavior of $\hat{\psi}_{[l]}$ for each fixed l; however, the latest result states that, under the information dominance, $\hat{\psi}_{[l]}$ becomes an asymptotically efficient estimator as $l \to \infty$. Still, the result does not imply the convergence of $\hat{\psi}_{[l]}$ itself to the MLE, or to any limiting estimator. This is a bit different from, for example, Theorems 3.2–3.4.

From a practical standpoint, the most important issue regarding MBP seems to be finding a "good" decomposition (3.106). In general, a condition for a good choice of l_w is called *information dominance*. In other words, $\ddot{l}_w$ needs to be larger than $\ddot{l}_e$ in a certain sense, where $\ddot{l}_w, \ddot{l}_e$ denote the matrices of second derivatives ([160], Theorem 2). However, because the second derivatives are difficult to evaluate, this condition is not easy to verify. On the other hand, the argument above suggests a potentially practical way to verify that one has had a good choice of l_w. Namely, one may let the procedure run for a few steps. If the sequence (3.111) shows sign of vanishing, it is an indication that a good choice has been made.

The MBP method is potentially applicable to at least some class of GLMMs. It is suggested that the h-likelihood of [155], described near the end of Section 3.4.1, may be used as l_w. However, if the random effects are normally distributed, this leads to a biased estimating equation. In fact, the solution to such an equation may not be consistent ([163]; also see Section 3.1). Another question is to what extent MBP helps in computing the MLE in GLMM. Jiang [2] (pp. 176–178) argues that, at least for certain types of GLMMs, such as those involving crossed random effects (see Section 3.3), computing the first derivatives of the log-likelihood is as difficult as computing the second derivatives. Thus, at least computationally, MBP does not seem to bring advantages in such cases.

Due to these concerns, we have chosen not to include MBP as part of Section 3.3, but rather a separate section. However, this view might be changed subject to solutions to some open problem related to this section (see Section 3.6.2).

3.6 GLMM diagnostics

In Section 2.5 we discussed LMM diagnostics. There are similar problems for GLMMs. In this section, we discuss some recent advance in this area. In the process, we introduce a useful, and interesting, method, called *tailoring*, in deriving a goodness-of-fit test statistic that is guaranteed to have an asymptotic χ^2 null distribution. The method is applied to GLMM diagnostics as a special case.

Unlike LMM diagnostics, the literature on GLMM diagnostics is much thinner. Gu [164] considered similar χ^2 tests to [76] and applied them to mixed logistic models, a special case of GLMMs. She considered both minimum χ^2 estimator (see below) and method of simulated moments (MSM) estimator [17] of the model parameters, and derived asymptotic null distributions of the test statistics, which are weighted χ^2. Tang [165] proposed a different χ^2-type goodness-of-fit test for GLMM diagnostics, which is not based on the cell frequencies. She proved that the asymptotic null distribution is χ^2. However, the test is based on the MLE, which is known to be computationally difficult to obtain. Furthermore, the test statistic involves the Moore-Penrose generalized inverse (G-inverse) of a normalizing matrix, which does not have an analytic expression. The interpretation of such a G-inverse may not be straightforward for a practitioner.

For cases of discrete responses, such as in typical situations of GLMMs, χ^2 tests based on cell frequencies, such as Pearson's χ^2 test, are much more natural than for cases of continuous observations. For example, if the responses are binomial, the range of the responses is a set of nonnegative integers. Thus, a natural choice of the cells are exactly those integers. In contrast, if the responses are continuous, one has to divide the range of the responses into intervals in order to apply Pearson's χ^2 test, and there are infinitely many ways of choosing the number of cells as well as the cells, given the number of cells. In fact, the choice of the cells in the latter case is a difficult, unsolved problem. See, for example, [76], for discussion on this issue and the reference therein. Also, a χ^2 asymptotic distribution is more desirable, in terms of simplicity, than the supremum of χ^2 of [77], or weighted χ^2 of [76] and [164]. Note that, in the latter cases, the weights are eigenvalues of some matrices, whose expressions are compicated, and involve unknown parameters. These parameters need to be estimated in order to obtain the critical values of the tests. Due to such a complication, [76] suggests to use a Monte-Carlo method to compute the critical value; but, by doing so, the usefulness of the asymptotic result may be undermined. Furthermore, it would be beneficial if the test statistic does not involve a G-inverse. In particular, the latter has no analytic expression; thus, it is difficult to see how the G-inverse is affected by the parameters, and sample sizes. As is well known, there may be multiple factors contributing to the sample size under a mixed effects model. Finally, computation has been a big issue in GLMM, especially for likelihood-based inference (e.g., [2], Section 3.4; [150]). Any diagnostic technique for GLMM has to be computationally efficient. For example, a test that requires computation of the MLE, as in [165], would be computationally less efficient than one that only requires a consistent estimator that is computationally simpler, such as the MSM estimator used in [164] (see [17]). In summary, it is desirable to (i) develop a Pearson type χ^2 test, based on cell frequencies, that (ii) results in a χ^2 asymptotic null distribution, (iii) does not involve a G-inverse, and (iv) is computationally attractive.

3.6.1 Tailoring

In this subsection, we describe a general approach to obtaining a test statistic that has an asymptotic χ^2 distribution under the null hypothesis. The original idea can be traced back to R. A. Fisher [166], who used the method to obtain an asymptotic χ^2 distribution for Pearson's χ^2-test, when the so-called minimum chi-square estimatoris used. However, Fisher did not put forward the method that he originated under a general framework, as we shall do here. Suppose that there is a sequence of s-dimensional random vectors, $B(\vartheta)$, which depend on a vector ϑ of unknown parameters such that, when ϑ is the true parameter vector, one has $E\{B(\vartheta)\} = 0$, $\text{Var}\{B(\vartheta)\} = I_s$, and, as the sample size increases,

$$|B(\vartheta)|^2 \xrightarrow{d} \chi_s^2, \tag{3.113}$$

where $|\cdot|$ denotes the Euclidean norm. However, because ϑ is unknown, one cannot use (3.113) for GoFT. What is typically done, such as in Pearson's χ^2-test, is to replace ϑ by an estimator, $\hat{\vartheta}$. Question is: what $\hat{\vartheta}$? The ideal scenario would be that, after replacing ϑ by $\hat{\vartheta}$ in (3.113), one has a reduction of degrees of freedom (d.f.), which leads to

$$|B(\hat{\vartheta})|^2 \xrightarrow{d} \chi_v^2, \tag{3.114}$$

where $v = s - r > 0$ and $r = \dim(\vartheta)$. This is the famous "subtract one degree of freedom for each parameter estimated" rule taught in many elementary statistics books (e.g., [71], p. 242). However, as is well known (e.g., [72]), depending on what $\hat{\vartheta}$ is used, (3.114) may or may not hold, regardless of what d.f. is actually involved. In fact, the only method that is known to achieve (3.114) without restriction on the distribution of the data is Fisher's minimum χ^2 method. In a way, the method allows one to "cut-down" the d.f. of (3.113) by r, and thus convert an asymptotic χ_s^2 to an asymptotic χ_v^2. For such a reason, we have coined the method, under the more general setting below, *tailoring*. We develop the method with a heuristic derivation, and refer the rigorous justification to [167].

The "right" estimator of ϑ for tailoring is supposed to be the solution to an estimating equation of the following form:

$$C(\vartheta) \equiv A(\vartheta)B(\vartheta) = 0, \tag{3.115}$$

where $A(\vartheta)$ is an $r \times s$ non-random matrix that plays the role of tailoring the s-dimensional vector, $B(\vartheta)$, to the r-dimensional vector, $C(\vartheta)$. The specification of A will become clear at the end of the derivation. Throughout the derivation, ϑ denotes the true parameter vector. For notation simplicity, we use A for $A(\vartheta)$, $\hat{A}$ for $A(\hat{\vartheta})$, etc. Under regularity conditions, one has the following expansions, which can be derived from the Taylor series expansion and large-sample theory (e.g., [1]):

$$\hat{\vartheta} - \vartheta \approx -\left\{E_\vartheta\left(\frac{\partial C}{\partial \vartheta'}\right)\right\}^{-1} C, \tag{3.116}$$

$$\hat{B} \approx B - E_\vartheta\left(\frac{\partial B}{\partial \vartheta'}\right)\left\{E_\vartheta\left(\frac{\partial C}{\partial \vartheta'}\right)\right\}^{-1} C. \tag{3.117}$$

Because $E_\vartheta\{B(\vartheta)\} = 0$ [see above (3.113)], one has

$$E_\vartheta\left(\frac{\partial C}{\partial \vartheta'}\right) = AE_\vartheta\left(\frac{\partial B}{\partial \vartheta'}\right). \tag{3.118}$$

Combining (3.117), (3.118), we get

$$\hat{B} \approx \{I_s - U(AU)^{-1}A\}B, \tag{3.119}$$

where $U = E_\vartheta(\partial B/\partial \vartheta')$. We assume that A is chosen such that

$$U(AU)^{-1}A \text{ is symmetric.} \tag{3.120}$$

Then, it is easy to verify that $I_s - U(AU)^{-1}A$ is symmetric and idempotent. If we further assume that the following limit exists:

$$I_s - U(AU)^{-1}A \longrightarrow P, \tag{3.121}$$

then P is also symmetric and idempotent. Thus, assuming that

$$B \xrightarrow{\text{d}} N(0, I_s), \tag{3.122}$$

which is typically the argument leading to (3.113), one has, by (3.119), $\hat{B} \xrightarrow{\text{d}} N(0, P)$, hence (e.g., [168], p. 58) $|\hat{B}|^2 \xrightarrow{\text{d}} \chi_\nu^2$, where $\nu = \text{tr}(P) = s - r$. This is exactly (3.114).

It remains to answer one last question: Is there such a non-random matrix $A = A(\vartheta)$ that satisfies (3.120) and (3.121)? We show that, not only the answer is yes, there is an optimal one. Let $A = N^{-1}U'W$, where W is a symmetric, non-random matrix to be determined, and N is a normalizing constant that depends on the sample size. By (3.116) and the fact that $\text{Var}_\vartheta(B) = I_s$ [see above (3.113)], we have

$$\text{var}_\vartheta(\hat{\vartheta}) \approx (U'WU)^{-1}U'W^2U(U'WU)^{-1} \geq (U'U)^{-1}, \tag{3.123}$$

by, for example, Lemma 5.1 of [1]. The equality on the right side of (3.123) holds when $W = I_s$, giving the optimal A:

$$A = A(\vartheta) = \frac{U'}{N} = \frac{1}{N}E_\vartheta\left(\frac{\partial B'}{\partial \vartheta}\right). \tag{3.124}$$

The A given by (3.124) clearly satisfy (3.120) [which is equal to $U(U'U)^{-1}U'$]. As for (3.121), suppose that $B(\vartheta)$ can be expressed as

$$B(\vartheta) = V(\vartheta)^{-1/2}\sum_{i=1}^{m}b_i(\vartheta), \tag{3.125}$$

where $b_i(\vartheta)$ is a random vector that depends on ϑ such that, for fixed θ, $b_i(\vartheta), 1 \leq$

$i \leq m$ are independent with $E_\vartheta\{b_i(\vartheta)\} = 0$, and $V(\vartheta) = \mathrm{Var}_\vartheta\{\sum_{i=1}^m b_i(\vartheta)\} = \sum_{i=1}^m \mathrm{Var}_\vartheta\{b_i(\vartheta)\}$. Then, we have, with $N = m$,

$$
\begin{aligned}
A(\vartheta)' &= \frac{1}{m}E_\vartheta\left(\frac{\partial B}{\partial \vartheta'}\right) \\
&= \frac{1}{m}V(\vartheta)^{-1/2}\sum_{i=1}^m E_\vartheta\left(\frac{\partial b_i}{\partial \vartheta'}\right) \\
&= \frac{1}{\sqrt{m}}\left\{\frac{1}{m}\sum_{i=1}^m V_i(\vartheta)\right\}^{-1/2}\left\{\frac{1}{m}\sum_{i=1}^m E_\vartheta\left(\frac{\partial b_i}{\partial \vartheta'}\right)\right\},
\end{aligned}
$$
(3.126)

where $V_i(\vartheta) = \mathrm{Var}_\vartheta\{b_i(\vartheta)\}$. From (3.126), it is seen that $\sqrt{m}A$ is expected to have a limit, hence $U(AU)^{-1}A = A'(AA')^{-1}A = (\sqrt{m}A)'\{(\sqrt{m}A)(\sqrt{m}A)'\}^{-1}(\sqrt{m}A)$ is expected to have a limit, under some non-singularity condition. Thus, (3.121) is expected to be satisfied. It should be noted that the solution to (3.115), $\hat{\vartheta}$, does not depend on the choice of N.

3.6.2 Application to GLMM

To apply the tailoring method to GLMMs, we obtain, as a first step, sufficient statistics at cluster levels. The idea is similar to [17]. Suppose that the observations come from m clusters and there are n_i observations, y_{ij}, $j = 1,\ldots,n_i$, for the ith cluster. Suppose that there is a vector of random effects, α_i, associated with the ith cluster such that, given $\alpha_1,\ldots,\alpha_m$, the observations y_{ij} are conditionally independent with conditional density

$$
f(y_{ij}|\theta_{ij},\phi_{ij}) = \exp\left\{\frac{y_{ij}\theta_{ij} - b(\theta_{ij})}{\phi_{ij}} + c(y_{ij},\phi_{ij})\right\},
$$
(3.127)

where ϕ_{ij} is known up to a dispersion parameter, ϕ, which is known in some cases. Furthermore, the natural parameters θ_{ij}s satisfy

$$
\theta_{ij} = x_{ij}'\beta + z_{ij}'\alpha_i,
$$
(3.128)

where h is a known function, and x_{ij}, z_{ij} are known vectors. A standard assumption for the random effects is that α_i, $1 \leq i \leq m$ are independent and distributed as $N(0,G)$, where the covariance matrix G depends on a vector, ψ, of variance components. (3.128) means that we are focusing on the case of canonical link. Let $p = \dim(\beta)$ and $d = \dim(\alpha_i)$. Suppose that the covariance matrix G depends on a q-dimensional vector, ψ, of variance components, that is, $G = G(\psi)$. Let $G = DD'$ be the Cholesky decomposition of G, where $D = D(\psi)$. Then, α_i can be expressed as $\alpha_i = D(\psi)\xi_i$, where $\xi_i \sim N(0,I_d)$. Furthermore, suppose that ϕ_{ij} has the following special form (e.g., [2], p. 191):

$$
\phi_{ij} = \phi/w_{ij},
$$
(3.129)

where ϕ is an unknown dispersion parameter, and w_{ij} is a known weight, for every i, j. Then, it can be shown that the conditional density of $y_i = (y_{ij})_{1 \leq j \leq n_i}$ given ξ_i (with respect to a σ-finite measure) can be expressed as

$$
f(y_i | \xi_i) = \exp \left\{ \left(\sum_{j=1}^{n_i} w_{ij} y_{ij} x_{ij} \right)' \left(\frac{\beta}{\phi} \right) + \left(\sum_{j=1}^{n_i} w_{ij} y_{ij} z_{ij} \right)' \left(\frac{D(\psi)}{\phi} \right) \xi_i \right.
$$
$$
\left. - \sum_{j=1}^{n_i} \left(\frac{w_{ij}}{\phi} \right) b(x_{ij}'\beta + z_{ij}'D(\psi)\xi_i) + \sum_{j=1}^{n_i} c \left(y_{ij}, \frac{\phi}{w_{ij}} \right) \right\}.
$$

Let $f(\cdot)$ denote the pdf of $N(0, I_d)$, and $\xi \sim N(0, I_d)$. It follows that

$$
f(y_i) = \int f(y_i | \xi_i) f(\xi_i) d\xi_i
$$
$$
= \exp \left\{ \left(\sum_{j=1}^{n_i} w_{ij} y_{ij} x_{ij} \right)' \left(\frac{\beta}{\phi} \right) \right\}
$$
$$
\mathrm{E} \left[\exp \left\{ \left(\sum_{j=1}^{n_i} w_{ij} y_{ij} z_{ij} \right)' \left(\frac{D(\psi)}{\phi} \right) \xi \right. \right.
$$
$$
\left. \left. - \sum_{j=1}^{n_i} \left(\frac{w_{ij}}{\phi} \right) b(x_{ij}'\beta + z_{ij}'D(\psi)\xi) \right\} \right]
$$
$$
\exp \left\{ \sum_{j=1}^{n_i} c \left(y_{ij}, \frac{\phi}{w_{ij}} \right) \right\}, \tag{3.130}
$$

where the expectation is with respect to ξ. From (3.130), it is clear that a set of sufficient statistics for all of the parameters involved, namely, β, ψ, and ϕ, are

$$
\sum_{j=1}^{n_i} w_{ij} y_{ij} x_{ij} \quad \text{and} \quad \sum_{j=1}^{n_i} w_{ij} y_{ij} z_{ij}. \tag{3.131}
$$

Note that the first summation in (3.131) is a $p \times 1$ vector, while the second summation is a $d \times 1$ vector. So, in all, there are $p + d$ components of those vectors; however, some of the components may be redundant, or functions of the other components. After removing the redundants, and functions of others, the remaining components form a vector, denoted by Y_i, so that the sufficient statistics in (3.131) are functions of Y_i. The Y_i's will be used for goodness-of-fit test of the following hypothesis:

$$
H_0 : \quad \text{The assumed GLMM holds} \tag{3.132}
$$

versus the alternative that there is a violation of the model assumption. In many cases, the null hypothesis is more specific about one particular part of the GLMM, such as the normality of the random effects, assuming that other parts of the model hold; the alternative thus also changes correspondingly.

If the values of y_{ij}'s belong to a finite subset of R, such as in the Binomial situation, the possible values of $Y_1, \ldots, Y_n$ is a finite subset $S \subset R^g$, where $g = \dim(Y_i)$, assuming that all of the Y_i's are of the same dimension. Let $C_1, \ldots, C_M$ be the different (vector) values in S. These are the cells under the general set-up. If the values of y_{ij}'s are not bounded, such as in the Poisson case, let K be a positive number such that the probability that $\max_i |Y_i| > K$ is small. Let $C_1, \ldots, C_{M-1}$ be the different (vector) values in $S \cap \{v \in R^g : |v| \leq K\}$, and $C_M = S \cap \{v \in R^g : |v| > K\}$, where S is the set of all possible values of the Y_i's. These are the cells under the general set-up. Now consider $O_i = [1_{(Y_i \in C_k)}]_{1 \leq k \leq M-1}$, and let $u_i(\theta) = E_\theta(O_i) = [P_\theta(Y_i \in C_k)]_{1 \leq k \leq M-1}$, where θ is the vector of parameters involved in the distribution of the observations. Let $b_i(\theta) = O_i - u_i(\theta)$. It is easy to show that $V_i(\theta) = \mathrm{Var}_\theta\{b_i(\theta)\} = \mathrm{Var}_\theta(O_i)$ has the following close-form expression:

$$V_i(\theta) = P_i(\theta) - p_i(\theta)p_i'(\theta), \tag{3.133}$$

where $P_i(\theta) = \mathrm{diag}\{p_{ik}(\theta), 1 \leq k \leq M-1\}$, with $p_{ik}(\theta) = P_\theta(Y_i \in C_k)$, $p_i(\theta) = [p_{ik}(\theta)]_{1 \leq k \leq M-1}$, and $p_i'(\theta) = \{p_i(\theta)\}'$. It follows, by a well-known formula of matrix inversion (e.g., [12], p. 275), that

$$\{V_i(\theta)\}^{-1} = \mathrm{diag}\left\{\frac{1}{p_{ik}(\theta)}, 1 \leq k \leq M-1\right\}$$
$$+ \frac{J_{M-1}}{1 - \sum_{k=1}^{M-1} p_{ik}(\theta)}, \tag{3.134}$$

assuming $\sum_{k=1}^{M-1} p_{ik}(\theta) < 1$, where J_a denotes the $a \times a$ matrix of 1's. We then consider $B(\theta) = V(\theta)^{-1/2} \sum_{i=1}^m b_i(\theta)$ with $V(\theta) = \sum_{i=1}^m V_i(\theta)$, as in (3.125). We illustrate with a simple example.

Example 3.12. Let y_{ij} be a binary outcome with $\mathrm{logit}\{P(y_{ij} = 1|\alpha)\} = \mu + \alpha_i$, $1 \leq i \leq n$, $1 \leq j \leq M-1$, where $\alpha_1, \ldots, \alpha_n$ are i.i.d. random effects with $\alpha_i \sim N(0, \sigma^2)$, $\alpha = (\alpha_i)_{1 \leq i \leq n}$, and μ, σ are unknown parameters with $\sigma \geq 0$. It is more convenient to use the following expression: $\alpha_i = \sigma \xi_i$, $1 \leq i \leq n$, where $\xi_1, \ldots, \xi_n$ are i.i.d. $N(0, 1)$ random variables. In this case, we have $x_{ij} = z_{ij} = w_{ij} = 1$, and $n_i = M-1, 1 \leq i \leq n$. Thus, both expressions in (3.131) are equal to $y_{i\cdot} = \sum_{j=1}^{M-1} y_{ij}$. It follows that the sufficient statistics are $Y_i = y_{i\cdot}, 1 \leq i \leq n$, which are i.i.d. The range of Y_i is $0, 1, \ldots, M-1$; thus, we have $C_k = \{k-1\}, 1 \leq k \leq M$. Let $\theta = (\mu, \sigma)'$. It is easy to show that

$$p_{ik}(\theta) = P_\theta(Y_i \in C_k) = \binom{M-1}{k-1} e^{(k-1)\mu} E\left[\frac{\exp\{(k-1)\sigma\xi\}}{\{1 + \exp(\mu + \sigma\xi)\}^{M-1}}\right],$$

$1 \leq k \leq M-1$, where the expectation is with respect to $\xi \sim N(0, 1)$. Note that, in this case, $p_{ik}(\theta)$ does not depend on i; and there is no need to compute $p_{iM}(\theta)$. Also, regarding computation of the A in (3.124), we have

$$\frac{\partial p_{ik}(\theta)}{\partial \mu} = \binom{M-1}{k-1} e^{(k-1)\mu} E\left[\frac{\exp\{(k-1)\sigma\xi\}}{\{1 + \exp(\mu + \sigma\xi)\}^{M-1}}\{(k-1)\right.$$
$$\left. - \frac{(M-1)\exp(\mu + \sigma\xi)}{1 + \exp(\mu + \sigma\xi)}\right\}\right],$$

$$\frac{\partial p_{ik}(\theta)}{\partial \sigma} = \binom{M-1}{k-1} e^{(k-1)\mu} \mathrm{E}\left[\frac{\exp\{(k-1)\sigma\xi\}}{\{1+\exp(\mu+\sigma\xi)\}^{M-1}}\left\{(k-1) \right.\right.$$
$$\left.\left. -\frac{(M-1)\exp(\mu+\sigma\xi)}{1+\exp(\mu+\sigma\xi)}\right\}\xi\right],$$

$1 \le k \le M - 1$. Again, there is no need to compute the derivatives for $k = M$.

In some cases, the range of Y_i may be different for different i. To avoid having zero cell probabilities, one strategy is to divide the data into (nonoverlapping) groups so that, within each group, the Y_i's have the same range. More specifically, let $Y_i, i \in I_l$ be the lth group whose corresponding cells are $C_{kl}, k = 1, \ldots, M_l$ with $p_{ikl}(\theta) = P_\theta(Y_i \in C_{kl}) > 0, i \in I_l, 1 \le k \le M_l, l = 1, \ldots, L$. The method described above can be applied to each group of the data, $Y_i, i \in I_l$, leading to the χ^2 test statistic, $\hat{\chi}_l^2$, that has the asymptotic $\chi^2_{M_l-r-1}$ distribution under the null hypothesis, $1 \le l \le L$. Then, because the groups are independent, the combined χ^2 statistic,

$$\hat{\chi}^2 = \sum_{l=1}^{L} \hat{\chi}_l^2, \qquad (3.135)$$

has the asymptotic χ^2 distribution with $\sum_{l=1}^{L}(M_l - r - 1) = M. - L(r+1)$ degrees of freedom, under the null hypothesis, where $M. = \sum_{l=1}^{L} M_l$. In conclusion, the goodness-of-fit test is carried out using (3.135) with the asymptotic $\chi^2_{M.-L(r+1)}$ null distribution.

3.6.3 A simulated example

We use a simulated example to verify the asymptotic theory and also study the finite-sample performance of the proposed test. The example is a special case of Example 3.12, with $n = 100$ or 200, and $M = 7$. The true parameters are $\mu = 1$ and $\sigma = 2$.

Consider testing the normality of the random effects, α_i, assuming other parts of the GLMM assumptions hold. Then, the null hypothesis, (3.132), is equivalent to

$$H_0: \quad \alpha_i \sim \text{Normal} \qquad (3.136)$$

versus the alternative that the distribution of α_i is not normal. Two specific alternatives are considered. Under the first alternative, $H_{1,1}$, the distribution of α_i is a centralized exponential distribution, namely, the distribution of $\zeta - 2$, where $\zeta \sim \text{Exponential}(0.5)$. Under the second alternative, $H_{1,2}$, the distribution of α_i is a normal-mixture, namely, the mixture of $N(-3, 0.5)$ with weight 0.2, $N(2/3, 0.5)$ with weight 0.3, and $N(4/5, 0.5)$ with weight 0.5. All simulation results are based on $K = 1000$ simulation runs.

First, we compare the empirical and asymptotic null distributions of the test statistic under different sample sizes, n. According to (3.114), the asymptotic null distribution of the test is χ_4^2. The left figure of Figure 3.1 shows the histogram of the simulated test statistics, for $n = 100$, under H_0, with the pdf of χ_4^2 plotted on top. It appears that the histogram matches the theoretical (asymptotic) distribution quite well. The corresponding cdf's are plotted in the right figure, and there is hardly

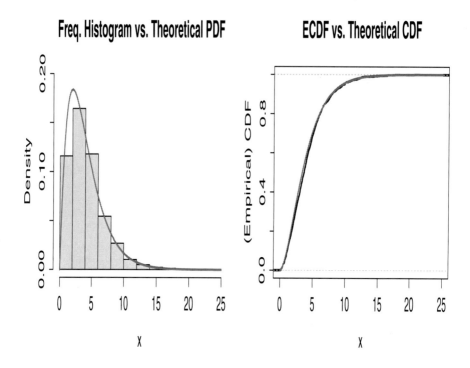

Figure 3.1 *Theoretical versus Empirical Distributions:* $n = 100$. *Left: pdfs; Right: cdfs.*

Table 3.5 *Empirical and Theoretical Quartiles*

Quartiles	$n = 100$	$n = 200$	χ_4^2
Q_1	2.094	1.848	1.923
Q_2	3.649	3.103	3.357
Q_3	5.493	5.203	5.385

any visible difference between the two. Figure 3.2 shows the corresponding plots for $n = 200$. Here, the matches are even better, more visibly in the histogram-pdf comparison. Some numerical summaries, in terms of the 1st (Q_1), 2nd (Q_2), and 3rd (Q_3) quartiles, are presented in Table 3.5.

Next, we consider size and power of the test at the levels of significance 0.01, 0.05, and 0.10, respectively. The simulated sizes and powers are presented in Table 3.6. It appears that the simulated sizes are closer to the nominal levels for $n = 100$; on the other hand, the simulated powers are much higher for $n = 200$.

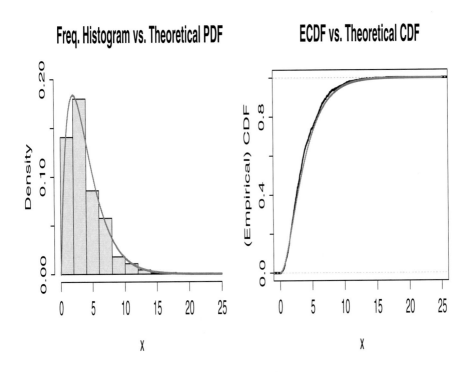

Figure 3.2 *Theoretical versus Empirical Distributions: n = 200. Left: pdfs; Right: cdfs.*

Table 3.6 *Simulated Size and Power*

Nominal Level	Size		Power against $H_{1,1}$		Power against $H_{1,2}$	
	$n = 100$	$n = 200$	$n = 100$	$n = 200$	$n = 100$	$n = 200$
0.01	0.010	0.007	0.148	0.408	0.329	0.766
0.05	0.051	0.043	0.345	0.650	0.577	0.912
0.10	0.104	0.080	0.483	0.768	0.704	0.952

3.7   Additional bibliographical notes and open problems

3.7.1   Bibliographical notes

Although PQL fails to produce consistent estimators for GLMM parameters, there is, at least, one occasion, in which the method is completely justified theoretically. This is when PQL is used in testing that all of the variance components are zero under the ANOVA GLMM, introduced above (3.38). See [137]. The reason is that, under the null hypothesis, the GLMM is reduced to a GLM [6]; thus, the MLE under the null hypothesis is the same as the MLE under a GLM, which is consistent and fairly easy to compute. Lin [137] showed that, under the null hypothesis that $\sigma_1 = \cdots = \sigma_q = 0$, a score test-statistic has an asymptotic χ_q^2 distribution; the test is locally asymptotically most powerful test if $q = 1$, and is a locally asymptotically most stringent test if $q > 1$ [169]. However, for testing the hypothesis that some, but not all, of the variances are zero, the test proposed by [137] is not asymptotically justified, because it is based on the second-order Laplace approximation as in [15] and [16]. As noted (see the second-to-last paragraph of Section 3.1), such an approximation leads to inconsistent estimator of the parameters not specified by the null hypothesis.

Sutradhar [170] proposed a quasi-likelihood (QL) method for estimation in GLMM as a way to improve the robust estimation in GLMM proposed by [19] (see Section 3.2.4). It considers data in a longitudinal setting and requires that the covariance matrix of $y_i = (y_{ij})_{1 \leq j \leq n_i}$, the vector of responses from the ith subject, as well as that of $s_i = (y_{ij}y_{ik})_{1 \leq j < k \leq n_i}$. The latter covariance matrix involves up to fourth moments of the conditional expectations of the data given the random effects. For example, in the case of binary responses, the covariance matrix of s_i involves expressions such as $E(p_{ia}p_{ib})$, $E(p_{ia}p_{ib}p_{ic})$, and $E(p_{ia}p_{ib}p_{ic}p_{id})$, where $p_{ij} = P(y_{ij} = 1 | \alpha)$ and α is the vector of random effects in the GLMM. These expectations can be evaluated by the simulated moment methods, as described in Section 3.2.3, or by numerical integration. Sutradhar [170] made a numerical comparison of the QL method and the first-step estimator of [19], described in Section 3.2.4, under the mixed logistic model $\text{logit}(p_{ij}) = \beta x_i + v_i$ with x_i being subject-level covariate and v_i independent $N(0, \sigma_v^2)$. The QL method turned out to be significantly more efficient than the first-step estimator in estimating σ_v^2. However, relative efficiency of QL and the second-step estimator has not been studied.

Following a similar approach to LMM diagnostics (see Section 2.5), [164] considered model diagnostics for GLMM, and established asymptotic distribution of a modified Pearson's χ^2-test statistic. For GLMM with clustered random effects, the minimum chi-square estimator is used in the test statistic; while, for GLMM with crossed random effects, the MSM estimator (see Section 3.2.3) is used. In a related work, [171] proposed a diagnostic test for checking the random-effect model misspecification in GLMM for clustered binary responses.

There has been growing interest in model selection in the context of mixed effects models, including GLMM. For a recent review on LMM selection, see [112]. Jiang [114] applied the fence methods to GLMM selection and proved consistency of the fence and its variations, including the *adaptive fence*. In particular, the authors applied a variation of the fence to a dataset from a survey conducted in Guatemala regarding the use of modern prenatal care for pregnancies. Using a different approach, [120] considered shrinkage selection/estimation of fixed and random effects in mixed effects models, including GLMM, and studied asymptotic properties. The approach involves the use of (Monte-Carlo) E-M algorithm (e.g., [147]). More recently, [172] considered predictive modeling in GLMM and studied asymptotic properties of several predictive modeling approaches. Hu et al. [173] proposed predictive shrinkage selection/estimation of fixed and random effects in GLMM that does not require the use of E-M algorithm, which is seen as a potential computational advantage.

There is a rich literature on GLMM asymptotics in the context of small area estimation. We leave these to Chapter 4 for detailed discussions.

3.7.2   Open problems

Regarding the GEE approach to inference about GLMM, so far there has been no theoretical results regarding the asymptotic relative efficiency between the second-step estimation of [19] and the QL method of [170]. Both methods were shown to improve the first-step estimation of [19] in empirical studies, and the second-step estimator was shown to be asymptotically more efficient than the first-step one asymptotically (see Section 3.2.4). Note that the second-step estimator is based quadratic functions of the data without the "diagonal terms" [see (3.40)], while the QL is based on quadratic functions of the data without such a restriction. Thus, intuitively, the QL should be more efficient than the second-step estimation, but it is unclear whether the difference will disappear asymptotically. Note that, asymptotically, the proportion of diagonal terms with respect to all the quadratic terms (diagonal terms and cross products) does disappear; and, by excluding the diagonal terms, one can avoid having to estimate an additional dispersion parameter, ϕ, that is sometimes involved in the GLMM [see (3.1)]. See, for example, [2] (pp. 192–193), for discussion.

As for maximum likelihood estimation, an apparent open problem, after the proof of consistency of MLE in GLMM (with crossed random effects [20]; see Section 3.3), is regarding asymptotic distribution of the MLE. Again, the problem is straightforward for GLMMs with clustered random effects, so the real challenge is to obtain the asymptotic distribution of the MLE for GLMMs with crossed random effects. It seems that the subset argument (see Section 3.3.2), although powerful in proving consistency of the MLE for such models, is not effective in obtain the asymptotic distribution. The reason is that the subset inequality [e.g., (3.68)] only gives an upper bound for the likelihood ratio, which is good enough to prove the consistency. However, to obtain the asymptotic distribution, this upper bound is not accurate enough; in other words, one also needs a lower bound that is effectively close to the upper bound. Perhaps, the subset argument can still play a role, at least in some part of the proof for the asymptotic distribution, but one just has not figured out a way to use it.

Another problem associated with the MLE is about the numerical convergence of the MBP (see Section 3.5). Here, a challenge is to prove the convergence under verifiable conditions, for example, under a GLMM with crossed random effects. The information dominance condition [see (3.112)], although having an intuitive interpretation, appears to be too general that it almost assumes what is needed to prove (the convergence property). Furthermore, it is desirable to study asymptotic properties of the limiting estimator of MBP, given the convergence of the latter. Provided that the convergence is proved, and the asymptotic distribution of the limiting estimator is established, then because the limiting estimator satisfies the likelihood equation (3.107), one would have obtained the asymptotic distribution of the MLE. However, this way of establishing the asymptotic distribution of the MLE may not be easier than a direct approach, at least for GLMMs with crossed random effects.

Compared to LMM diagnostics (e.g., [76], [77]; see Section 2.5), the literature on GLMM diagnostics is relatively sparse, especially in terms of rigorous theoretical developments. On the other hand, Pearson's χ^2-test, or similar ideas, for goodness-of-fit appear to be more natural at least for some GLMMs than for LMM for continuous responses. For example, suppose that, given the random effects, α, responses $y_i, i = 1, \ldots, n$ are independent and distributed as Binomial$(6, p_i)$, where the conditional probability p_i depends on the random effects (and some fixed effects as well). In this case, it is entirely natural to consider a Pearson-type test involving 6 cells. In contrast, for a LMM with continuous responses, it is not so clear how many cells, or intervals, should be used for a Pearson-type test. See the last paragraph of Section 2.6.2 for some discussion on this issue.

The consistency results for PGWLS and QIF are more satisfactory for the estimation of fixed effects than for random effects. While this is not totally surprising because, as noted, in most cases where mixed effects models are used, there is not sufficient information for consistently estimating each individual random effect, something more reasonable can be expected. For example, in some cases there is sufficient information for estimating at least some of the random effects individually. In fact, this is typical in small area estimation (see the next chapter), where the sample sizes for different small areas are inhomogeneous due to the inhomogeneity of their corresponding (sub)populations. As a more specific example, in the BRFSS data discussed in Example 3.9, there were HSAs; the sample sizes corresponding to those HSAs ranged from 4 to 2301 [152]. A reasonable asymptotic framework for such situations would be that $n_i \to \infty$ for some but not all of $1 \le i \le m$, where i corresponds to the cluster associated with α_i, the ith component of α, and $m = \dim(\alpha)$, and $m \to \infty$. Under such an assumption, one would expect that the estimator of α_i is consistent as long as $n_i \to \infty$. So far, no such a result has been established for any method that jointly estimates the fixed and random effects.

Another problem associated with PGWLS or QIF is the selection of the constant λ in (3.92) or (3.97). Although, asymptotically, many choices of λ would do the same, there may be a difference in terms of finite-sample performance. So, preferably, one would like to choose λ in a way that is data-driven, and establish the associated asymptotic properties for estimating the fixed and random effects.

In spite of several attempts being made (e.g., [14], [174]), there is, so far, no esay way to effectively extend the REML idea to GLMM. This is, of course, partially due to the computational difficulty associated with GLMM, but more importantly, whatever the extension, one expects to see the kind of asymptotic theory (see Section 2.2) that distinguishes REML from straight maximum likelihood; in other words, the asymptotic superiority of REML in the context of GLMM. So far, no such a theory exists for GLMM.

Finally, the Tailoring method, introduced in Section 3.6.1, offers a unified approach to goodness-of-fit tests. However, so far the method has only been applied to

GLMMs with clustered random effects. It is of practical interest to develop model diagnostic tools for GLMMs with crossed random effects. In particular, it is of interest to know how to apply tailoring to the latter case; apparently, the application is not straightforward.

Chapter 4

Small Area Estimation

Small area estimation (SAE) has become a very active area of statistical research and applications. Here the term small area typically refers to a population for which reliable statistics of interest cannot be produced based on direct sampling from the population due to certain limitations of the available data. Examples of small areas include a geographical region (e.g., a state, county, municipality, etc.), a demographic group (e.g., a specific age $\times$ sex $\times$ race group), a demographic group within a geographic region, etc. For example, in a statewide telephone survey of sample size 4300 in the state of Nebraska, only 14 observations are available to estimate the prevalence of alcohol abuse in Boone county, a small county in Nebraska. The problem is even more severe for a direct survey estimation of the prevalence for white females in the age group 25–44 in this county, because only one observation is available from the survey. See [175] for details. In absence of adequate direct samples from the small areas, statistical methods have been developed in order to "borrow strength." See [176] for an updated account of various methods used in SAE.

A defining characteristic of SAE problems is that the numbers of observations for different small areas are typically small, or at least some of these numbers of observations are small. On the other hand, it is fairly common that there are many small areas that are involved. In fact, there has been extensive asymptotic analys in SAE assuming that the number of small areas, m, is large, while the numbers of observations from the ith small area, n_i, is bounded, $1 \leq i \leq m$. As an area of applications of the mixed effects model asymptotics discussed in the last two chapters, we present a comprehensive coverage of asymptotic methods in SAE. It should be noted that the current chapter is not merely applications of the results and techniques of the previous chapters. In fact, many new asymptotic methods have been introduced in the studies of SAE that are potentially beneficial to other subject fields.

4.1 The Prasad–Rao method

Typically, the borrowing of strength is done via statistical modeling. An explicit (linear) model for composite SAE may be expressed as:

$$Y_i = X_i\beta + Z_iv_i + e_i, \quad i = 1,\ldots,m, \tag{4.1}$$

where m is the number of small areas; Y_i represents the vector of observations from the ith small area; X_i is a matrix of known covariates for the ith small area, and β is a vector of unknown regression coefficients (the fixed effects); Z_i is a known matrix. Furthermore, v_i is a vector of small-area specific random effects; and e_i represents a vector of sampling errors. It is assumed that Y_i is $n_i \times 1$, X_i is $n_i \times p$, β is $p \times 1$, Z_i is $n_i \times b_i$, v_i is $b_i \times 1$ and e_i is $n_i \times 1$. Also assumed is that the v_i's and e_i's are independent such that $E(v_i) = 0$, $\text{Var}(v_i) = G_i$; $E(e_i) = 0$ and $\text{Var}(e_i) = R_i$. Here, G_i and R_i usually depend on a vector ψ of unknown variance components. It is clear that (3.1) is a special case of LMM. Two special cases of the above model are the Fay-Herriot model (Example 2.12) and nested-error regression model (Example 2.6).

A problem of main interest in SAE is estimation, or prediction, of small-area means. A small-area mean may be expressed, at least approximately, as a mixed effect, $\eta = b'\beta + a'v$, where a and b are known vectors, and β and $v = (v_i)_{1 \leq i \leq m}$ are the vectors of fixed and random effects, respectively. The estimation of η is thus typically done via the EBLUP, as described in Section 2.4. Although the EBLUP is fairly easy to obtain, assessing its uncertainty is quite challenging. A measure of the uncertainty that is commonly used is the MSPE. However, unlike the BLUP, the MSPE of the EBLUP does not, in general, have a closed-form expression. This is because once the variance components ψ are replaced by their estimators, the predictor is no longer linear in Y. A naive approach to estimation of the MSPE of EBLUP would be to first obtain the MSPE of BLUP, which can be expressed in closed-form as a function of ψ (see below), and then replace ψ by $\hat{\psi}$ in the expression of the MSPE of BLUP, where $\hat{\psi}$ is the consistent estimator of ψ. However, as will be seen, this approach underestimates the MSPE of EBLUP, as it does not take into account the additional variation associated with the estimation of ψ.

Prasad and Rao [29] proposed a method based on the Taylor series expansion to produce the second-order unbiased MSPE estimator for EBLUP. Here, the term "second-order unbiased" is with respect to the above naive MSPE estimator, which is first-order unbiased. The latter property is because, roughly speaking, the difference between the BLUP and EBLUP is of the order $O(m^{-1/2})$. To see this, note that the BLUP can be expressed as (2.45), where $\tilde{\beta}$ is given by (2.46). Denote the BLUP by $\tilde{\eta} = \tilde{\eta}(\psi)$, which depends on ψ. Then, the EBLUP is simply $\hat{\eta} = \tilde{\eta}(\hat{\psi})$. By the Taylor expansion, we have $\tilde{\eta}(\hat{\psi}) - \tilde{\eta}(\psi) \approx (\partial\tilde{\eta}/\partial\psi')(\hat{\psi} - \psi) = O_P(m^{-1/2})$ under some regularity conditions (see [1], Section 3.4). Therefore, $E(\hat{\eta} - \tilde{\eta})^2$ is typically of the order $O(m^{-1})$. On the other hand, [64] showed that, under normality,

$$
\begin{aligned}
\text{MSPE}(\hat{\eta}) &= E(\hat{\eta} - \eta)^2 \\
&= E(\tilde{\eta} - \eta)^2 + E(\hat{\eta} - \tilde{\eta})^2 \\
&= \text{MSPE}(\tilde{\eta}) + E(\hat{\eta} - \tilde{\eta})^2.
\end{aligned}
\tag{4.2}
$$

Equation (4.2) clearly suggests that the naive MSPE estimator underestimates the true MSPE, because it only takes into account the first term on the right side. Furthermore, if one replaces ψ by $\hat{\psi}$ in the expression of $\text{MSPE}(\tilde{\eta})$, it introduces a bias of the order $O(m^{-1})$ [not $O(m^{-1/2})$]. Thus, the bias of the naive MSPE estimator is $O(m^{-1})$. By second-order unbiasedness of the Prasad–Rao method, it means that

$$
E\left(\widehat{\text{MSPE}} - \text{MSPE}\right) = o(m^{-1}),
\tag{4.3}
$$

where $\mathrm{MSPE} = \mathrm{MSPE}(\hat{\eta})$ and $\widehat{\mathrm{MSPE}}$ represents the Prasad–Rao estimator of MSPE that is derived below. Furthermore, we have the closed-form expression

$$\mathrm{MSPE}(\tilde{\eta}) \;\; = \;\; a'(G - GZ'V^{-1}ZG)a + d'(X'V^{-1}X)^{-1}d, \qquad (4.4)$$

where $d = b - X'V^{-1}ZGa$. Here, we assume that X is of full rank p. Note that, typically, the first term on the right side of (4.4) is $O(1)$ and the second-term is $O(m^{-1})$. An implication is that $\mathrm{MSPE}(\tilde{\eta}) = O(1)$. In view of (4.2) and (4.4), a main goal of the Prasad–Rao method is therefore to derive an approximation to $\mathrm{E}(\hat{\eta} - \tilde{\eta})^2$. Assume that suitable regularity conditions are satisfied. Then we have, by the Taylor expansion and (2.45),

$$
\begin{aligned}
\hat{\eta} - \tilde{\eta} \;\; &= \;\; \tilde{\eta}(\hat{\psi}) - \tilde{\eta}(\psi) \\
&= \;\; \frac{\partial \tilde{\eta}}{\partial \psi'}(\hat{\psi} - \psi) + \frac{1}{2}(\hat{\psi} - \psi)'\frac{\partial^2 \tilde{\eta}(\tilde{\psi})}{\partial \psi \partial \psi'}(\hat{\psi} - \psi),
\end{aligned}
$$

where $\tilde{\psi}$ lies between ψ and $\hat{\psi}$. Suppose that $\hat{\psi}$ is a $\sqrt{m}$-consistent estimator in the sense that $\sqrt{m}(\hat{\psi} - \psi) = O_P(1)$ (again, see [1], Section 3.4) and the following hold:

$$\left|\frac{\partial \tilde{\eta}}{\partial \psi}\right| = O_P(1), \qquad \sup_{|\tilde{\psi}-\psi|\leq|\hat{\psi}-\psi|} \left\|\frac{\partial^2 \tilde{\eta}(\tilde{\psi})}{\partial \psi \partial \psi'}\right\| = O_P(1).$$

Then, by the method of formal derivation ([1], Section 4.3), we have

$$\mathrm{E}(\hat{\eta} - \tilde{\eta})^2 \;\; = \;\; \mathrm{E}\left\{\frac{\partial \tilde{\eta}}{\partial \psi'}(\hat{\psi} - \psi)\right\}^2 + o(m^{-1}). \qquad (4.5)$$

Now, suppose the first term on the right side of (4.5) can be expressed as

$$\mathrm{E}\left\{\frac{\partial \tilde{\eta}}{\partial \psi'}(\hat{\psi} - \psi)\right\}^2 \;\; = \;\; \frac{a(\psi)}{m} + o(m^{-1}), \qquad (4.6)$$

where $a(\cdot)$ is a known differentiable function. Also, let $b(\psi)$ denote the right side of (4.4). By (4.5) and (4.6), to obtain a second-order unbiased estimator of $\mathrm{E}(\hat{\eta} - \tilde{\eta})^2$, all one needs to do is to replace ψ in $a(\psi)$ by $\hat{\psi}$ because the resulting bias is $o(m^{-1})$ (why?). However, one cannot use the same strategy to estimate $\mathrm{MSPE}(\tilde{\eta}) = b(\psi)$, because the resulting bias is $O(m^{-1})$ rather than $o(m^{-1})$. In order to reduce the latter bias to $o(m^{-1})$, we use the following bias correction procedure. Note that, by the Taylor expansion, we have

$$b(\hat{\psi}) \;\; = \;\; b(\psi) + \frac{\partial b}{\partial \psi'}(\hat{\psi} - \psi) + \frac{1}{2}(\hat{\psi} - \psi)'\frac{\partial^2 b}{\partial \psi \partial \psi'}(\hat{\psi} - \psi) + o(m^{-1});$$

hence, by the method of formal derivation (again, see [1], Section 4.3), we have

$$
\begin{aligned}
\mathrm{E}\{b(\hat{\psi})\} \;\; &= \;\; b(\psi) + \frac{\partial b}{\partial \psi'}\mathrm{E}(\hat{\psi} - \psi) + \frac{1}{2}\mathrm{E}\left\{(\hat{\psi} - \psi)'\frac{\partial^2 b}{\partial \psi \partial \psi'}(\hat{\psi} - \psi)\right\} \\
&\quad + o(m^{-1}) \\
&= \;\; b(\psi) + \frac{c(\psi)}{m} + o(m^{-1}).
\end{aligned}
$$

Here, we make the assumption that $E(\hat{\psi} - \psi) = O(m^{-1})$, which holds under regularity conditions. Now, we can apply the same plug-in technique used above for estimating $a(\psi)$ to the estimation of $c(\psi)$. In other words, we estimate $b(\psi)$ by $b(\hat{\psi}) - c(\hat{\psi})/m$ because the bias of this estimator is

$$
\begin{aligned}
E\left\{b(\hat{\psi}) - \frac{c(\hat{\psi})}{m}\right\} - b(\psi) &= b(\psi) + \frac{c(\psi)}{m} + o(m^{-1}) - \frac{E\{c(\hat{\psi})\}}{m} - c(\psi) \\
&= \frac{E\{c(\psi) - c(\hat{\psi})\}}{m} + o(m^{-1}) \\
&= o(m^{-1}),
\end{aligned}
$$

provided that $c(\cdot)$ is a smooth (e.g., differentiable) function.

In conclusion, if we define the Prasad–Rao estimator as

$$
\widehat{\text{MSPE}} = b(\hat{\psi}) + \frac{a(\hat{\psi}) - c(\hat{\psi})}{m}, \tag{4.7}
$$

then we have

$$
\begin{aligned}
E(\widehat{\text{MSPE}}) &= E\{b(\hat{\psi})\} + \frac{E\{a(\hat{\psi})\}}{m} - \frac{E\{c(\hat{\psi})\}}{m} \\
&= b(\psi) + \frac{c(\psi)}{m} + o(m^{-1}) \\
&\quad + \frac{a(\psi)}{m} + \frac{E\{a(\hat{\psi}) - a(\psi)\}}{m} \\
&\quad - \frac{c(\psi)}{m} - \frac{E\{c(\hat{\psi}) - c(\psi)\}}{m} \\
&= b(\psi) + \frac{a(\psi)}{m} + o(m^{-1}) \\
&= \text{MSPE}(\tilde{\eta}) + E(\hat{\eta} - \tilde{\eta})^2 + o(m^{-1}) \\
&= \text{MSPE} + o(m^{-1}),
\end{aligned}
$$

using (4.2), (4.4), (4.5) and (4.6) near the end. Therefore, (4.3) holds.

Prasad and Rao (1990) obtained a detailed expression of (4.6) and, hence, (4.7) for the two special cases discussed earlier—that is, the Fay–Herriot model (Example 2.12) and the nested-error regression model (Example 2.6), assuming normality and using the method of moments (MoM) estimators of ψ. The main advantag of the MoM estimators is that they have closed-form expressions. On the other hand, these estimators are less commonly used these days than the ML and REML estimators (see Chapter 2). Datta and Lahiri [177] extended the Prasad–Rao method to cover cases where ML or REML estimators of ψ are used. Their results were obtained under the longitudinal model of Example 2.4, which includes the Fay–Herriot model and nested-error regression model as special cases. See the next section for details.

Equation (4.7) is called the Prasad–Rao MSPE estimator. A more popular form of it is the following. Denote the two terms on the right side of (4.4) by $g_1(\psi)$ and

$g_2(\psi)$, respectively. Also denote the right side of (4.6) without the $o(m^{-1})$ by $g_3(\psi)$. Then, the Prasad–Rao MSPE estimator can be expressed as

$$\widehat{\text{MSPE}} = g_1(\hat{\psi}) + g_2(\hat{\psi}) + g_3(\hat{\psi}) + \Delta(\hat{\psi}), \qquad (4.8)$$

where $\Delta(\hat{\psi})$ corresponds to a bias-correction term to $g_1(\hat{\psi})$. Although (4.8) is derived under the normality assumption, [178] showed that it has certain robustness with respect to non-normal distribution of the random effects, v_i, but not with respect to non-normal distribution of the errors, e_i, provided that the MoM estimator of ψ is used. A number of other extensions of the Prasad–Rao method have been given, which we shall discuss in the sequel. But before we do that, first, in the next section, let us focus on a special class of small area models, for which more explicit expressions of the g and Δ functions in (4.8) can be given.

4.2 The Fay–Herriot model

The Fay–Herriot model was first introduced in Example 2.12. It is one of the most popular models for small-area estimation that have been practically used. Yet, the model is surprisingly simple and therefore very useful in introducing the basic as well as advanced asymptotic techniques for SAE. Recall that a Fay–Herriot model can be expressed as

$$y_i = x_i'\beta + v_i + e_i, \qquad (4.9)$$

$i = 1, \ldots, m$, with the assumptions detailed in Example 2.12. The small-area mean under the Fay–Herriot model can be expressed as $\theta_i = x_i'\beta + v_i$. The EBLUP of θ_i is given by the right side of (2.44), with A replaced by $\hat{A}$ and β replaced by the right side of (2.47), again with A replaced by $\hat{A}$. Here, $\hat{A}$ is assumed to be a consistent estimator of A. A slightly different expression is given by

$$\hat{\theta}_i = y_i - B_i(\hat{A})\{y_i - x_i'\tilde{\beta}(\hat{A})\}, \qquad (4.10)$$

where $B_i(A) = D_i/(A + D_i)$ and $\tilde{\beta}(A)$ is the BLUE of β, given by (2.47), where A is the true variance of the v_i's.

The $\hat{A}$ in (4.10) is supposed to be a consistent, or more precisely, $\sqrt{m}$-consistent estimator of A. Here, by $\sqrt{m}$-consistent it means $\hat{A} - A = O_P(m^{-1/2})$. There are several choices for $\hat{A}$. Prasad and Rao [29] used MoM, which we call P-R estimator:

$$\hat{A}_{\text{PR}} = \frac{y'P_{X\perp}y - \text{tr}(P_{X\perp}D)}{m - p}, \qquad (4.11)$$

where $P_{X\perp} = I_m - P_X$ with $P_X = X(X'X)^{-1}X'$ and $X = (x_i')_{1 \le i \le m}$. Without loss of generality, we assume that X has full rank p. [177] considered ML and REML estimators of A, denoted by $\hat{A}_{\text{ML}}$ and $\hat{A}_{\text{RE}}$, respectively, which do not have closed-form expressions. Another estimator of A, proposed by [61], is given below. Let

$$Q = Q(A) = \Gamma^{-1} - \Gamma^{-1}X(X'\Gamma^{-1}X)^{-1}X'\Gamma^{-1}$$

with $\Gamma = \mathrm{Var}(y) = \mathrm{diag}(A + D_i, 1 \le i \le m)$. The Fay–Herriot (F-H) estimator of A, denoted by $\hat{A}_{\mathrm{FH}}$, is obtained by solving iteratively the equation

$$\frac{y'Q(A)y}{m-p} = 1 \tag{4.12}$$

for A. It can be shown that (4.12) is unbiased in the sense that the expectation of the left side is equal to the right side if A is the true variance. Note that

$$y'Q(A)y = \sum_{i=1}^{m} \frac{\{y_i - x_i'\tilde{\beta}(A)\}^2}{A + D_i}. \tag{4.13}$$

All of the estimators, $\hat{A}_{\mathrm{PR}}$, $\hat{A}_{\mathrm{ML}}$, $\hat{A}_{\mathrm{RE}}$, and $\hat{A}_{\mathrm{FH}}$, are $\sqrt{m}$-consistent. In particular, the $\sqrt{m}$-consistency of $\hat{A}_{\mathrm{ML}}$ and $\hat{A}_{\mathrm{RE}}$ follows from the results of [25], which do not require normality (see Sections 2.1 and 2.2). The $\sqrt{m}$-consistency of of $\hat{A}_{\mathrm{PR}}$ and $\hat{A}_{\mathrm{FH}}$ may be implied by the general theory of estimating equations (see Section 3.2).

Furthermore, all of the estimators possess the following properties: (i) They are even functions of the data; that is, the estimators are unchanged when y is replaced by $-y$; (ii) they are translation invariant; that is, the estimators are unchanged when y is replaced by $y + Xb$ for any $b \in R^p$). Harville [179] showed that for any estimator $\hat{A}$ that satisfies (i) and (ii), the MSPE of the corresponding EBLUP has the following decomposition:

$$\begin{aligned}\mathrm{MSPE}(\hat{\theta}_i) &= \mathrm{E}(\tilde{\theta}_{\mathrm{B},i} - \theta_i)^2 + \mathrm{E}(\tilde{\theta}_i - \tilde{\theta}_{\mathrm{B},i})^2 + \mathrm{E}(\hat{\theta}_i - \tilde{\theta}_i)^2 \\ &= g_{1i}(A) + g_{2i}(A) + \mathrm{E}(\hat{\theta}_i - \tilde{\theta}_i)^2, \end{aligned} \tag{4.14}$$

where $\tilde{\theta}_{\mathrm{B},i}$ is the BP of θ_i (in terms of minimizing the MSPE), given by (4.10) with $\hat{A}$ and $\tilde{\beta}(\hat{A})$ replaced by A and β, the true parameters, and $\tilde{\theta}_i$ is the BLUP of θ_i, given by (4.10) with $\hat{A}$ replaced by A, the true A. To compute the BP, both A and β have to be known; to compute the BLUP, A has to be known. Both BP and BLUP are not available in typical situations; so the EBLUP is the only one among the three that is computable. However, decomposition (4.14) is very useful in deriving a second-order unbiased estimator of the MSPE of EBLUP. Further analytic expressions can be obtained for the first two terms, namely,

$$g_{1i}(A) = \frac{AD_i}{A + D_i}, \tag{4.15}$$

$$g_{2i}(A) = \left(\frac{D_i}{A+D_i}\right)^2 x_i' \left(\sum_{j=1}^{m} \frac{x_j x_j'}{A+D_j}\right)^{-1} x_i. \tag{4.16}$$

Because these terms as functions of A have nothing to do with the estimator, they remain the same regardless of what estimator of A is used as long as (i) and (ii) are satisfied. Also, note that $g_{1i}(A) = O(1)$ and $g_{2i}(A) = O(m^{-1})$, under regularity conditions. On the other hand, the third term on the right side of (4.14) depends on the estimator of A. Under regularity conditions, it can be shown (e.g., [177]) that

$$\mathrm{E}(\hat{\theta}_i - \tilde{\theta}_i)^2 = \frac{D_i^2}{(A+D_i)^3}\mathrm{var}(\hat{A}) + o(m^{-1})$$

$$= g_{3i}(A) + o(m^{-1}), \qquad (4.17)$$

where $\mathrm{var}(\hat{A})$ is the asymptotic variance of $\hat{A}$. It remains to evaluate $\mathrm{var}(\hat{A})$, and thus $g_{3i}(A)$, for the different estimators of A. Prasad and Rao [29] showed that for $\hat{A}_{\mathrm{PR}}$,

$$g_{3i}(A) = \frac{2D_i^2}{(A+D_i)^3 m^2} \sum_{j=1}^{m} (A+D_j)^2. \qquad (4.18)$$

Datta and Lahiri [177] showed that for both $\hat{A}_{\mathrm{ML}}$ and $\hat{A}_{\mathrm{RE}}$,

$$g_{3i}(A) = \frac{2D_i^2}{(A+D_i)^3} \left\{ \sum_{j=1}^{m} (A+D_j)^{-2} \right\}^{-1} \qquad (4.19)$$

(here it is assumed that p, the rank of X, is bounded; otherwise, the asymptotic variance of $\hat{A}_{\mathrm{ML}}$ and $\hat{A}_{\mathrm{RE}}$ may be different; see Section 12.2). Finally, [180] showed that, for $\hat{A}_{\mathrm{FH}}$, we have

$$g_{3i}(A) = \frac{2D_i^2 m}{(A+D_i)^3} \left\{ \sum_{j=1}^{m} (A+D_j)^{-1} \right\}^{-2}. \qquad (4.20)$$

Despite the different expressions, it is seen from (4.18)–(4.20) that $g_{3i}(A) = O(m^{-1})$, which makes sense in view of (4.17). In conclusion, the leading term in the MSPE decomposition is $g_{1i}(A)$, which is $O(1)$, followed by two $O(m^{-1})$ terms, $g_{2i}(A)$ and $g_{3i}(A)$, so that

$$\mathrm{MSPE}(\hat{\theta}_i) = g_{1i}(A) + g_{2i}(A) + g_{3i}(A) + o(m^{-1}) \qquad (4.21)$$

with $g_{3i}(A)$ depending on the method of estimation of A.

It can be shown that

$$\text{right side of } (4.19) \leq \text{right side of } (4.20)$$
$$\leq \text{right side of } (4.18).$$

This means that in terms of the asymptotic (predictive) efficiency, ML and REML estimators are the best, followed by the F-H estimator, and then by the P-R estimator. These theoretical results are confirmed by the results of simulation studies carried out by [180], who showed that the EBLUPs based on ML, REML, and F-H estimators of A perform similarly in terms of the MSPE (ML and REML perform slightly better under normality), and all three perform significantly better than the EBLUP with the P-R estimator of A when there is a large variability in the D_i's among the small areas. Note that in the balanced case (i.e., $D_i = D, 1 \leq i \leq m$), the P-R, REML, and F-H estimators are identical (provided that the estimator is nonnegative), whereas the ML estimator is different, although the difference is expected to be small when m is large.

Based on the MSPE approximation (4.21) and using a similar bias-correction technique as in the previous section, second-order unbiased estimators of the MSPE can be obtained. Datta and Lahiri [177] showed that if $b(A)$ is the asymptotic bias of $\hat{A}$ up to the order $o(m^{-1})$, then the estimator

$$
\begin{aligned}
\widehat{\mathrm{MSPE}}(\hat{\theta}_i) &= g_{1i}(\hat{A}) + g_{2i}(\hat{A}) + 2g_{3i}(\hat{A}) \\
&\quad - \left(\frac{D_i}{\hat{A}+D_i}\right)^2 b(\hat{A})
\end{aligned}
\tag{4.22}
$$

is second-order unbiased; that is,

$$
\mathrm{E}\{\widehat{\mathrm{MSPE}}(\hat{\theta}_i) - \mathrm{MSPE}(\hat{\theta}_i)\} = o(m^{-1}).
\tag{4.23}
$$

The authors further showed that for $\hat{A}_{\mathrm{PR}}$ and $\hat{A}_{\mathrm{RE}}$, $b(A) = 0$, so that for EBLUP with P-R estimator of A, we have

$$
\widehat{\mathrm{MSPE}}(\hat{\theta}_i) = g_{1i}(\hat{A}_{\mathrm{PR}}) + g_{2i}(\hat{A}_{\mathrm{PR}}) + 2g_{3i,\mathrm{PR}}(\hat{A}_{\mathrm{PR}}),
\tag{4.24}
$$

where $g_{3i,\mathrm{PR}}(A)$ denotes the $g_{3i}(A)$ given by (4.18); similarly, for EBLUP with the REML estimator of A, $\widehat{\mathrm{MSPE}}(\hat{\theta}_i)$ is given by (4.24) with PR replaced by RE, where $g_{3i,\mathrm{RE}}(A)$ denotes the $g_{3i}(A)$ given by (4.19). As for EBLUP with ML estimator of A, [177] showed that

$$
\begin{aligned}
&\widehat{\mathrm{MSPE}}(\hat{\theta}_i) \\
&= \text{(4.24) with PR replaced by ML} \\
&\quad - \left(\frac{D_i}{\hat{A}_{\mathrm{ML}}+D_i}\right)^2 \left\{ \sum_{j=1}^{m} (\hat{A}_{\mathrm{ML}}+D_j)^{-2} \right\}^{-1} \\
&\quad \times \mathrm{tr}\left\{ \left(\sum_{j=1}^{m} \frac{x_j x_j'}{\hat{A}_{\mathrm{ML}}+D_j} \right)^{-1} \sum_{j=1}^{m} \frac{x_j x_j'}{(\hat{A}_{\mathrm{ML}}+D_j)^2} \right\},
\end{aligned}
\tag{4.25}
$$

where $g_{3i,\mathrm{ML}}$ is the same as $g_{3i,\mathrm{RE}}$. Finally, for EBLUP with the F-H estimator of A, [180] showed that

$$
\begin{aligned}
&\widehat{\mathrm{MSPE}}(\hat{\theta}_i) \\
&= \text{(4.24) with PR replaced by FH} \\
&\quad - 2\left(\frac{D_i}{\hat{A}_{\mathrm{FH}}+D_i}\right)^2 \left\{ \sum_{j=1}^{m} (\hat{A}_{\mathrm{FH}}+D_j)^{-1} \right\}^{-3} \\
&\quad \times \left[m \sum_{j=1}^{m} (\hat{A}_{\mathrm{FH}}+D_j)^{-2} - \left\{ \sum_{j=1}^{m} (\hat{A}_{\mathrm{FH}}+D_j)^{-1} \right\}^2 \right],
\end{aligned}
\tag{4.26}
$$

where $g_{3i,\mathrm{FH}}(A)$ denotes $g_{3i}(A)$ given by (4.20).

The (unconditional) MSPE is one way to measure the uncertainty of the EBLUP. There are other measures of uncertainties associated with conditional expectations. First, note that one can write

$$
\mathrm{MSPE}(\hat{\theta}_i) = \mathrm{E}[\mathrm{E}\{(\hat{\theta}_i - \theta_i)^2 | \theta\}],
\tag{4.27}
$$

where the outside conditional expectation is with respect to the distribution of $\theta = (\theta_i)_{1 \le i \le m}$ and the inside conditional expectation is with respect to the sampling distribution of y, known as the design-based (conditional) MSPE. If one considers the inside conditional expectation in (4.27), it leads to

$$\text{MSPE}_1(\hat{\theta}_i|\theta) = \text{E}\{(\hat{\theta}_i - \theta_i)^2|\theta\} \tag{4.28}$$

as a measure of uncertainty conditional on the small-area means. A good thing about MSPE_1 is that it has an exactly unbiased estimator. More generally, consider a predictor of the form $\hat{\theta}_i = y_i + h_i(y)$, where $h_i(\cdot)$ is a differentiable function. Rivest and Belmonte [181] showed

$$\text{MSPE}_1(\hat{\theta}_i|\theta) = D_i + 2\text{E}\{(y_i - \theta_i)h_i(y)|\theta\} + \text{E}\{h_i^2(y)|\theta\}. \tag{4.29}$$

Furthermore, applying the well-known Stein's identity (e.g., [182], p. 124), we have

$$\text{E}\{(y_i - \theta_i)h_i(y)|\theta\} = D_i\text{E}\left(\frac{\partial h_i}{\partial y_i}\bigg|\theta\right). \tag{4.30}$$

Equations (4.29) and (4.30) lead to the expression

$$\begin{aligned}
\text{MSPE}_1(\hat{\theta}_i|\theta) &= D_i + 2D_i\text{E}\left(\frac{\partial h_i}{\partial y_i}\bigg|\theta\right) + \text{E}\{h_i^2(y)|\theta\} \\
&= \text{E}\left\{D_i\left(1 + 2\frac{\partial h_i}{\partial y_i}\right) + h_i^2(y)\bigg|\theta\right\}.
\end{aligned} \tag{4.31}$$

Thus, an exactly (design-)unbiased estimator of MSPE_1 is

$$\widehat{\text{MSPE}}_1(\hat{\theta}_i) = D_i\left(1 + 2\frac{\partial h_i}{\partial y_i}\right) + h_i^2(y). \tag{4.32}$$

In particular, for the EBLUP, we have

$$h_i(y) = -\frac{D_i}{\hat{A} + D_i}\{y_i - x_i'\tilde{\beta}(\hat{A})\}, \tag{4.33}$$

where $\hat{A}$ is a specified estimator of A and $\tilde{\beta}(A)$ is given below (4.10). Thus, for the EBLUP, we have

$$\begin{aligned}
\frac{\partial h_i}{\partial y_i} &= \frac{D_i}{(\hat{A} + D_i)^2}\{y_i - x_i'\tilde{\beta}(\hat{A})\}\frac{\partial \hat{A}}{\partial y_i} \\
&\quad - \frac{D_i}{\hat{A} + D_i}\left(1 - x_i'\frac{\partial \tilde{\beta}}{\partial y_i}\right),
\end{aligned} \tag{4.34}$$

$$\begin{aligned}
\frac{\partial \tilde{\beta}}{\partial y_i} &= \left(\sum_{j=1}^m \frac{x_j x_j'}{\hat{A} + D_j}\right)^{-1} \\
&\quad \times \left\{\frac{x_i}{\hat{A} + D_i} - \sum_{j=1}^m \frac{y_j - x_j'\tilde{\beta}(\hat{A})}{(\hat{A} + D_j)^2}x_j\frac{\partial \hat{A}}{\partial y_i}\right\}
\end{aligned} \tag{4.35}$$

(note that the latter expression is a vector). It remains to obtain expressions for $\partial \hat{A}/\partial y_i$ for the different estimators of A. For the P-R estimator, the expression is simple, because $\hat{A}_{PR}$ has the closed-form expression (4.11), namely,

$$\frac{\partial \hat{A}_{PR}}{\partial y_i} = \frac{2}{m-p}\{y_i - x_i'(X'X)^{-1}X'y\}. \tag{4.36}$$

For the other estimators of A, the expressions for $\partial \hat{A}/\partial y_i$ are more complicated and the implicit function theorem in calculus needs to be used (see [183]). Despite the exact unbiasedness, the estimator (4.32) may be unstable. The reason is that, unlike $\widehat{MSPE}(\hat{\theta}_i)$, the estimator (4.32) involves the term y_i, the direct estimator from the ith small area. This term is subject to a large sampling variation compared to an estimator, such as $\hat{A}$, that is based on observations from all of the small areas.

Alternatively, we can express the MSPE as

$$MSPE(\hat{\theta}_i) = E[E\{(\hat{\theta}_i - \theta_i)^2|y_i\}]. \tag{4.37}$$

This leads to another measure of uncertainty:

$$MSPE_2(\hat{\theta}_i|y_i) = E\{(\hat{\theta}_i - \theta_i)^2|y_i\}. \tag{4.38}$$

Note that, unlike (4.28), here the conditioning is on y_i, the direct estimator from the ith small area [184]. Also, note that if the direct estimator, y_i, is used instead of $\hat{\theta}_i$, then we have

$$MSPE_2(y_i|y_i) = \{y_i - E(\theta_i|y_i)\}^2 + var(\theta_i|y_i)$$
$$= g_{1i}(A) + \left(\frac{D_i}{A+D_i}\right)^2 (y_i - x_i'\beta)^2, \tag{4.39}$$

where $g_{1i}(A)$ is given by (4.15). The question then is whether one can do better than y_i with the EBLUP $\hat{\theta}_i$. Datta et al. [183] obtained the following approximation:

$$MSPE_2(\hat{\theta}_i|y_i) = g_{1i}(A) + g_{2i}(A) + \frac{(y_i - x_i'\beta)^2}{A+D_i}g_{3i}(A)$$
$$+ o_P(m^{-1}), \tag{4.40}$$

where $g_{ri}(A)$, $r = 1,2,3$, are the same as before. Comparing (4.39) and (4.40) and noting that both $g_{2i}(A)$ and $g_{3i}(A)$ are $O(m^{-1})$, we see that a term of order $O(1)$ is replaced by something of the order $O_P(m^{-1})$, when y_i is replaced by $\hat{\theta}_i$. Thus, the EBLUP is doing better, as expected.

Using a similar bias-correction technique as in Section 4.1, [183] obtained a second-order unbiased estimator of $MSPE_2$. The estimator has the general form

$$\widehat{MSPE}_2(\hat{\theta}_i|y_i) = g_{1i}(\hat{A}) + g_{2i}(\hat{A}) + g_{3i}(\hat{A})\left[1 + \frac{\{y_i - x_i'\tilde{\beta}(\hat{A})\}^2}{\hat{A}+D_i}\right]$$
$$- \left(\frac{D_i}{\hat{A}+D_i}\right)^2 b_i(\hat{A}), \tag{4.41}$$

where, like $g_{3i}(A)$, $b_i(A)$ depends on which estimator of A is used. For example,

$$
\begin{aligned}
b_i(A) \;=\; & \left\{ (A+D_i)\sum_{j=1}^{m}(A+D_i)^{-2} \right\}^{-1} \\
& \times \left[\frac{\{y_i - x_i'\tilde{\beta}(A)\}^2}{A+D_i} - 1 \right]
\end{aligned}
\tag{4.42}
$$

for the REML estimator of A, and

$$
\begin{aligned}
b_i(A) \;=\; & 2\left\{ \sum_{j=1}^{m}(A+D_j)^{-1} \right\}^{-3} \\
& \times \left[m\sum_{j=1}^{m}(A+D_j)^{-2} - \left\{ \sum_{j=1}^{m}(A+D_j)^{-1} \right\}^{2} \right] \\
& + \left\{ \sum_{j=1}^{m}(A+D_j)^{-1} \right\}^{-1} \left[\frac{\{y_i - x_i'\tilde{\beta}(A)\}^2}{A+D_i} - 1 \right]
\end{aligned}
\tag{4.43}
$$

for the F-H estimator of A.

Finally, we consider hierarchical Bayes estimation of the small-area means. Note that the Fay–Herriot model can be formulated as a hierarchical model, once a prior distribution, $p(\beta, A)$, is specified on β and A. In this regard, we assume the following:

(i) Given the θ_i's, $y_i, i = 1, \ldots, m$, are independent such that $y_i \sim N(\theta_i, D_i)$.

(ii) Given β and A, $\theta_i, i = 1, \ldots, m$, are independent such that $\theta_i \sim N(x_i'\beta, A)$.

(iii) The prior $p(\beta, A) \propto \pi(A)$, where $\pi(A)$ is a distribution for A ($\propto$ means "proportional to").

Note that without (iii), this is the same as the Fay–Herriot model that we have been considering. The hierarchical Bayes estimator of θ_i is the posterior mean, $E(\theta_i|y)$; a measure of uncertainty for the Bayes estimator is the posterior variance, $\text{var}(\theta_i|y)$. It can be shown that the conditional distribution of θ_i given A and y is normal with mean equal to the right side of (4.10) with $\hat{A}$ replaced by A and variance equal to $g_{1i}(A) + g_{2i}(A)$. It follows that

$$
\begin{aligned}
E(\theta_i|y) \;&=\; E\{E(\theta_i|A,y)|y\} \\
&=\; y_i - E\left[\left(\frac{D_i}{A+D_i}\right)\{y_i - x_i'\tilde{\beta}(A)\}\,\middle|\, y \right],
\end{aligned}
\tag{4.44}
$$

$$
\begin{aligned}
\text{var}(\theta_i|y) \;&=\; E\{\text{var}(\theta_i|A,y)|y\} + \text{var}\{E(\theta_i|A,y)\} \\
&=\; E\{g_{1i}(A)|y\} + E\{g_{2i}(A)|y\} \\
&\quad + \text{var}\left[\left(\frac{D_i}{A+D_i}\right)\{y_i - x_i'\tilde{\beta}(A)\}\,\middle|\, y \right].
\end{aligned}
\tag{4.45}
$$

The conditional expectations involved in (4.44) and (4.45) do not have closed-form expressions. To obtain second-order approximations to (4.44) and (4.45), [180] used

Laplace approximation (e.g., [1], Section 4.6). Let $c(A)$ denote a function for which $E\{c(A)|y\}$ exists. Define $h(A)$ by

$$\exp\{-mh(A)\} \quad = \quad |\Gamma(A)|^{-1/2}|X'\Gamma^{-1}(A)X|^{-1/2}\exp\left\{-\frac{1}{2}y'Q(A)y\right\},$$

where $\Gamma(A)$ and $Q(A)$ are the same as in (4.12) [where we used the notation Γ instead of $\Gamma(A)$]. Note that $mh(A)$ is the same as the negative of the restrictive log-likelihood for estimating A (see Section 2.2). The minimizer of $h(A)$ is therefore the REML estimator of A, $\hat{A}_{RE}$. To express the Laplace approximation to $E\{c(A)|y\}$ in a neat way, we need to introduce some notation. Write $\hat{c} = c(\hat{A}_{RE})$, $\hat{c}_r = \partial^r c(A)/\partial A^r|_{\hat{A}_{RE}}$, $r = 1, 2$, $\hat{h}_r = \partial^r h(A)/\partial A^r|_{\hat{A}_{RE}}$, $r = 2, 3$, and $\hat{\rho}_1 = \partial \log \pi(A)/\partial A|_{\hat{A}_{RE}}$. Then we have

$$E\{c(A)|y\} \quad = \quad \hat{c} + \frac{1}{2m\hat{h}_2}\left(\hat{c}_2 - \frac{\hat{h}_3}{\hat{h}_2}\hat{c}_1\right) + \frac{\hat{c}_1}{m\hat{h}_2}\hat{\rho}_1$$
$$+ O_P(m^{-2}). \tag{4.46}$$

Similarly, provided that $var\{c(A)|y\}$ exists, we have

$$var\{c(A)|y\} \quad = \quad \frac{\hat{c}_1^2}{m\hat{h}_2} + O_P(m^{-2}). \tag{4.47}$$

By applying (4.46) and (4.47) to (4.44) and (4.45), second-order approximations of $E(\theta_i|y)$ and $var(\theta_i|y)$ can be obtained. See [180] for the results and the Ph. D. dissertation of D. D. Smith (2001; University of Georgia) for the details of the proofs.

Note that, unlike the Laplace approximations we have previously seen [e.g., (3.5)], here we have the orders of the approximation errors. For example, the approximation error in (4.46) is $O_P(m^{-2})$ and $E(c(A)|y)$ is $O_P(1)$. We call such an approximation *asymptotically accurate*. In contrast, the Laplace approximation in (3.5) is not asymptotically accurate in that the approximation error does not go to zero (e.g., in probability) as the sample size increases. One may wonder why there is such a difference. For the most part, the Laplace approximation becomes accurate if, as the sample size increases, the underlying distribution becomes concentrated near the mode of the distribution. The posterior distribution of A becomes concentrated near its mode as $m \to \infty$, as more information becomes available about A. On the other hand, in typical situations of GLMM discussed in Chapter 3, the distribution of the random effects does not become concentrated near its mode, even as the sample size increases, as the information about individual random effects remains limited. For example, in the mixed logistic model of Section 3.1, the n_i's (which correspond to the information about the individual α_i's) remain bounded even as $m \to \infty$.

4.3 Extension of Prasad–Rao method

The Prasad–Rao method was developed under the assumption that the data are clustered with independence between clusters. This includes the majority of the previous studies in SAE, where data from different small areas are considered independent (e.g., [176]). More recently, there have been growing interest in nonparametric SAE (e.g., [185]), and SAE with spatially correlated area-specific random effects (e.g., [186]). For such cases, the Prasad–Rao method described in Section 4.1 may not apply, because the between-cluster independence may not hold.

Das et al. [63] extended the Prasad–Rao method to general linear mixed models, which may not have a block-diagonal covariance structure. Their results are potentially applicable to the cases mentioned above. First, note that the Kackar–Harville identity (4.2) holds for normal linear mixed models in general, with or without the block-diagonal covariance structure, where η is any mixed effects that can be expressed as $\eta = \phi'\beta + \psi'v$, where v is the vector of all of the random effects involved. Secondly, expression (4.4) also holds for normal linear mixed models in general; so this term has a closed-form expression. It remains to approximate the second term on the right side of (4.2) to the second order. [185] used the following results due to [63] to obtain the second-order approximation to this term. Note that the results do not require normality (but the Kackar–Harville identity does). Let $l(\theta) = l(\theta, y)$ be a function, where $\theta = (\theta_i)_{1 \leq i \leq s}$ is a parameter with the parameter space Θ and y is the vector of observations. For example, l may be the restricted log-likelihood function that leads to the REML estimator of θ, the vector of variance components (see Section 2.2).

Theorem 4.1. Suppose that the following hold:
(i) $l(\theta, y)$ is three-times continuously differentiable with respect to θ;
(ii) the true $\theta \in \Theta^o$, the interior of Θ;
(iii) $-\infty < \limsup_{n\to\infty} \lambda_{\max}(D^{-1}AD^{-1}) < 0$, where $A = \mathrm{E}\{\partial^2 l/\partial\theta^2\}$ with the second derivative evaluated at the true θ and $D = \mathrm{diag}(d_1, \ldots, d_s)$, with d_i's being positive constants satisfying $d_* = \min_{1 \leq i \leq s} d_i \to \infty$, as $n \to \infty$;
(iv) the gth moments of the following are bounded for some $g > 0$:

$$\frac{1}{d_i}\left|\frac{\partial l}{\partial\theta_i}\right|, \quad \frac{1}{\sqrt{d_i d_j}}\left|\frac{\partial^2 l}{\partial\theta_i \partial\theta_j} - \mathrm{E}\left(\frac{\partial^2 l}{\partial\theta_i \partial\theta_j}\right)\right|, \quad \frac{d_*}{d_i d_j d_k}M_{ijk}, \quad 1 \leq i,j,k \leq s,$$

where the first and second derivatives are evaluated at the true θ and

$$M_{ijk} = \sup_{\tilde{\theta}\in S_\delta(\theta)}\left|\frac{\partial^3 l}{\partial\theta_i \partial\theta_j \partial\theta_k}\bigg|_{\theta=\tilde{\theta}}\right|$$

with $S_\delta(\theta) = \{\tilde{\theta} : |\tilde{\theta}_i - \theta_i| \leq \delta d_*/d_i, 1 \leq i \leq s\}$ for some $\delta > 0$. Then the following results hold:
(I) There exists $\hat{\theta}$ such that for any $0 < \rho < 1$, there is an event set B satisfying for large n and on B, $\hat{\theta} \in \Theta$, $\partial l/\partial\theta|_{\theta=\hat{\theta}} = 0$, $|D(\hat{\theta}-\theta)| < d_*^{1-\rho}$, and

$$\hat{\theta} = \theta - A^{-1}a + r,$$

where θ is the true θ, $a = \partial l/\partial\theta$, evaluated at the true θ, and $|r| \leq d_*^{-2\rho}u$ with $\mathrm{E}(u^g) = O(1)$.
(II) $\mathrm{P}(B^c) \leq cd_*^{\tau g}$ with $\tau = (1/4) \wedge (1-\rho)$ and c being a constant.

Theorem 4.1 plays a key role in the proof of the following theorem. Write the BLUP of η as $\tilde{\eta} = \tilde{\eta}(\theta, y) = \phi'\tilde{\beta} + \psi'\tilde{v}$, where $\tilde{\beta}$ is the BLUE of β and $\tilde{v}$ the BLUP of v, both depending on the unknown true θ. Define a truncated estimator $\hat{\theta}_t$ of θ as follows: $\hat{\theta}_t = \hat{\theta}$ if $|\hat{\theta}| \leq L_n$, and $\hat{\theta}_t = \theta^*$ otherwise, where $\hat{\theta}$ is the estimator in Theorem 4.1, θ^* is a known vector in Θ, and L_n is a sequence of positive numbers such that $L_n \to \infty$ as $n \to \infty$. Consider the EBLUP $\hat{\eta} = \tilde{\eta}(\hat{\theta}_t, y)$. Define $s_0 = \sup_{|\theta| \leq L_n} |\tilde{\eta}(\theta, y)|$ and $s_2 = \sup_{|\theta| \leq L_n} \|\partial^2 \tilde{\eta}/\partial\theta\partial\theta'\|$, where $\|M\| = \lambda_{\max}^{1/2}(M'M)$ is the spectral norm of matrix M and $b = \partial\tilde{\eta}/\partial\theta$, evaluated at the true θ.

Theorem 4.2. Suppose that the conditions of Theorem 4.1 are satisfied. Furthermore, suppppose that there is $h > 0$, $g > 8$, and nonnegative constants g_j, $j = 0, 1, 2$, such that $E(s_0^{2h}) = O(d_*^{g_0})$, $E\{|b|^{2g/(g-2)}\} = O(d_*^{g_1})$, and $E(s_2^2) = O(d_*^{g_2})$. If the inequalities

$$g_0 < \frac{g}{4}(h-1) - 2h, \; g_1 < \left(\frac{g}{g-2}\right)\left\{\frac{1}{2} \wedge \left(\frac{g}{4} - 2\right)\right\}, \; g_2 < \frac{1}{2}$$

hold, then we have

$$E(\hat{\eta} - \tilde{\eta})^2 = E(b'A^{-1}a)^2 + o(d_*^{-2}), \qquad (4.48)$$

where A and a are the same as in Theorem 4.1.

For a mixed ANOVA satisfying (2.1) and (2.3), the first term on the right side of (4.48) can be specified up to the order $o(d_*^{-2})$. Note that in this case, $\theta = (\tau^2, \sigma_1^2, \ldots, \sigma_s^2)$. Define $S(\theta) = V^{-1}ZG\psi$, where $V = \text{Var}(y) = \tau^2 I_n + \sum_{i=1}^s \sigma_i^2 Z_i Z_i'$, $Z = (Z_1 \ldots Z_s)$ and $G = \text{Var}(\alpha) = \text{diag}(\sigma_1^2 I_{m_1}, \ldots, \sigma_s^2 I_{m_s})$. Then [note that a minus sign is missing in front of the trace on the right side of (3.4) of [63]] we have

$$E\{b'A^{-1}a\}^2 = -\text{tr}\left(\frac{\partial S'}{\partial\theta}V\frac{\partial S}{\partial\theta}A^{-1}\right) + o(d_*^{-2}). \qquad (4.49)$$

Combining (4.48) and (4.49) with the Kackar-Harville identity, we obtain a second-order approximation to the MSPE of EBLUP as

$$\text{MSPE}(\hat{\eta}) = g_1(\theta) + g_2(\theta) + g_3(\theta) + o(d_*^{-2}), \qquad (4.50)$$

where $g_1(\theta) = \psi'(G - GZ'V^{-1}ZG)\psi$, $g_2(\theta) = \{\phi - X'S(\theta)\}'(X'V^{-1}X)^{-1}\{\phi - X'S(\theta)\}$, and $g_3(\theta)$ is the first term on the right side of (4.49). By (4.50) and using a similar bias-correction technique as in Section 4.2, a second-order unbiased estimator of the MSPE can be obtained. See [63].

Unlike (4.21), the remaining term on the right side of (4.50) is expressed as $o(d_*^{-2})$. For mixed ANOVA models, [63] showed that d_i^2 may be chosen as $\text{tr}\{(Z_i'PZ_i)^2\}$, $0 \leq i \leq s$, where P is given by (2.16). Of course, in the case of Fay–Herriot model, we has $d_* = \sqrt{m}$, so (4.50) reduced to (4.21).

As an application of Theorem 4.2, we consider a problem regarding nonparametric SAE. Nonparametric models has received considerable attention in recent literature on SAE. In particular, [185] proposed a spline-based nonparametric model.

For simplicity, let us first consider a nonparametric regression model, which can be expressed as

$$y_i = f(x_i) + \varepsilon_i, \quad i = 1, \ldots, n, \tag{4.51}$$

where $f(\cdot)$ is an unknown function and the errors ε_i are independent with mean 0 and constant variance. The linear regression model is a special case of (4.51) with $f(x) = x'\beta$, a vector of unknown regression coefficients. In the latter case, f is unknown only up to a vector of parameters, β. In other words, the form of f is known. It is clear that a nonparametric regression model offers more flexibility than the linear regression model in modeling the mean function of the observations. On the other hand, it is difficult to make an inference about the model if f is completely unknown. A common strategy in practice is to approximate f by a function that can be chosen from a rich family of parametric functions. One such family is called *P-splines*. For simplicity, consider the case of univariate x. Then a P-spline can be expressed as

$$\tilde{f}(x) = \beta_0 + \beta_1 x + \cdots + \beta_p x^p \\ + \gamma_1 (x - \kappa_1)_+^p + \cdots + \gamma_q (x - \kappa_q)_+^p, \tag{4.52}$$

where p is the degree of the spline, q is the number of knots, κ_j, $1 \le j \le q$ are the knots, and $x_+ = x 1_{(x>0)}$. Graphically, a P-spline is pieces of (pth degree) polynomials smoothly connected at the knots. The P-spline model, which is (4.51) with f replaced by $\tilde{f}$, is fitted by *penalized least squares*—that is, by minimizing

$$|y - X\beta - Z\gamma|^2 + \lambda |\gamma|^2, \tag{4.53}$$

with respect to β and γ, where $y = (y_i)_{1 \le i \le n}$, the ith row of X is $(1, x_i, \ldots, x_i^p)$, the ith row of Z is $[(x_i - \kappa_1)_+^p, \ldots, (x_i - \kappa_q)_+^p]$, $1 \le i \le n$, and λ is a penalty, or smoothing, parameter. To determine λ, [187] used the following interesting connection to a linear mixed model. Suppose that the ε_i's are distributed as $N(0, \tau^2)$. Then, if the γ's are treated as independent random effects with the distribution $N(0, \sigma^2)$, the minimizer of (4.53) is the same as the BLUE for β and the BLUP for γ (see Section 2.4), provided that λ is identical to the ratio τ^2/σ^2. Thus, the P-spline model is fitted the same way as the linear mixed model

$$y = X\beta + Z\gamma + \varepsilon \tag{4.54}$$

(e.g., [2], Section 2.3.1). It should be pointed out that the connection is asymptotically valid only if the true underlying function f is not a P-spline. To see this, consider the following example.

Example 4.1. Suppose that the true underlying function is a quadratic spline with two knots, given by

$$f(x) = 1 - x + x^2 - 2(x-1)_+^2 + 2(x-2)_+^2, \quad 0 \le x \le 3$$

(the shape is a half-circle between 0 and 1 facing up, a half-circle between 1 and 2 facing down, and a half-circle between 2 and 3 facing up). Note that this function is smooth in that it has a continuous derivative. Obviously, the best approximating spline is f itself, for which $q = 2$. However, if one uses the above linear mixed model connection, the ML (or REML) estimator of σ^2 is consistent only if $q \to \infty$ (i.e., the number of appearances of the spline random effects goes to infinity). This can be seen from the asymptotic theory in Chapter 2 (see Sections 2.1 and 2.2); but, intuitively, it make sense even without the theory. The seeming inconsistency has two worrisome consequences: (i) The meaning of λ may be conceptually difficult to interpret and (ii) the behavior of the estimator of λ may be unpredictable.

Nevertheless, in most applications of P-splines, the unknown function f is unlikely to be a P-spline, so, in a way, (4.52) is only used as an approximation, with the approximation error vanishing as $q \to \infty$. So, from now on this is the case that we focus on. [185] incorporated the spline model with the small-area random effects. By making use of the spline–mixed-model connection, their approximating model simply has the term $W\alpha$ added to the right side of (4.54); that is,

$$ y = X\beta + Z\gamma + W\alpha + \varepsilon, \qquad (4.55) $$

where α is the vector of small-area random effects and W is a known matrix. It is assumed that γ, α, and ε are uncorrelated with means 0 and covariance matrices Σ_γ, Σ_α, and Σ_ε, respectively. The BLUE and BLUP are given by

$$ \tilde{\beta} = (X'V^{-1}X)^{-1}X'V^{-1}y, $$
$$ \tilde{\gamma} = \Sigma_\gamma Z'V^{-1}(y - X\tilde{\beta}), $$
$$ \tilde{\alpha} = \Sigma_\alpha W'V^{-1}(y - X\tilde{\beta}), $$

where $V = \text{Var}(y) = Z\Sigma_\gamma Z' + W\Sigma_\alpha W' + \Sigma_\varepsilon$ (see Section 2.4). The EBLUE and EBLUP, denoted by $\hat{\beta}$, $\hat{\gamma}$, and $\hat{\alpha}$, are obtained by replacing V by $\hat{V}$ in the corresponding expressions. Here, we assume that $V = V(\theta)$, where θ is a vector of variance components, so that $\hat{V} = V(\hat{\theta})$. The REML estimator is used for $\hat{\theta}$, as suggested by [185]. The problem of main interest is the estimation of small-area means. More precisely, we may treat (4.55) as a super-population model, which can be expressed as

$$ Y_{ij} = x'_{ij}\beta + z'_{ij}\gamma + w'_{ij}\alpha + \varepsilon_{ij}, \qquad (4.56) $$

$j = 1,\ldots,N_i$, where N_i is the population size for the ith small area or subpopulation. Then, the ith small area mean is approximately equal to the mixed effect

$$ \theta_i = \bar{x}'_{i,\text{P}}\beta + \bar{z}'_{i,\text{P}}\gamma + \bar{w}'_{i,\text{P}}\alpha, \qquad (4.57) $$

where $\bar{x}_{i,\text{P}}$, $\bar{z}_{i,\text{P}}$, and $\bar{w}_{i,\text{P}}$ are the population means of x_{ij}, z_{ij}, and w_{ij}, respectively, over $j = 1,\ldots,N_i$, and the approximation error is $O_\text{P}(N_i^{-1/2})$. Thus, by ignoring the approximation error, we may treat θ_i as the small-area mean. Also, note that, in most applications, we have $w'_{ij}\alpha = \alpha_i$, the random effect associated with the ith small area, so that $\bar{w}'_{i,\text{P}}\alpha = \alpha_i$. We will focus on this case in the sequel. The EBLUP of θ_i is

$$ \hat{\theta}_i = \bar{x}'_{i,\text{P}}\hat{\beta} + \bar{z}'_{i,\text{P}}\hat{\gamma} + \hat{\alpha}_i, \qquad (4.58) $$

where $\hat{\alpha}_i$ is the ith component of $\hat{\alpha}$.

The MSPE of the EBLUP is, again, of interest. Here, however, we cannot apply the results of Prasad–Rao in Section 4.1. The reason is that the latter results apply only to the case where the covariance matrix of y, V, is block-diagonal. It is easy to see that V is not block-diagonal in this case. However, the result of Theorem 4.2 is still valid. More specifically, for (4.55), assuming $\Sigma_\gamma = \sigma_\gamma^2 I_q$, $\Sigma_\alpha = \sigma_\alpha^2 I_m$, and $\Sigma_\varepsilon = \sigma_\varepsilon^2 I_n$, where σ_γ^2 and σ_ε^2 are positive, we have $d_0^2 = \mathrm{tr}(P^2)$, $d_1^2 = \mathrm{tr}(PZZ'PZZ')$ and $d_2^2 = \mathrm{tr}(PWW'PWW')$ (see the second paragraph below Theorem 4.2). For (4.50) to be meaningful, we need to show that

$$d_*^2 = d_0^2 \wedge d_1^2 \wedge d_2^2 \to \infty \tag{4.59}$$

as $m \to \infty$, at the very least. We assume $W = \mathrm{diag}(1_{n_i}, 1 \le i \le m)$, as is often the case, where n_i is the sample size for the ith small area, which are assumed to be bounded. Then we have $\lambda_{\max}(W\Sigma_\alpha W') = \sigma_\alpha^2 \lambda_{\max}(WW') = \sigma_\alpha^2 \lambda_{\max}(W'W) = \sigma_\alpha^2 \max_{1 \le i \le m} n_i$. It follows that

$$
\begin{aligned}
V &= \Sigma_\varepsilon + Z\Sigma_\gamma Z' + W\Sigma_\alpha W' \\
&\le \sigma_\varepsilon^2 I_n + \lambda_{\max}(W\Sigma_\alpha W')I_n + \sigma_\gamma ZZ' \\
&= bI_n + \sigma_\gamma^2 ZZ',
\end{aligned}
\tag{4.60}
$$

where $b = \sigma_\varepsilon^2 + \sigma_\alpha^2 \max_{1 \le i \le m} n_i$ is positive and bounded. Let $\lambda_1, \ldots, \lambda_q$ be the eigenvalues of $Z'Z$. Then the eigenvalues of ZZ' are $\lambda_1, \ldots, \lambda_q, 0, \ldots, 0$ (there are $n - q$ zeros after λ_q). Therefore, the eigenvalues of V^{-2} are $(b + \sigma_\gamma^2 \lambda_j)^{-2}$, $1 \le j \le q$, and $b^{-2}, \ldots, b^{-2}$ ($n - q$ b^{-2}'s). It follows that

$$\mathrm{tr}(V^{-2}) = \frac{n-q}{b^2} + \sum_{j=1}^q (b + \sigma_\gamma^2 \lambda_j)^{-2}. \tag{4.61}$$

Also, we have $V \ge \sigma_\varepsilon^2 I_n$; hence, by, for example, (i) of [1], Section 5.3.1, we have $V^{-1} \le \sigma_\varepsilon^{-2} I_n$. It follows, by (iii) of [1], Section 5.3.1, that $\|V^{-1}\|^2 = \lambda_{\max}(V^{-2}) = \{\lambda_{\max}(V^{-1})\}^2 \le \{\lambda_{\max}(\sigma_\varepsilon^{-2} I_n)\}^2 = \sigma_\varepsilon^{-4}$. By the triangle inequality [e.g., [1], (5.43)], we have

$$
\begin{aligned}
\|V^{-1}\|_2 &= \|P + V^{-1}X(X'V^{-1}X)^{-1}X'V^{-1}\|_2 \\
&\le \|P\|_2 + \|V^{-1}X(X'V^{-1}X)^{-1}X'V^{-1}\|_2 \\
&\le \|P\|_2 + \|V^{-1}\|\sqrt{p+1} \\
&\le \|P\|_2 + \sigma_\varepsilon^{-2}\sqrt{p+1}.
\end{aligned}
$$

Therefore, by (4.61), we have $d_0 = \|P\|_2 \ge \|V^{-1}\|_2 - \sigma_\varepsilon^2 \sqrt{p+1} \to \infty$, provided, for example, that $n - q \to \infty$.

Next, consider d_1. Note that $d_1^2 = \|Z'QZ\|_2^2$, and $Z'QZ = Z'V^{-1}Z - Z'V^{-1}X(X'V^{-1}X)^{-1}X'V^{-1}Z$. Thus, by a similar argument, we have

$$
\begin{aligned}
\|Z'V^{-1}Z\|_2 &= \|Z'QZ + Z'V^{-1}X(X'V^{-1}X)^{-1}X'V^{-1}Z\|_2 \\
&\le \|Z'QZ\|_2 + \|Z'V^{-1}X(X'V^{-1}X)^{-1}X'V^{-1}Z\|_2.
\end{aligned}
$$

Again, by (i), (ii) of [1], Section 5.3.1 and the fact that $V \geq \sigma_\varepsilon^2 I_n + \sigma_\gamma^2 ZZ'$, we have

$$Z'V^{-1}Z \leq Z'(\sigma_\varepsilon^2 I_n + \sigma_\gamma^2 ZZ')^{-1}Z.$$

Note that the nonzero eigenvalues of $Z'(\sigma_\varepsilon^2 I_n + \sigma_\gamma^2 ZZ')^{-1}Z$ are the same as the nonzero eigenvalues of $(\sigma_\varepsilon^2 I_n + \sigma_\gamma^2 ZZ')^{-1/2} ZZ' (\sigma_\varepsilon^2 I_n + \sigma_\gamma^2 ZZ')^{-1/2}$, which are $(\sigma_\varepsilon^2 + \sigma_\gamma^2 \lambda_j)^{-1} \lambda_j$, $1 \leq j \leq q$, followed by $n - q$ zeros. It follows by (iii) of [1], Section 5.3.1, that $\|Z'V^{-1}Z\| \leq \sigma_\gamma^{-2}$. On the other hand, it can be shown that $\|Z'V^{-1}X(X'V^{-1}X)^{-1}X'V^{-1}Z\|_2 \leq \|Z'V^{-1}Z\|\sqrt{p+1}$. Therefore, we have $\|Z'QZ\|_2 \geq \|Z'V^{-1}Z\|_2 - \sigma_\gamma^{-2}\sqrt{p+1}$. Now use, again, the inequality (4.60) and similar arguments as above to show that

$$\|Z'V^{-1}Z\|_2^2 \geq \sum_{j=1}^{q} \left(\frac{\lambda_j}{b + \sigma_\gamma^2 \lambda_j} \right)^2. \tag{4.62}$$

Thus, $d_1 \to \infty$, provided, for example, that the right side of (4.62) goes to infinity.

Finally, let us consider d_2. By a similar argument, it can be shown that

$$\|W'QW\|_2 \geq \|W'V^{-1}W\|_2 - c\sqrt{p+1} \tag{4.63}$$

for some constant c. Now, again, using (4.60) and an inverse matrix identity (e.g., [1], Appendix A.1), we have

$$V^{-1} \geq (bI_n + \sigma_\gamma^2 ZZ')^{-1}$$
$$= b^{-1}\{I_n - \delta Z(I_q + \delta Z'Z)^{-1}Z'\},$$

where $\delta = \sigma_\gamma^2 / b$. Thus, we have

$$W'V^{-1}W \geq b^{-1}\{W'W - \delta W'Z(I_q + \delta Z'Z)^{-1}Z'W\}.$$

Write $B = \delta Z(I_q + \delta Z'Z)^{-1}Z'$. Then, we have

$$\lambda_{\max}(B) = \lambda_{\max}\{\delta(I_q + \delta Z'Z)^{-1/2}Z'Z(I_q + \delta Z'Z)^{-1/2}\}$$
$$= \max_{1 \leq j \leq q} \delta \lambda_j (1 + \delta \lambda_j)^{-1}$$
$$\leq 1.$$

It follows that $\|W'BW\|_2 \leq (\max_{1 \leq i \leq m} n_i)q$. By (iii) of [1], Section 5.3.1, we have

$$\|W'W\|_2 \leq \|bW'V^{-1}W + W'BW\|_2$$
$$\leq b\|W'V^{-1}W\|_2 + (\max_{1 \leq i \leq m} n_i)q.$$

Thus, by (4.63) and the fact that $\|W'W\|_2 = (\sum_{i=1}^{m} n_i^2)^{1/2}$, we have

$$d_2 = \|W'QW\|_2$$
$$\geq b^{-1}\left\{ \left(\sum_{i=1}^{m} n_i^2 \right)^{1/2} - \left(\max_{1 \leq i \leq m} n_i \right)q \right\} - c\sqrt{p+1}$$
$$\longrightarrow \infty,$$

provided that, for example, $n_i \geq 1$, $1 \leq i \leq m$, and $q/\sqrt{m} \to 0$.

In the above arguments that lead to (4.59), there is a single assumption whose meaning is not very clear. This is the assumption that the right side of (4.62) goes to ∞. Note that the assumption does not have to hold, even if $q \to \infty$. Below we consider a specific example and show that the assumption holds in this case, provided that $q \to \infty$ at a certain slower rate than n.

Example 4.2. Consider the following special case: $n_i = 1$, $1 \leq i \leq m$ (hence $n = m$), $x_i = i/n$, $1 \leq i \leq n = qr$, where r is a positive integer, and $\kappa_u = (u-1)/q$, $1 \leq u \leq q$. We first show that for any fixed q, there is a positive integer $n(q)$ such that

$$\lambda_{\min}(Z'Z) \;\geq\; 1, \quad n \geq n(q). \tag{4.64}$$

Note that $Z'Z = [\sum_{i=1}^{n}(x_i - \kappa_u)_+^p (x_i - \kappa_v)_+^p]_{1 \leq u,v \leq q}$. We have

$$\frac{1}{n}\sum_{i=1}^{n}(x_i - \kappa_u)_+^p (x_i - \kappa_v)_+^p \;\longrightarrow\; \int_0^1 (x - \kappa_u)_+^p (x - \kappa_v)_+^p \, dx$$

$$\equiv \quad b_{uv}, \quad 1 \leq u,v \leq q,$$

as $n \to \infty$. The matrix $B = (b_{uv})_{1 \leq u,v \leq q}$ is positive definite. To see this, let $\xi = (\xi_u)_{1 \leq u \leq q} \in R^q$. Then

$$\xi'B\xi \;=\; \int_0^1 \left\{ \sum_{u=1}^{q} \xi_u (x - \kappa_u)_+^p \right\}^2 dx \geq 0$$

and the equality holds if and only if

$$\sum_{u=1}^{q} \xi_u (x - \kappa_u)_+^p \;=\; 0, \quad x \in [0,1]. \tag{4.65}$$

Let $x \in (0, \kappa_2]$; then by (4.65), we have $\xi_1 x^p = 0$ and hence $\xi_1 = 0$. Let $x \in (\kappa_2, \kappa_3]$; then by (4.65) and the fact that $\xi_1 = 0$, we have $\xi_2(x - \kappa_2)^p = 0$, hence $\xi_2 = 0$; and so on. This implies $\xi_u = 0$, $1 \leq u \leq q$. It follows, by Weyl's eigenvalue perturbation theorem [e.g., [1], (5.51)], that

$$\frac{1}{n}\lambda_{\min}(Z'Z) \;=\; \lambda_{\min}(B) > 0;$$

hence, there is $n(q) \geq 1$ such that (4.64) holds.

Without loss of generality, let $n(q)$, $q = 1, 2, \ldots$, be strictly increasing. Define the sequence $q(n)$ as $q(n) = 1$, $1 \leq n < n(1)$, and $q(n) = j$, $n(j) \leq n < n(j+1)$, $j \geq 1$. By the definition, it is seen that, with $q = q(n)$, (4.64) holds as long as $n \geq n(1)$. It follows that $\lambda_j \geq 1$, $1 \leq j \leq q$; hence, the right side of (4.62) is bounded from below by $q/(b + \sigma_\gamma^2)^2$, for $q = q(n)$ with $n \geq n(1)$, which goes to infinity as $n \to \infty$.

Another potential application of Theorem 4.2 is SAE with spatially correlated random effects. In many cases, especially if the small areas are adjacent or nearby in geological locations, it makes sense to assume that there is some correlation between the adjacent or nearby small areas, and the magnitude of the correlation decreases as the distance between the small areas increases. For example, [188] considered a conditionally autoregressive (CAR) model for the correlations among the area-specific random effects. Here, the random effects, $v_1, \ldots, v_m$ for the small areas is assumed to be multivariate normal with mean being the zero vector. The covariance matrix is assumed to have the expression

$$\Sigma = (D - C)^{-1}, \tag{4.66}$$

where D is a diagonal matrix; C is a symmetric matrix whose diagonal elements are zero, so that $D - C$ is positive definite. Under such a spatial-correlation model, the covariance matrix of the data is, once again, not (necessarily) block-diagonal. Therefore, the Prasad–Rao method of Section 4.1 cannot be applied. See Section 4.6 for further discussion.

4.4   Empirical best prediction with binary data

As discussed in Section 4.1, the Prasad–Rao method is based on the Taylor series expansion. It is appealing to a practitioner for its simplicity of implementation. In many cases, the data collected from the small areas are binary (indicators). Following [189], we consider a mixed logistic model for binary data y_{ij}, the jth observation from the ith small area. Let α_i be a random effect associated with the ith small area, $1 \leq i \leq m$. It is assumed that given $\alpha = (\alpha_i)_{1 \leq i \leq m}$, y_{ij}, $i = 1, \ldots, m$, $j = 1, \ldots, n_i$, are conditionally independent binary such that

$$\text{logit}\{P(y_{ij} = 1 | \alpha)\} = x'_{ij}\beta + \alpha_i, \tag{4.67}$$

where $x_{ij} = (x_{ijk})_{1 \leq k \leq p}$ is a vector of auxiliary variables that are available and $\beta = (\beta_k)_{1 \leq k \leq p}$ is a vector of unknown parameters. Furthermore, the random effects $\alpha_1, \ldots, \alpha_m$ are independent and distributed as $N(0, \sigma^2)$ with σ^2 being an unknown variance. It is clear that the mixed logistic model is a special case of GLMM discussed in the Chapter 3.

Our main interest is to estimate the small-area means. Let N_i be the population size for the ith small area. Then the small-area mean is equal to the proportion $p_i = N_i^{-1} \sum_{k=1}^{N_i} Y_{ik}$. The difference between Y_{ik} and y_{ij} is that Y_{ik} is the kth value of interest in the population, whereas y_{ij} is the jth sampled value (so $y_{ij} = Y_{ik}$ for some k). Now, consider (13.1) as a *superpopulation model*, with j replaced by k and y by Y, in the sense that the finite population $Y_{ik}, k = 1, \ldots, N_i$, are realizations of random variables satisfying the model. Then we have

$$\frac{1}{N_i} \sum_{k=1}^{N_i} \{Y_{ik} - P(Y_{ik} = 1 | \alpha_i)\} = O_P(N_i^{-1/2}). \tag{4.68}$$

By (4.67) (with j replaced by k and y by Y), the left side of (13.2) is equal to $p_i - \zeta_i(\beta, \alpha_i)$, where $\zeta_i(\beta, \alpha_i) = N_i^{-1} \sum_{k=1}^{N_i} h(x'_{i,k}\beta + \alpha_i)$ with $h(x) = e^x/(1 + e^x)$. It follows that, to the order of $O_P(N_i^{-1/2})$, the small-area mean p_i can be approximated by the mixed effect $\zeta_i(\beta, \alpha_i)$. Here, a mixed effect is a (possibly nonlinear) function of the fixed effects β and random effect α_i. The population size N_i is usually quite large. Note that the small areas are not really "small" in terms of their population sizes. For example, a (U.S.) county is often considered a small area, even though there are tens of thousands of people living in the county; so N_i is large. However, N_i is much smaller compared to the population of the United States, and this is why n_i is small (because the sample size is proportional to the population size). Therefore, the approximation error on the right side of (4.68) is often ignored, and we therefore consider $\xi_i(\beta, \alpha_i)$ as the small-area mean.

The estimation problem now becomes a prediction problem, and we can treat the problem more generally without restricting to a specific function $\zeta_i(\cdot, \cdot)$. It turns out that the functional form of ζ_i does not create much complication as long as it is smooth. Therefore, for the sake of simplicity, we focus on a special case, that is, the prediction of the area specific random effect

$$\zeta_i = \alpha_i. \tag{4.69}$$

If β and σ are known, the best predictor (BP) of α_i, in the sense of minimum MSPE, is the conditional expectation $\tilde{\alpha}_i = \mathrm{E}(\alpha_i|y) = \mathrm{E}(\alpha_i|y_i)$, where $y_i = (y_{ij})_{1\leq j\leq n_i}$. It can be shown that this can be expressed as

$$\tilde{\alpha}_i = \sigma\frac{\mathrm{E}[\xi\exp\{\phi_i(y_{i\cdot}, \sigma\xi, \beta)\}]}{\mathrm{E}[\exp\{\phi_i(y_{i\cdot}, \sigma\xi, \beta)\}]}$$
$$= \psi_i(y_{i\cdot}, \theta), \tag{4.70}$$

where $y_{i\cdot} = \sum_{j=1}^{n_i} y_{ij}$, $\theta = (\beta', \sigma)'$, $\phi_i(k, u, v) = ku - \sum_{j=1}^{n_i}\log\{1+\exp(x_{ij}'v+u)\}$, and the expectation is taken with respect to $\xi \sim N(0, 1)$. Unlike the linear mixed models (see Section 2.4), here the BP does not have a closed-form expression. Nevertheless, only one-dimensional integrals are involved in the expression, which can be evaluated numerically, either by numerical integration or by the Monte-Carlo method. On the other hand, the lack of analytic expression does not keep us from studing the asymptotic behavior of the BP. For example, the following hold:

(i) $\tilde{\alpha}_i/\sigma \to \mathrm{E}(\xi) = 0$ as $\sigma \to 0$. In other words, $\tilde{\alpha}_i = o(\sigma)$ as $\sigma \to 0$.
(ii) For any $1 \leq k \leq n_i - 1$, as $\sigma \to \infty$,

$$\psi_i(k, \theta) \longrightarrow \frac{\int u\exp\{\phi_i(k, u, \beta)\}\,du}{\int \exp\{\phi_i(k, u, \beta)\}\,du}.$$

(iii) $\psi_i(0, \theta) \to -\infty$ and $\psi_i(n_i, \theta) \to \infty$ as $\sigma \to \infty$.
(iv) Suppose that $x_{ij} = x_i$ (i.e., the auxiliary variables are at area level). Then, as $\sigma \to \infty$, we have

$$\psi_i(k, \theta) \longrightarrow \sum_{l=1}^{k-1} l^{-1} - \sum_{l=1}^{n_i-k-1} l^{-1} - x_i'\beta, \quad 1 \leq k \leq n_i - 1, n_i \geq 2,$$

where $\sum_1^0(\cdot)$ is understood as zero. Note that $\sum_{l=1}^{k-1} l^{-1} \sim \log(k) + C$, where C is Euler's constant. Therefore, as $\sigma \to \infty$, we have

$$\tilde{\alpha}_i \approx \mathrm{logit}(\bar{y}_{i\cdot}) - x_i'\beta, \tag{4.71}$$

where $\bar{y}_{i.} = n_i^{-1} y_{i.}$, provided that $1 \leq y_{i.} \leq n_i - 1$ and $n_i \geq 2$. Note that (4.71) holds even if $y_{i.} = 0$ or n_i.

In practice, the parameters θ are usually unknown. It is then customary to replace θ by $\hat{\theta}$, a consistent estimator. The result is called EBP, given by

$$\hat{\alpha}_i = \psi_i(y_{i.}, \hat{\theta}). \tag{4.72}$$

We first study the asymptotic behavior of the EBP when $m \to \infty$ and $n_i \to \infty$. The first limiting process is reasonable because the total number of small areas, m, is usually quite large. For example, in the NHIS problem discussed in [190], the number of small areas of interest is about 600. The second limiting process, however, is unreasonable because the sample size n_i is typically small. On the other hand, it is important to know what would happen in the "ideal" situation, so that one can be sure that the approach is fundamentally sound, at least in large sample. Write

$$\hat{\alpha}_i - \alpha_i = |\psi_i(y_{i.}, \hat{\theta}) - \psi_i(y_{i.}, \theta)| + |\psi_i(y_{i.}, \theta) - \alpha_i^*| + |\alpha_i^* - \alpha_i|, \tag{4.73}$$

where α_i^* is the maximizer of $\phi_i(y_{i.}, u, \beta)$ over u. The idea is to show that the three terms on the right side of (4.73) have the orders $O_P(|\hat{\theta} - \theta|)$, $O_P(n_i^{-1})$, and $O_P(n_i^{-1/2})$, respectively. The order of the last term is derived using virtually the same technique as for the MLE (e.g., [1], Section 4.7). The order of the second term is the result of Laplace approximation to integrals (e.g., [1], Section 4.6). More details about the approximation error in the Laplace approximation can be found in Chapter 4 of [191]. As for the first term, let us point out how the idea of the Laplace approximation is, again, useful here. By the Taylor expansion, it suffices to show that $\partial \psi_i / \partial \theta$ is bounded in probability. Write $g_i(v) = -n_i^{-1} \phi_i(y_{i.}, \sigma v, \beta)$. For any function $f = f(y_{i.}, v, \theta)$, define

$$T_i(f) = \frac{\int f \exp(-n_i g_i) \phi \, dv}{\int \exp(-n_i g_i) \phi \, dv},$$

where ϕ is the pdf of $N(0, 1)$. Let $\varphi_i(y_{i.}, \theta) = T_i(v)$, where v represents the identity function; that is, $f(v) = v$. Then we have $\psi_i(y_{i.}, \theta) = \sigma T_i(v)$. It is easy to show that

$$\frac{\partial \varphi_i}{\partial \theta_k} = n_i \left\{ T_i(v) T_i \left(\frac{\partial g_i}{\partial \theta_k} \right) - T_i \left(v \frac{\partial g_i}{\partial \theta_k} \right) \right\}, \tag{4.74}$$

$1 \leq k \leq p + 1$, where $\theta_{p+1} = \sigma$. The expression on the right side might suggest that our goal is hopeless: As $n_i \to \infty$, how can the right side of (13.8) be bounded? However, the difference inside the curly brackets has a special form. By the Laplace approximation (e.g., [191], Chapter 4), we have

$$T_i(f) = f(\tilde{v}_i) + O(n_i^{-1}), \tag{4.75}$$

where $\tilde{v}_i$ is the minimizer of g_i with respect to v. Using (4.75), we have a cancellation of the leading term inside the curly brackets, or square brackets in the following expression, for the right side of (4.74) (verify this):

$$n_i^{-1}\left[\{\tilde{v}_i+O(n_i^{-1})\}\left\{\left.\frac{\partial g_i}{\partial\theta_k}\right|_{\tilde{v}_i}+O(n_i^{-1})\right\}-\left\{\tilde{v}_i\left.\frac{\partial g_i}{\partial\theta_k}\right|_{\tilde{v}_i}+O(n_i^{-1})\right\}\right]$$

$$=\quad\left.\frac{\partial g_i}{\partial\theta_k}\right|_{\tilde{v}_i}O(1)+\tilde{v}_iO(1)-O(1)+O(n_i^{-1}).$$

The expression can now be expected to be bounded.

A consequence of (4.73) plus the orders of the terms is that the EBP is consistent in the sense that $\hat{\alpha}_i-\alpha_i\xrightarrow{\text{P}}0$ as $n_i\to\infty$ (recall that $\hat{\theta}$ is a consistent estimator). However, as noted, the assumption $n_i\to\infty$ is impractical for small-area estimation. So from now on we will forget about the consistency and focus on the estimation of the MSPE of the EBP given that n_i is bounded. It is easy to verify the following decomposition of the MSPE:

$$\begin{aligned}\text{MSPE}(\hat{\alpha}_i)&=\text{E}(\hat{\alpha}_i-\alpha_i)^2\\&=\text{E}(\hat{\alpha}_i-\tilde{\alpha}_i)^2+\text{E}(\tilde{\alpha}_i-\alpha_i)^2,\end{aligned}\tag{4.76}$$

where $\tilde{\alpha}_i=\text{E}(\alpha_i|y)$. Using expression (4.70) for $\tilde{\alpha}_i$, the second term has a closed-form expression, namely,

$$\text{E}(\tilde{\alpha}_i-\alpha_i)^2\quad=\quad\sigma^2-b_i(\theta),\tag{4.77}$$

where $b_i(\theta)=\text{E}\{\text{E}(\alpha_i|y_i)\}^2=\sum_{k=0}^{n_i}\psi_i^2(k,\theta)p_i(k,\theta)$ and

$$\begin{aligned}p_i(k,\theta)&=\quad\text{P}(y_{i\cdot}=k)\\&=\quad\sum_{z\in S(n_i,k)}\exp\left(\sum_{j=1}^{n_i}z_jx'_{ij}\beta\right)\text{E}[\exp\{\phi_i(z_{\cdot},\sigma\xi,\beta)\}]\end{aligned}$$

with $S(n,k)=\{z=(z_1,\ldots,z_n)\in\{0,1\}^n:z_1+\cdots+z_n=k\}$, $z_{\cdot}=\sum_{j=1}^{n_i}z_j$, and the expectation is taken with respect to $\xi\sim N(0,1)$. As for the first term, we use a formal derivation technique (see [1], Section 4.3). By the Taylor expansion, we have

$$\begin{aligned}\hat{\alpha}_i-\tilde{\alpha}_i&=\quad\psi_i(y_{i\cdot},\hat{\theta})-\psi_i(y_{i\cdot},\theta)\\&=\quad\frac{\partial\psi_i}{\partial\theta'}(\hat{\theta}-\theta)+\frac{1}{2}(\hat{\theta}-\theta)'\left(\frac{\partial^2\psi_i}{\partial\theta\partial\theta'}\right)(\hat{\theta}-\theta)+o_{\text{P}}(|\hat{\theta}-\theta|^2).\end{aligned}$$

Suppose that

$$\hat{\theta}-\theta\quad=\quad O_{\text{P}}(m^{-1/2}).\tag{4.78}$$

It follows that

$$\text{E}(\hat{\alpha}_i-\tilde{\alpha}_i)^2\quad=\quad m^{-1}\text{E}\left\{\frac{\partial\psi_i}{\partial\theta'}\sqrt{m}(\hat{\theta}-\theta)\right\}^2+o(m^{-1}).\tag{4.79}$$

To obtain a further approximation, let us first make it simpler by assuming that $\hat{\theta} = \hat{\theta}_{-i}$, an estimator that does not depend on $y_{i\cdot}$. We know that by making such an assumption, we may draw criticisms right the way: How practical is this assumption? Well, before we worry about it, let us first see how it works, if the assumption holds. The bottom line is that, as we will argue later, the special form $\hat{\theta} = \hat{\theta}_{-i}$ does not really matter. Note that if $\hat{\theta} = \hat{\theta}_{-i}$, then the estimator is independent of $y_{i\cdot}$. If we write $V_i(\theta) = m\mathrm{E}(\hat{\theta}_{-i} - \theta)(\hat{\theta}_{-i} - \theta)'$, then we have

$$
\begin{aligned}
&\mathrm{E}\left\{ \frac{\partial \psi_i}{\partial \theta'} \sqrt{m}(\hat{\theta}_{-i} - \theta) \right\}^2 \\
&= \mathrm{E}[\mathrm{E}\{(\cdots)^2 | y_{i\cdot} = k\}|_{k=y_{i\cdot}}] \\
&= \mathrm{E}\left[\left\{ \frac{\partial}{\partial \theta'} \psi_i(k, \theta) \right\} V_i(\theta) \left\{ \frac{\partial}{\partial \theta} \psi_i(k, \theta) \right\} \Big|_{k=y_{i\cdot}} \right] \\
&= \mathrm{E}\left\{ \frac{\partial \psi_i}{\partial \theta'} V_i(\theta) \frac{\partial \psi_i}{\partial \theta} \right\} \\
&= \sum_{k=0}^{n_i} \left\{ \frac{\partial}{\partial \theta'} \psi_i(k, \theta) \right\} V_i(\theta) \left\{ \frac{\partial}{\partial \theta} \psi_i(k, \theta) \right\} p_i(k, \theta) \\
&\equiv a_i(\theta).
\end{aligned} \tag{4.80}
$$

If we combined (4.76), (4.77), (4.79), and (4.80), we get

$$
\mathrm{MSPE}(\hat{\alpha}_{i,-i}) = \sigma^2 - b_i(\theta) + a_i(\theta)m^{-1} + o(m^{-1}). \tag{4.81}
$$

Here, $\hat{\alpha}_{i,-i}$ indicates that θ is estimated by $\hat{\theta}_{-i}$ in the EBP.

To show that we can replace $\hat{\alpha}_{i,-i}$ by $\hat{\alpha}_i$, an EBP with θ estimated by $\hat{\theta}$, we need to show that the difference made by such a replacement is $o(m^{-1})$. Suppose that (4.78) holds and, in addition,

$$
\hat{\theta}_{-i} - \hat{\theta} = o_{\mathrm{P}}(m^{-1/2}). \tag{4.82}
$$

The motivation for (4.82) is the following. Consider, for simplicity, the sample mean $\hat{\mu} = m^{-1} \sum_{j=1}^{m} X_j$ based on i.i.d. observations $X_1, \ldots, X_m$. Then $\hat{\mu}_{-i} = (m-1)^{-1} \sum_{j \neq i} X_j$; so we have

$$
\hat{\mu}_{-i} - \hat{\mu} = \frac{1}{m(m-1)} \sum_{j \neq i} X_j - \frac{1}{m} X_i = O_{\mathrm{P}}(m^{-1}),
$$

provided that $\mathrm{E}(|X_i|) < \infty$. Hence (4.82) holds. Later we consider a class of estimators of θ that satisfy (4.82). Suppose that (4.78) and (4.82) hold; then we have $\hat{\theta}_{-i} - \theta = \hat{\theta} - \theta + \hat{\theta}_{-i} - \hat{\theta} = O_{\mathrm{P}}(m^{-1/2})$. Therefore, by the Taylor expansion, we have

$$
\begin{aligned}
\hat{\alpha}_{i,-i} - \tilde{\alpha}_i &= \frac{\partial \psi_i}{\partial \theta'}(\hat{\theta}_{-i} - \theta) + o_{\mathrm{P}}(|\hat{\theta}_{-i} - \theta|) \\
&= O_{\mathrm{P}}(m^{-1/2}),
\end{aligned}
$$

$$\hat{\alpha}_i - \hat{\alpha}_{i,-i} = (\hat{\alpha}_i - \tilde{\alpha}_i) - (\hat{\alpha}_{i,-i} - \tilde{\alpha}_i)$$

$$= \left\{ \frac{\partial \psi_i}{\partial \theta'}(\hat{\theta} - \theta) + o_P(|\hat{\theta} - \theta|) \right\} - \left\{ \frac{\partial \psi_i}{\partial \theta'}(\hat{\theta}_{-i} - \theta) + o_P(|\hat{\theta}_{-i} - \theta|) \right\}$$

$$= \frac{\partial \psi_i}{\partial \theta'}(\hat{\theta} - \hat{\theta}_{-i}) + o_P(m^{-1/2})$$

$$= o_P(m^{-1/2}).$$

It follows that

$$\mathrm{MSPE}(\hat{\alpha}_i) = \mathrm{MSPE}(\hat{\alpha}_{i,-i}) + 2\mathrm{E}(\hat{\alpha}_i - \hat{\alpha}_{i,-i})(\hat{\alpha}_{i,-i} - \alpha_i) + \mathrm{E}(\hat{\alpha}_i - \hat{\alpha}_{i,-i})^2$$

$$= \mathrm{MSPE}(\hat{\alpha}_{i,-i}) + o(m^{-1}). \tag{4.83}$$

Note that $\mathrm{E}(\hat{\alpha}_i - \hat{\alpha}_{i,-i})(\hat{\alpha}_{i,-i} - \alpha_i) = \mathrm{E}(\hat{\alpha}_i - \hat{\alpha}_{i,-i})(\hat{\alpha}_{i,-i} - \tilde{\alpha}_i)$. To obtain the final approximation, note that, again by (4.78) and (4.82), the $V_i(\theta)$ involved in $a_i(\theta)$ [see (4.81)] can be approximated by

$$V_i(\theta) = m\{\mathrm{E}(\hat{\theta} - \theta)(\hat{\theta} - \theta)' + \mathrm{E}(\hat{\theta}_{-i} - \hat{\theta})(\hat{\theta} - \theta)'$$

$$+ \mathrm{E}(\hat{\theta} - \theta)(\hat{\theta}_{-i} - \hat{\theta})' + \mathrm{E}(\hat{\theta}_{-i} - \hat{\theta})(\hat{\theta}_{-i} - \hat{\theta})'\}$$

$$= m\{\mathrm{E}(\hat{\theta} - \theta)(\hat{\theta} - \theta)' + o(m^{-1})\}$$

$$= V(\theta) + o(1),$$

where $V(\theta) = m\mathrm{E}(\hat{\theta} - \theta)(\hat{\theta} - \theta)'$. It follows that

$$a_i(\theta) = \sum_{k=0}^{n_i} \left\{ \frac{\partial}{\partial \theta'} \psi_i(k,\theta) \right\} V(\theta) \left\{ \frac{\partial}{\partial \theta} \psi_i(k,\theta) \right\} p_i(k,\theta)$$

$$+ o(1). \tag{4.84}$$

Denoting the summation on the right side of (4.84) by $c_i(\theta)$, we have, by (4.83), (4.81), and (4.84),

$$\mathrm{MSPE}(\hat{\alpha}_i) = \sigma^2 - b_i(\theta) + c_i(\theta)m^{-1} + o(m^{-1}). \tag{4.85}$$

In other words, we have obtained a similar approximation as (4.81) without assuming that $\hat{\theta} = \hat{\theta}_{-i}$. Once again, the method of formal derivation is used here without justifying each step rigorously. For example, it is not necessarily true that $\mathrm{E}\{o_P(m^{-1})\} = o(m^{-1})$. However, these steps can be justified, with rigor, under suitable regularity conditions (see [189]).

It remains to find estimators that satisfy (4.78) and (4.82). Recall the method of moments estimators for GLMM discussed in Section 3.2.3. It can be shown that, subject to some regularity conditions, these estimators satisfy (4.78) and (4.82). In the current case, the estimating equations (3.39) and (3.40) take the following form:

$$\sum_{i=1}^{m} \sum_{j=1}^{n_i} x_{ijk} y_{ij} = \sum_{i=1}^{m} x_{ijk} \mathrm{E}_\theta(y_{ij}), \quad 1 \le k \le p, \tag{4.86}$$

$$\sum_{i=1}^{m} \sum_{j_1 \ne j_2} y_{ij_1} y_{ij_2} = \sum_{i=1}^{m} \sum_{j_1 \ne j_2} \mathrm{E}_\theta(y_{ij_1} y_{ij_2}) \tag{4.87}$$

with $E_\theta(y_{ij}) = E\{h(x'_{ij}\beta + \sigma\xi)\}$ and

$$E_\theta(y_{ij_1}y_{ij_2}) \;=\; E\{h(x'_{ij_1}\beta + \sigma\xi)h(x'_{ij_2}\beta + \sigma\xi)\}, \quad j_1 \neq j_2$$

with $h(x) = e^x/(1+e^x)$, where the expectations are taken with respect to $\xi \sim N(0,1)$.

Based on the MSPE approximation (4.85), we can derive a second-order unbiased estimator of the MSPE. This means finding $\widehat{\mathrm{MSPE}}(\hat{\alpha}_i)$ such that

$$E\{\widehat{\mathrm{MSPE}}(\hat{\alpha}_i) - \mathrm{MSPE}(\hat{\alpha}_i)\} \;=\; o(m^{-1}). \tag{4.88}$$

Write $d_i(\theta) = \sigma^2 - b_i(\theta)$. For the $c_i(\theta)$ in (4.85), we can simply replace θ by $\hat{\theta}$, a consistent estimator (e.g., the method of moment estimator). This is because the difference is $m^{-1}\{c_i(\hat{\theta}) - c_i(\theta)\} = o_P(m^{-1})$, provided that $c_i(\cdot)$ is continuous. However, we cannot simply replace $d_i(\theta)$ by $d_i(\hat{\theta})$, because the difference is $d_i(\hat{\theta}) - d_i(\theta) = O_P(m^{-1/2})$, in the typical situations, provided that $\hat{\theta}$ satisfies (4.78). Therefore, we have to do a *bias correction* in order to reduce the bias to $o(m^{-1})$. Let $M(\theta)$ denote the vector of the difference between the two sides of (4.86) and (4.87) and let $\hat{\theta}$ be the solution to these equations [i.e., $M(\hat{\theta}) = 0$]. We have, by the Taylor expansion,

$$
\begin{aligned}
0 \;=\;& M(\hat{\theta}) \\
=\;& M(\theta) + \frac{\partial M}{\partial \theta'}(\hat{\theta} - \theta) + \frac{1}{2}\left[(\hat{\theta} - \theta)'\frac{\partial^2 M_k}{\partial\theta\partial\theta'}(\hat{\theta} - \theta)\right]_{1\leq k\leq p+1} \\
& + \frac{1}{6}\left[\sum_{a,b,c}\frac{\partial^3 \tilde{M}_k}{\partial\theta_a\partial\theta_b\partial\theta_c}(\hat{\theta}_a - \theta_a)(\hat{\theta}_b - \theta_b)(\hat{\theta}_c - \theta_c)\right]_{1\leq k\leq p+1},
\end{aligned}
$$

where M_k represents the kth component of M and $\tilde{M}_k$ denotes M_k evaluated at $\tilde{\theta}_k$, which lies between θ and $\hat{\theta}$ (but depends on k), $1 \leq k \leq p+1$. In typical situations, we have $(\partial M/\partial\theta')^{-1} = O_P(m^{-1})$, and the second and third derivatives of M_k are $O_P(m)$. These, plus (4.78), imply

$$
\begin{aligned}
0 \;=\;& M(\theta) + \frac{\partial M}{\partial \theta'}(\hat{\theta} - \theta) + \frac{1}{2}\left[(\hat{\theta} - \theta)'\frac{\partial^2 M_k}{\partial\theta\partial\theta'}(\hat{\theta} - \theta)\right]_{1\leq k\leq p+1} \\
& + O_P(m^{-1/2}) \tag{4.89} \\
=\;& M(\theta) + \frac{\partial M}{\partial \theta'}(\hat{\theta} - \theta) + O_P(1). \tag{4.90}
\end{aligned}
$$

Equation (4.90) implies that

$$\hat{\theta} - \theta \;=\; -\left(\frac{\partial M}{\partial \theta'}\right)^{-1}M(\theta) + O_P(m^{-1}). \tag{4.91}$$

We now bring (4.91) back to (4.89) to replace $\hat{\theta} - \theta$ in the quadratic form, leading to

$$0 \;=\; M(\theta) + \frac{\partial M}{\partial \theta'}(\hat{\theta} - \theta) + \frac{1}{2}\hat{Q} + O_P(m^{-1/2}), \tag{4.92}$$

where $\hat{Q} = [M'(\theta)(\partial M'/\partial\theta)^{-1}(\partial^2 M_k/\partial\theta\partial\theta')(\partial M/\partial\theta')^{-1}M(\theta)]_{1\le k\le p+1}$. This leads to the following expansion:

$$\hat{\theta} - \theta = -\left(\frac{\partial M}{\partial\theta'}\right)^{-1}\left\{M(\theta) + \frac{1}{2}\hat{Q}\right\} + O_P(m^{-3/2}) \qquad (4.93)$$

$$= -\left(\frac{\partial M}{\partial\theta'}\right)^{-1}M(\theta) + O_P(m^{-1}). \qquad (4.94)$$

The second equation is due to the fact that $\hat{Q} = O_P(1)$.

Now use another Taylor expansion, this time for $d_i(\hat{\theta})$:

$$d_i(\hat{\theta}) = d_i(\theta) + \frac{\partial d_i}{\partial\theta'}(\hat{\theta} - \theta) + \frac{1}{2}(\hat{\theta} - \theta)'\frac{\partial^2 d_i}{\partial\theta\partial\theta'}(\hat{\theta} - \theta) + O_P(m^{-3/2}).$$

By plugging (4.93) into the first-order term and (4.94) into the second-order term, we obtain the following approximation:

$$d_i(\hat{\theta}) = d_i(\theta) - \frac{\partial d_i}{\partial\theta'}\left(\frac{\partial M}{\partial\theta'}\right)^{-1}\left\{M(\theta) + \frac{1}{2}\hat{Q}\right\} + \frac{1}{2}\hat{R} + O_P(m^{-3/2})$$

$$= d_i(\theta) + m^{-1}\hat{B}_i(\theta) + O_P(m^{-3/2}),$$

where $\hat{R}$ is the same as the components of $\hat{Q}$ except replacing $\partial^2 M_k/\partial\theta\partial\theta'$ by $\partial^2 d_i(\theta)/\partial\theta\partial\theta'$ and

$$\hat{B}_i(\theta) = -m\left[\frac{\partial d_i}{\partial\theta'}\left(\frac{\partial M}{\partial\theta'}\right)^{-1}\left\{M(\theta) + \frac{1}{2}\hat{Q}\right\} - \frac{1}{2}\hat{R}\right]$$

$$= -m\frac{\partial d_i}{\partial\theta'}\left(\frac{\partial M}{\partial\theta'}\right)^{-1}M(\theta) + O_P(1)$$

$$= -m\frac{\partial d_i}{\partial\theta'}\left\{E\left(\frac{\partial M}{\partial\theta'}\right)\right\}^{-1}M(\theta) + O_P(1).$$

The second equation holds because $\hat{R} = O_P(m^{-1})$, $(\partial M/\partial\theta')^{-1} = O_P(m^{-1})$, and $\hat{Q} = O_P(1)$; the third equation is due to the facts that

$$\left(\frac{\partial M}{\partial\theta'}\right)^{-1} - \left\{E\left(\frac{\partial M}{\partial\theta'}\right)\right\}^{-1}$$

$$= -\left\{E\left(\frac{\partial M}{\partial\theta'}\right)\right\}^{-1}\left\{\frac{\partial M}{\partial\theta'} - E\left(\frac{\partial M}{\partial\theta'}\right)\right\}\left(\frac{\partial M}{\partial\theta'}\right)^{-1}$$

$$= O(m^{-1})O_P(m^{1/2})O_P(m^{-1})$$

$$= O_P(m^{-3/2}),$$

and $M(\theta) = O_P(m^{1/2})$. Now, use the fact that $E\{M(\theta)\} = 0$ to get

$$B_i(\theta) \equiv E\{\hat{B}_i(\theta)\} = O(1) \qquad (4.95)$$

[note that $M(\theta) = O_P(m^{1/2})$ is not enough to get (4.95)]. The left side of (4.95), multiplied by m^{-1}, is the leading term for the bias of $d_i(\hat{\theta})$. To bias-correct the latter, we subtract an estimator of the bias, $B_i(\hat{\theta})$ [note that $B_i(\hat{\theta})$ is different from $\hat{B}_i(\theta)$]. This time, the plugging in of $\hat{\theta}$ results in an overall difference of $o(m^{-1})$. The bias-corrected MSPE estimator is therefore

$$\widehat{\text{MSPE}}(\hat{\alpha}_i) = d_i(\hat{\theta}) + \frac{c_i(\hat{\theta}) - B_i(\hat{\theta})}{m}. \tag{4.96}$$

In practice, $B_i(\hat{\theta})$ may be evaluated using a parametric bootstrap method; namely, the bootstrap samples are generated under the mixed logistic model with $\hat{\theta}$ treated as the true parameters. For each bootstrap sample, the expression $\hat{B}_i(\hat{\theta})$ is evaluated. The sample mean of these evaluations over the different bootstrap samples is then an approximation to $B_i(\hat{\theta})$. See the next subsection.

It can be shown that the estimator given by (4.96) satisfies (4.88). Note that (4.96) has a form similar to the Prasad–Rao MSPE estimator (4.7). Therefore, the result obtained in this section may be viewed as an extension of the Prasad–Rao method to mixed logistic models for small-area estimation. Once again, the method of formal derivation is used in deriving the second-order unbiased MSPE estimator. However, the end result [i.e., (4.88) holds for the MSPE estimator (4.96)], can be rigorously justified (see [189]).

4.5 Resampling methods in SAE

In the previous sections we discussed the Prasad–Rao method, also known as *linearization* method in MSPE estimation. An advantage of the linearization method is that it often leads to analytic expressions for the MSPE estimator, or something that is close to analytic expressions, whose computation are relatively fast. A disadvantage is that such a method often involves tedious derivations, and the final analytic expression is likely to be complicated. More importantly, errors often occur in the process of derivations as well as computer programming based on the lengthy expressions.

An alternative to the linearization method is resampling methods, which include *jackknife* and *bootstrap*. These methods usually require much less algebraic derivations, thus reducing the chance of errors. Moreover, the methods are often "one-formula-for-all"; in other words, one needs not to re-derive the formula, as in Prasad–Rao type methods, every time there is a new problem. For example, [177] showed that the Prasad–Rao estimator of the MSPE of the EBLUP needs to be adjusted in order to maintain second-order unbiasedness, depending on whether the ML or REML estimators of ψ is used. See Section 4.2.

4.5.1 Jackknifing the MSPE of EBP

The prediction problem in Sections 4.1, 4.2, and 4.4 may be formulated more generally as estimating the MSPE of an empirical predictor. Suppose that we are interested in predicting an unobservable random vector $\zeta = (\zeta_l)_{1 \leq l \leq t}$. For example, ζ may be a vector of random effects, or mixed effects, that are associated with a linear or generalized linear mixed model. The prediction will be based on independent, vector-valued observations, $y_1, \ldots, y_m$, whose joint distribution depends on a vector $\psi = (\psi_k)_{1 \leq k \leq s}$ of unknown parameters. We consider an example.

Example 4.3. Consider the nested error regression model of Example 2.6. For a given small area i, let $\zeta = \bar{X}'_{i,\mathrm{P}}\beta + v_i$, where $\bar{X}_{i,\mathrm{P}}$ is the population mean for the x_{ij}'s. By a similar argument as in Section 4.4, it can be shown that ζ_i is approximately equal to the population mean for the ith small area under the assumed model, and the approximation error is $O_\mathrm{P}(N_i^{-1/2})$, where N_i is the population size for the small area. Furthermore, let $y_{i'} = (y_{i'j})_{1 \leq j \leq n_{i'}}$, the vector of observations from the i'th small area, $1 \leq i' \leq m$. The prediction of ζ will be based on $y_1, \ldots, y_m$, which are independent, with their joint distribution depending on $\psi = (\beta', \sigma_v^2, \sigma_e^2)'$.

The BP, in the sense of the minimum MSPE, is given by

$$
\begin{aligned}
\tilde{\zeta} &= \mathrm{E}(\zeta | y_1, \ldots, y_m) \\
&= \pi(y_S, \psi),
\end{aligned}
\tag{4.97}
$$

a vector-valued function of $y_S = (y_i)_{i \in S}$, where S is a subset of $\{1, \ldots, m\}$, and ψ. Since ψ is typically unknown, the BP is not computable. It is then customary to replace ψ by $\hat{\psi}$, a consistent estimator. The result is what we call empirical best predictor, or EBP, given by

$$
\hat{\zeta} = \pi(y_S, \hat{\psi}).
\tag{4.98}
$$

Typically, the MSPE of the EBP is much more difficult to evaluate than the MSPE of BP. We use an example to illustrate.

Example 4.3 (continued). Under the normality assumption, the conditional distribution of ζ given $y_1, \ldots, y_m$ is the same as the conditional distribution of ζ given y_i. The latter is normal with mean

$$
\begin{aligned}
\mathrm{E}(\zeta | y_i) &= \bar{X}'_{i,\mathrm{P}}\beta + \mathrm{E}(v_i | y_i) \\
&= \bar{X}'_{i,\mathrm{P}}\beta + \frac{n_i \sigma_v^2}{\sigma_e^2 + n_i \sigma_v^2}(\bar{y}_{i\cdot} - \bar{x}'_{i\cdot}\beta)
\end{aligned}
\tag{4.99}
$$

and variance $\sigma_v^2 \sigma_e^2 / (\sigma_e^2 + n_i \sigma_v^2)$. Thus, $\tilde{\zeta}$ is given by the right side of (4.99). It can be shown that $\mathrm{MSPE}(\tilde{\zeta}) = \mathrm{E}(\tilde{\zeta} - \zeta)^2 = \sigma_v^2 \sigma_e^2 / (\sigma_e^2 + n_i \sigma_v^2) = \mathrm{var}(\zeta | y_i)$. On the other hand, the MSPE of $\hat{\zeta}$ does not have a closed-form expression, in general, and may depend on what estimator of ψ is used.

The jackknife method provides an attractive alternative for the MSPE estimation. Note that, unlike in the previous section, here the main interest is prediction, rather than estimation. Note that the MSPE has the following decomposition. Consider, for now, the case of univariate ζ. Then we have

$$
\begin{aligned}
\mathrm{MSPE}(\hat{\zeta}) &= \mathrm{E}(\hat{\zeta} - \zeta)^2 \\
&= \mathrm{E}(\hat{\zeta} - \tilde{\zeta})^2 + \mathrm{E}(\tilde{\zeta} - \zeta)^2 \\
&= \mathrm{MSAE}(\hat{\zeta}) + \mathrm{MSPE}(\tilde{\zeta}).
\end{aligned}
\tag{4.100}
$$

The first term on the right side of (4.100) corresponds to mean squared approximation error (MSAE) of $\hat{\zeta}$ to $\tilde{\zeta}$, whereas the second term is the MSPE of $\tilde{\zeta}$ (considered as a predictor). A jackknife estimator of the MSAE is given by

$$\widehat{\mathrm{MSAE}}(\hat{\zeta})_{\mathrm{J}} \quad = \quad \frac{m-1}{m}\sum_{i=1}^{m}(\hat{\zeta}_{-i}-\hat{\zeta})^2, \tag{4.101}$$

where $\hat{\zeta}_{-i}$ is some kind of jackknife replication of $\hat{\zeta}$ (the exact definition will be given after some discussion). Furthermore, suppose that $\mathrm{MSPE}(\tilde{\zeta}) = b(\psi)$, where $b(\cdot)$ is a known function. We consider a bias-corrected jackknife estimator of the second term:

$$\widehat{\mathrm{MSPE}}(\tilde{\zeta})_{\mathrm{J}} \quad = \quad b(\hat{\psi}) - \frac{m-1}{m}\sum_{i=1}^{m}\{b(\hat{\psi}_{-i})-b(\hat{\psi})\}, \tag{4.102}$$

where $\hat{\psi}$ is a suitable estimator of ψ and $\hat{\psi}_{-i}$ is the jackknife replication of $\hat{\psi}$. We then combine the two to obtain a jackknife estimator of the MSPE of $\hat{\zeta}$:

$$\widehat{\mathrm{MSPE}}(\hat{\zeta})_{\mathrm{J}} \quad = \quad \widehat{\mathrm{MSAE}}(\hat{\zeta})_{\mathrm{J}} + \widehat{\mathrm{MSPE}}(\tilde{\zeta})_{\mathrm{J}}. \tag{4.103}$$

The subject that is new here is $\hat{\zeta}$, which is a predictor, not estimator. The question is: What is an appropriate definition of $\hat{\zeta}_{-i}$? The following example shows that a naive definition may not work.

Example 4.4 (James–Stein estimator and naive jackknife). Let the observations $y_1,\ldots,y_m$ be independent such that $y_i \sim N(\theta_i, 1)$, $1 \le i \le m$. In the context of simultaneous estimation of $\theta = (\theta_1,\ldots,\theta_m)'$, it is well known that, for $m \ge 3$, the James–Stein estimator dominates the maximum likelihood estimator, which is simply $y = (y_1,\ldots,y_m)'$, in terms of the frequentist risk under a sum of squares error loss function (e.g., [192], p. 302). Efron and Morris [193] provided an empirical Bayes justification of the James–Stein estimator. Their Bayesian model may be equivalently written as a simple random effects model, $y_i = v_i + e_i, i = 1,\ldots,m$, where the random effects v_i and sampling errors e_i are independent with $v_i \sim N(0,A)$ and $e_i \sim N(0,1)$, where A is unknown. Write $B = (1+A)^{-1}$. Then the James–Stein estimator can be interpreted as an EBP under the random effects model. For example, the BP of $\zeta = v_1$ is $\tilde{\zeta} = (1-B)y_1$. The estimator of B proposed by [193] is $\hat{B} = (m-2)/\sum_{i=1}^{m}y_i^2$. Alternatively, the MLE of B is $\hat{B} = m/\sum_{i=1}^{m}y_i^2$. By plugging in $\hat{B}$, we get the EBP, $\hat{\zeta} = (1-\hat{B})y_1$. A straightforward extension of the jackknife for estimating the MSPE of $\hat{\zeta}$ would define the ith jackknife replication of $\hat{\zeta}$ as one derived the same way except without the observation y_i. Denote this naive jackknife replication by $\hat{\zeta}^*_{-i}$. The question is: What is $\hat{\zeta}^*_{-1}$? To be more specific, suppose that the MLE of B is used. If we follow the derivation of of the EBP, then we must have $\tilde{\zeta}_{-1} = 0$; hence, $\hat{\zeta}^*_{-1} = 0$. On the other hand, for $i \ge 2$, we have $\tilde{\zeta}_{-i} = (1-B)y_1$ (same as $\tilde{\zeta}$) and $\hat{B}_{-i} = (m-1)/\sum_{j\ne i}y_j^2$; hence, $\hat{\zeta}^*_{-i} = (1-\hat{B}_{-i})y_1$. It can be shown that $\mathrm{MSPE}(\hat{\zeta}) = 1 - B + 2B/m + o(m^{-1})$. On the other hand, we have $\mathrm{E}(\hat{\zeta}^*_{-1} - \hat{\zeta})^2 = \mathrm{E}(\hat{\zeta}^2) = A(1-B) + o(1)$ and $\mathrm{E}(\hat{\zeta}^*_{-i} - \hat{\zeta})^2 = 2B/m^2 + o(m^{-2})$, $i \ge 2$.

With these, it can be shown that the expectation of the right side of (4.103), with $\hat{\zeta}_{-i}$ replaced by $\zeta^*_{-i}, 1 \leq i \leq m$, is equal to $\mathrm{MSPE}(\hat{\zeta}) + A(1-B) + o(1)$. In other words, the bias of the naive jackknife MSPE estimator does not even go to zero as $m \to \infty$.

The problem with the naive jackknife can be seen clearly from the above example. The observation y_1 plays a critical role in the prediction of $\zeta = v_1$. This observation cannot be removed no matter what. Any jackknife replications should only be with respect to $\hat{\psi}$, not y_S [see (4.98)]. Therefore, we define the ith jackknife replication of $\hat{\zeta}$ as

$$\hat{\zeta}_{-i} = \pi(y_S, \hat{\psi}_{-i}), \tag{4.104}$$

where y_S is the same as in (4.98) and $\hat{\psi}_{-i}$ is the ith jackknife replication of $\hat{\psi}$, described below.

So far, the presence of $\hat{\psi}$ is, more or less, just a notation—we have not given the specific form of $\hat{\psi}$. In many applications, the estimator $\hat{\psi}$ belongs to a class of M-estimators. Here, an M-estimator is associated with a solution, ψ, to the following equation:

$$F(\psi) = \sum_{j=1}^{m} f_j(\psi, y_j) + a(\psi) = 0, \tag{4.105}$$

where $f_j(\cdot, \cdot)$ are vector-valued functions satisfying $\mathrm{E}\{f_j(\psi, y_j)\} = 0, 1 \leq j \leq m$, if ψ is the true parameter vector, and $a(\cdot)$ is a vector-valued function which may depend on the joint distribution of $y_1, \ldots, y_m$. When $a(\psi) \neq 0$, it plays the role of a modifier, or penalizer. We consider some examples.

Example 4.5 (ML estimation). Consider the longitudinal linear mixed model (2.4). Let $\psi = (\beta', \theta')'$. Under regularity conditions, the MLE of ψ satisfies (4.105) with $a(\psi) = 0$, $f_j(\psi, y_j) = [f_{j,k}(\psi, y_j)]_{1 \leq k \leq p+q}$, where p is the dimension of β and q the dimension of θ;

$$[f_{j,k}(\psi, y_j)]_{1 \leq k \leq p} = X_j' V_j^{-1}(\theta)(y_j - X_j \beta);$$

$$f_{j,p+l}(\psi, y_j) = (y_j - X_j \beta)' V_j^{-1}(\theta) \frac{\partial V_j}{\partial \theta_l} V_j^{-1}(\theta)(y_j - X_j \beta)$$

$$- \mathrm{tr}\left\{V_j^{-1}(\theta) \frac{\partial V_j}{\partial \theta_l}\right\}, \quad 1 \leq l \leq q,$$

where $V_j(\theta) = R_j + Z_j G_j Z_j'$.

Example 4.6 (REML estimation). Continue with the previous example. Similarly, the REML estimator of θ is defined as a solution to the REML equation (see Section 2.2). The REML estimator of β is defined as $\hat{\beta} = \{\sum_{j=1}^{m} X_j' V_j^{-1}(\hat{\theta}) X_j\}^{-1} \sum_{j=1}^{m} X_j' V_j^{-1}(\hat{\theta}) y_j$, where $\hat{\theta}$ is the REML estimator of θ. It can be shown that the REML estimator of ψ satisfies (4.105), where the f_j's are the same as in Example 14.6; $a(\psi) = [a_k(\psi)]_{1 \leq k \leq p+q}$ with $a_k(\psi) = 0, 1 \leq k \leq p$, and

$$a_{p+l}(\psi) = \sum_{j=1}^{m} \mathrm{tr}\left[V_j^{-1}(\theta) X_j \{X' V^{-1}(\theta) X\}^{-1} X_j' V_j^{-1}(\theta) \frac{\partial V_j}{\partial \theta_l}\right],$$

$1 \le l \le q$. Here, $X = (X_j)_{1 \le j \le m}$ and $V(\theta) = \text{diag}\{V_j(\theta), 1 \le j \le m\}$.

Consider a jackknife replication of (4.105); that is,

$$F_{-i}(\psi) = \sum_{j \ne i} f_j(\psi, y_j) + a_{-i}(\psi) = 0, \qquad (4.106)$$

$1 \le i \le m$. The ith jackknife replication of ψ, ψ_{-i}, is defined as a solution to (4.106). Sometimes, a solution to (4.105) may not exist, or exist but not within the parameter space (e.g., negative values for variances). Therefore, we define an M-estimator of ψ as $\hat{\psi} = \psi$ if the solution to (4.105) exists within the parameter space, and $\hat{\psi} = \psi^*$ otherwise, where ψ^* is a known vector within the parameter space. Similarly, we define $\hat{\psi}_{-i} = \psi_{-i}$ if the solution to (4.106) exists within the parameter space, and $\hat{\psi}_{-i} = \psi^*$ otherwise. [30], [194] showed that the jackknife MSPE estimator, defined by (4.103) with $\hat{\psi}, \hat{\psi}_{-i}, 1 \le i \le m$, being the M-estimator and its jackknife replications, has a bias reduction property, known as asymptotic unbiasedness. For simplicity, here we consider a very special case. See the next section for a multivariate extension.

Let ψ be a scalar parameter and $S = \{1\}$ in (4.98). Also, assume that $f_j = f$ (i.e., not dependent on j) and $a(\psi) = a_{-i}(\psi) = 0$ in (4.105) and (4.106). Let $y_1, \ldots, y_m$ be independent with the same distribution as Y. Before we study the asymptotic unbiasedness of the jackknife MSPE estimator, let us first consider an important property of the M-estimator. For notation convenience, write $\hat{\psi}_{-0} = \hat{\psi}$ and $F_{-0} = F$. The M-estimators $\hat{\psi}_{-i}, 0 \le i \le m$, are said to be consistent uniformly (c.u.) at rate m^{-d} if for any $b > 0$, there is a constant B (possibly depend on b) such that $P(A_{i,b}^c) \le Bm^{-d}, 0 \le i \le m$, where $A_{i,b} = \{F_{-i}(\hat{\psi}_{-i}) = 0, |\hat{\psi}_{-i} - \psi| \le b\}$ and ψ is the true parameter. The M-estimating equations are said to be *standard* if $f(\psi, Y) = (\partial/\partial \psi) l(\psi, Y)$ for some function $l(\psi, u)$ that is three times continuously differentiable with respect to ψ and satisfies

$$E\left\{\frac{\partial^2}{\partial \psi^2} l(\psi, Y)\right\} > 0.$$

The ML and REML equations (see Examples 4.5 and 4.6) are multivariate extensions of the standard M-estimating equations. The following theorem is given in [30] as a proposition.

Theorem 4.3. Suppose that the M-estimating equations are standard and the $2d$th moments of

$$\left|\frac{\partial^r}{\partial \psi^r} l(\psi, Y)\right|, \ r = 1, 2, \quad \sup_{|\psi' - \psi| \le b_0} \left|\frac{\partial^3}{\partial \psi^3} l(\psi', Y)\right|$$

are finite for some $d \ge 1$ and $b_0 > 0$, where ψ is the true parameter. Then there exist M-estimators $\hat{\psi}_{-i}, 0 \le i \le m$, that are c.u. at rate m^{-d}.

The next thing we do is to establish the asymptotic unbiasedness properties for the jackknife estimators of $\text{MSAE}(\hat{\zeta})$ and $\text{MSPE}(\tilde{\theta})$ separately and then combine them [see (4.100) and (4.103)]. To do so we also need some regularity conditions on the EBP (14.98) (note that now $S = \{1\}$). We assume that

$$|\pi(Y_1, \psi)| \le \omega(Y_1)(1 \vee |\psi|^\lambda) \qquad (4.107)$$

for some constant $\lambda > 0$ and measurable function $\omega(\cdot)$ such that $\omega(\cdot) \geq 1$ $[a \vee b = \max(a,b)]$. The c.u. property can now be generalized. Let $\mathscr{A} = \sigma(y_1, \ldots, y_m)$, the σ-field generated by the y_i's. Define a measure μ_ω on $\mathscr{A}$ as

$$\mu_\omega(A) \quad = \quad \mathrm{E}\{\omega^2(Y) 1_A\}, \quad A \in \mathscr{A}. \tag{4.108}$$

The M-estimators $\hat{\psi}_{-i}, 0 \leq i \leq m$, are said to be c.u. with respect to μ_ω (c.u. μ_ω) at rate m^{-d} if for any $b > 0$, there is a constant B (possibly dependent on b) such that $\mu_\omega(A_{i,b}^c) \leq B m^{-d}, 0 \leq i \leq m$, where $A_{i,b}$ is the same as above. Below are some remarks regarding c.u. μ_ω and its connection to c.u.

Remark 1. Because $\omega(\cdot) \geq 1$, that the M-estimators are c.u. μ_ω at rate m^{-d} implies that they are c.u. at rate m^{-d}. Conversely, if there is $\tau > 2$ such that $\mathrm{E}(|\omega(Y)|^\tau) < \infty$, then that the M-estimators are c.u. at rate m^{-d} implies that they are c.u. μ_ω at rate $m^{-d(1-2/\tau)}$. In particular, if $\mathrm{E}\{\omega^4(Y)\} < \infty$, then if the M-estimators are c.u. at rate m^{-2d}, they are c.u. μ_ω at rate m^{-d}. This is useful in checking the c.u. μ_ω property because, under a suitable moment condition, it reduces to checking the c.u. property.

Remark 2. In practice, the function ω may be chosen in the following way: Find a positive number λ such that $\omega(y) = \sup_\psi \{|\pi(y, \psi)|/(1 \vee |\psi|^\lambda)\} < \infty$ for every y and use this ω.

The following theorem states the asymptotic unbiasedness of the jackknife MSAE estimator.

Theorem 4.4. Suppose that (i) $\mathrm{E}\{(\partial/\partial\psi)f(\psi,Y)\} \neq 0$; (ii) for some constants $d > 2$ and $b_0 > 0$, the expectations of the following are finite, where ψ is the true parameter:

$$|f(\psi,Y)|^{2d}, \quad \left|\frac{\partial}{\partial\psi}f(\psi,Y)\right|^{2d}, \quad \sup_{|\psi'-\psi|\leq b_0}\left|\frac{\partial^2}{\partial\psi^2}f(\psi',Y)\right|^{2d},$$

$$\sup_{|\psi'-\psi|\leq b_0}\left\{\frac{\partial^3}{\partial\psi^3}f(\psi',Y)\right\}^4;$$

$$\omega^4(Y), \quad \left\{\frac{\partial^2}{\partial\psi^2}\pi(Y,\psi)\right\}^4, \quad \sup_{|\psi'-\psi|\leq b_0}\left\{\frac{\partial^3}{\partial\psi^3}\pi(Y,\psi')\right\}^2,$$

$$\sup_{|\psi'-\psi|\leq b_0}\left|\frac{\partial}{\partial\psi}\pi(Y,\psi')\right|^{2d};$$

and (iii) $\hat{\psi}_{-i}, 0 \leq i \leq m$ are c.u. μ_ω at rate m^{-d}. Then we have

$$\mathrm{E}\{\widehat{\mathrm{MSAE}}(\hat{\zeta})_J\} - \mathrm{MSAE}(\hat{\zeta}) \quad = \quad o(m^{-1-\varepsilon})$$

for any $0 < \varepsilon < (d-2)/(2d-1)$.

The next theorem states the asymptotic unbiasedness of the jackknife bias-corrected estimator for the MSPE of $\tilde{\zeta}$.

Theorem 4.5. Suppose that (i) $\mathrm{E}\{(\partial/\partial\psi)f(\psi,Y)\} \neq 0$; (ii) for some $d > 2$ and

$b_0 > 0$, the $2d$th moments of the following are finite:

$$\left| f(\psi, Y) \right|, \quad \left| \frac{\partial^r}{\partial \psi^r} f(\psi, Y) \right|, \quad r = 1, 2, \quad \sup_{|\psi' - \psi| \le b_0} \left| \frac{\partial^3}{\partial \psi^3} f(\psi, Y) \right|,$$

and $\sup_{|\psi' - \psi| \le b_0} |b^{(4)}(\psi')|$ is bounded, where ψ is the true parameter and $b^{(4)}$ the fourth derivative; and (iii) $\hat{\psi}_{-i}, 0 \le i \le m$ are c.u. at rate m^{-d}. Then we have

$$E\{\widehat{\mathrm{MSPE}}(\tilde{\zeta})_J\} - \mathrm{MSPE}(\tilde{\zeta}) = o(m^{-1-\varepsilon})$$

for any $0 < \varepsilon < (d-2)/(2d+1)$.

By combining Theorems 4.4 and 4.5 and in view of (14.100) and (14.103), we obtain the asymptotic unbiasedness of the jackknife MSPE estimator.

Theorem 4.6. Suppose that (i)–(iii) of Theorem 4.4 and (ii) of Theorem 4.5 hold; then we have

$$E\{\widehat{\mathrm{MSPE}}(\hat{\zeta})_J\} - \mathrm{MSPE}(\hat{\zeta}) = o(m^{-1-\varepsilon})$$

for any $0 < \varepsilon < (d-2)/(2d+1)$.

Note that, because $\varepsilon > 0$, the result of Theorem 4.6 implies that $E\{\widehat{\mathrm{MSPE}}(\hat{\zeta})_J\} - \mathrm{MSPE}(\hat{\zeta}) = o(m^{-1})$. Thus, the jackknife MSPE estimator is second-order unbiased according to our earlier definition (see Sections 4.1 and 4.2).

At this point, we would like to discuss some of the ideas used in the proofs. A basic technique is Taylor series expansions (see below). This allows us to approximate the EBP by something simpler. Typically, the approximation error in the Taylor expansion is expressed in terms of O_P or o_P. We need to "convert" such a result to convergence in expectation. In this regard, the following lemma is found very useful.

Lemma 4.1. Let ξ_n, η_n, and ζ_n be sequences of random variables and let $\mathscr{A}_n$ be a sequence of events. Suppose that $\xi_n = \eta_n + \zeta_n$ on $\mathscr{A}_n$ and the following hold: $E(\xi_n^2 1_{\mathscr{A}_n^c}) \le cn^{-a_1}$, $E(\eta_n^2 1_{\mathscr{A}_n^c}) \le cn^{-a_2}$, $E(\eta_n^2) \le c$, and $|\zeta_n| \le n^{-a_3} v_n$ with $E(v_n^2) \le c$, where the a's are positive constants. Then, for any $0 < \varepsilon \le a_1 \wedge a_2 \wedge a_3$, we have

$$\left| E(\xi_n^2) - E(\eta_n^2) \right| \le cn^{-\varepsilon},$$

where c depends only on the a's and the (unspecified) c's.

We now give an outline of the Taylor expansions. The details can be found in [194]. First, consider expansions of the M-estimators. It is fairly straightforward to do it for $\hat{\psi} - \psi$. On the other hand, it is more challenging to obtain an expansion for $\hat{\psi}_{-i} - \hat{\psi}$, the main reason being that we need to carry out the expansion for $\hat{\psi}_{-i} - \hat{\psi}$ to a higher order (than for $\hat{\psi} - \psi$) [because we need to consider the sum of $(\hat{\psi}_{-i} - \hat{\psi})^2$ over i]. The idea is to do the Taylor expansion twice, or do it in two steps, first obtaining a rough approximation and then using it for a more accurate result. Write $f_j = f_j(\psi, y_j)$, $g_j = (\partial / \partial \psi) f_j(\psi, y_j)$, $f_. = \sum_j f_j$, $f_{-i} = \sum_{j \ne i} f_j$, $\bar{f} == m^{-1} f_.$, and so forth. Define $D_i = \{\hat{\psi}_{-i}$ satisfies $F_{-i} = 0$ and $|\hat{\psi}_{-i} - \psi| \le m^{\delta - 1/2}\}, 0 \le i \le m$, and $G = \{|\bar{g} - E(\bar{g})| \le m^{-\Delta} |E(\bar{g})|\}$ for some $\delta, \Delta > 0$. Let $C_i = D_0 \cap D_i, i \ge 1$. For expanding $\hat{\psi} - \psi$, we have

$$f_j(\hat{\psi}, y_j) = f_j + g_j(\hat{\psi} - \psi) + \frac{1}{2} h_j(\hat{\psi} - \psi)^2 + r_j,$$

where $h_j = (\partial^2/\partial\psi^2)f_j(\psi_j^*, y_j)$ and ψ_j^* lies between ψ and $\hat{\psi}$ [note that ψ_j^* depends on j (why?)], and $|r_j| \leq m^{3\delta - 3/2}u$ and $E(u^d)$ is bounded. Hereafter, you do not need to verify the bounds for the r's, as the goal here is to illustrate the main idea, but you may want to think about why, as always. Also, the r's are not necessarily the same, even if the same notation is used (this is similar to the unspecified c; see [1], Section 14.2). If we sum over j, then we have, on C_0 (why does the first equation hold?),

$$
\begin{aligned}
0 &= \sum_j f_j(\hat{\psi}, y_j) \\
&= f_{\cdot} + g_{\cdot}(\hat{\psi} - \psi) + \frac{1}{2}h_{\cdot}(\hat{\psi} - \psi)^2 + r. & (4.109) \\
&= f_{\cdot} + E(g_{\cdot})(\hat{\psi} - \psi) + r_1 + r, & (4.110)
\end{aligned}
$$

where $r_1 = \{g_{\cdot} - E(g_{\cdot})\}(\hat{\psi} - \psi)$, $|r| \leq m^{2\delta}u$, and $E(u^d)$ is bounded. This implies that

$$
\hat{\psi} - \psi = -\{E(g_{\cdot})\}^{-1}f_{\cdot} + r \qquad (4.111)
$$

with $|r| \leq m^{2\delta - 1}u$ and $E(u^d)$ bounded.

The two equation numbers in (4.109) and (4.110) are not left by mistake (normally one would need just one equation number for such a series of equations), as it will soon become clear. For expanding $\hat{\psi}_{-i} - \hat{\psi}$, we have, on C_i,

$$
f_j(\hat{\psi}_{-i}, y_j) = f_j + g_j(\hat{\psi}_{-i} - \psi) + \frac{1}{2}h_{-i,j}(\hat{\psi}_{-i} - \psi)^2 + r_{-i,j},
$$

where $h_{-i,j} = (\partial^2/\partial\psi^2)f_j(\psi_{-i,j}^*, y_j)$, $\psi_{-i,j}^*$ is between ψ and $\hat{\psi}_{-i}$, and $|r_{-i,j}| \leq m^{3\delta - 3/2}u_j$ with $E(u_j^d)$ bounded. We then sum over i to have, on C_i,

$$
\begin{aligned}
0 &= \sum_{j \neq i} f_j(\hat{\psi}_{-i}, y_j) \\
&= f_{\cdot -i} + g_{\cdot -i}(\hat{\psi}_{-i} - \psi) + \frac{1}{2}h_{-i,-i}(\hat{\psi}_{-i} - \psi)^2 + r_{-i,-i} & (4.112) \\
&= f_{\cdot} + E(g_{\cdot})(\hat{\psi}_{-i} - \psi) + r_i & (4.113)
\end{aligned}
$$

with $|r_i| \leq m^{2\delta}u_i$ and $E(u_i^d)$ bounded. Again, the two equations numbers, (4.112) and (4.113) are not left without a purpose. We now subtract (4.113) from (14.110) to get $0 = E(g_{\cdot})(\hat{\psi} - \hat{\psi}_{-i}) + r_i$, where $|r_i| \leq m^{2\delta}u_i$ and $E(u_i^d)$ is bounded. This impies a (rough) bound

$$
|\hat{\psi}_{-i} - \hat{\psi}| \leq m^{2\delta - 1}u_i \qquad (4.114)
$$

with $E(u_i^d)$ bounded. Using (4.114), it can be shown that

$$
h_{-i,-i}(\hat{\psi}_{-i} - \psi)^2 = E(h_{-i,-i})(\hat{\psi} - \psi)^2 + r_i \qquad (4.115)
$$

with $|r_i| \leq m^{3\delta - 1/2}u_i$ and $E(u_i^d$ bounded. We now subtract (14.112) from (14.109) and observe that $h_{\cdot}(\hat{\psi} - \psi)^2 = E(h_{\cdot})(\hat{\psi} - \psi)^2 + r$ with $|r| \leq m^{2\delta - 1/2}u$ and $E(u^d)$

bounded, to get $0 = f_i + g.(\hat{\psi} - \hat{\psi}_{-i}) + r_i$ with $|r_i| \leq m^{3\delta - 1/2} u_i$ and $\mathrm{E}(u_i^d)$ bounded. This gives us a more accurate expansion:

$$\begin{aligned} \hat{\psi}_{-i} - \hat{\psi} &= g_.^{-1}(f_i + r_i) \\ &= \{\mathrm{E}(g.)\}^{-1} f_i + r_i \end{aligned} \tag{4.116}$$

with $|r_i| \leq m^{-1-\varepsilon} u_i$ for $\varepsilon = (1/2 - 3\delta) \wedge \Delta$ and $\mathrm{E}(u_i^d)$ bounded.

The expansions for $\hat{\psi}_{-i} - \hat{\psi}$ lead to those for $b(\hat{\psi}_{-i}) - b(\hat{\psi}), 1 \leq i \leq m$ [see (14.26)], and similar techniques are used in obtaining expansions for the EBPs, $\hat{\zeta}_{-i} - \hat{\zeta}, 1 \leq i \leq m$, which are the keys to the proofs.

To conclude this section, we would like to make a note on the extension of Theorem 4.6 beyond the simple case that we have considered. This has much to do with the role of the a's in (4.105) and 14.106). In the simple case we assumed that the a's are zero (functions), so there is no such a problem. In general, [30] had the following restriction on the a's in order to obtain the asymptotic unbiasedness of the jackknife MSPE estimator,

$$\sum_{i=1}^{m} \{a(\psi) - a_{-i}(\psi)\} = O(m^{-\nu}) \tag{4.117}$$

for some constant $\nu > 0$, where ψ is the true parameter vector. To see that (4.117) is not a serious restriction, we illustrate it with an example.

Example 4.6 (continued). Write $\Delta_i = a(\psi) - a_{-i}(\psi)$, $V_j = V_j(\theta)$, and $V = V(\theta)$. From the definition of a we have $a_{-i,k}(\psi) = 0, 1 \leq k \leq p$, and $a_{-i,p+l}(\psi)$ is defined the same way as $a_{p+l}(\psi)$ except with $\sum_{j=1}^{m}$ replaced by $\sum_{j \neq i}, 1 \leq l \leq q$. Note that $X'V^{-1}X = \sum_{j=1}^{m} A_j$, with $A_j = X_j'V_j^{-1}X_j$, also needs to be adjusted. Thus, $\Delta_{i,k} = 0, 1 \leq k \leq p$, and, for $1 \leq l \leq q$,

$$\Delta_{i,p+l} = \mathrm{tr}(A_.^{-1} B_{i,l}) + \sum_{j \neq i} \mathrm{tr}\{(A_.^{-1} - A_{.-i}^{-1}) B_{j,l}\},$$

where $B_{j,l} = X_j V_j^{-1}(\partial V_j / \partial \theta_l) V_j^{-1} X_j$, $A_. = A_{.-0} = \sum_{j=1}^{m} A_j$, and $A_{.-i} = \sum_{j \neq i} A_j$ (verify the expression for $\Delta_{i,p+l}$). Under regularity conditions, we have $A_.^{-1} - A_{.-i}^{-1} = -A_.^{-1} A_i A_.^{-1} + O(m^{-3})$. Thus, we have

$$\begin{aligned} \sum_{i=1}^{m} \Delta_{i,p+l} &= \sum_{i=1}^{m} \mathrm{tr}(A_.^{-1} B_{i,l}) - \sum_{i=1}^{m} \sum_{j \neq i} \{\mathrm{tr}(A_.^{-1} A_i A_.^{-1} B_{j,l}) + O(m^{-3})\} \\ &= \sum_{i=1}^{m} \mathrm{tr}(A_.^{-1} B_{i,l}) - \sum_{i=1}^{m} \sum_{j=1}^{m} \mathrm{tr}(A_.^{-1} A_i A_.^{-1} B_{j,l}) \\ &\quad + \sum_{i=1}^{m} \mathrm{tr}(A_.^{-1} A_i A_.^{-1} B_{i,l}) + O(m^{-1}) \\ &= \sum_{i=1}^{m} \mathrm{tr}(A_.^{-1} B_{i,l}) - \sum_{j=1}^{m} \mathrm{tr}\left\{A_.^{-1} \left(\sum_{i=1}^{m} A_i\right) A_.^{-1} B_{j,l}\right\} + O(m^{-1}) \\ &= O(m^{-1}). \end{aligned}$$

Thus, (4.117) is satisfied with $\nu = 1$.

4.5.2 Monte-Carlo jackknife for SAE after model selection

One difficulty in obtaining nearly unbiased estimator of the MSPE, which so far has not been discussed, is that the latter is a positive quantity, which has to be taken into account. Typically, it is relatively easy to obtain a positive MSPE estimator that is first-order unbiased, that is, its bias is of the order $O(m^{-1})$. To achieve the second-order unbiasedness, either analytical (e.g., [29]) or resampling (e.g., [30], [31]) methods are used. However, with very few exceptions [29], [195], these techniques do not produce MSPE estimators that are guaranteed positive, in spite of achieving the second-order unbiasedness. To ensure that the MSPE estimator is positive, some modification of the (second-order unbiased) MSPE estimator is often made. For example, [31] suggested the following strategy. Let $\widehat{\text{MSPE}}_1$ and $\widehat{\text{MSPE}}_2$ be two estimators of the same MSPE, for example, the former being an MSPE estimator with a additive bias-correction, and the latter one with a multiplicative bias-correction. Both MSPE estimators have some types of problems. For example, $\widehat{\text{MSPE}}_1$ can take negative values, and $\widehat{\text{MSPE}}_2$ can be unreliable ([31]) The idea is to combine the two estimators by letting $\widehat{\text{MSPE}} = \widehat{\text{MSPE}}_1$ if something happens, and $\widehat{\text{MSPE}} = \widehat{\text{MSPE}}_2$ otherwise. This strategy takes care of the positivity issue, but it does not necessarily preserve the second-order unbiasedness, even if $\widehat{\text{MSPE}}_1$ and $\widehat{\text{MSPE}}_2$ are both second-order unbiased. The reason is simple. Suppose that we have

$$E(\widehat{\text{MSPE}}_j) \quad = \quad \text{MSPE} + o(m^{-1}), \quad j = 1, 2. \qquad (4.118)$$

Let $\mathscr{A}$ be the event set such that the combined MSPE estimator can be expressed as

$$\widehat{\text{MSPE}} \quad = \quad (\widehat{\text{MSPE}}_1) 1_{\mathscr{A}} + (\widehat{\text{MSPE}}_2) 1_{\mathscr{A}^c},$$

hence $E(\widehat{\text{MSPE}}) = E\{(\widehat{\text{MSPE}}_1) 1_{\mathscr{A}}\} + E\{(\widehat{\text{MSPE}}_2) 1_{\mathscr{A}^c}\}$. However, it is not necessarily true that either $E\{(\widehat{\text{MSPE}}_1) 1_{\mathscr{A}}\} = E\{(\text{MSPE}) 1_{\mathscr{A}}\} + o(m^{-1})$, or $E\{(\widehat{\text{MSPE}}_2) 1_{\mathscr{A}^c}\} = E\{(\text{MSPE}) 1_{\mathscr{A}^c}\} + o(m^{-1})$, or both, even though (1) hold. One may go back, again, to correct the bias of $\widehat{\text{MSPE}}$ to make it second-order unbiased, but then it may, again, lose the property of being positive, and so on. In fact, it is very difficult to achieve this "double-goal," that is, produce an MSPE estimator that is both second-order unbiased and guaranteed positive. So far, only the P-R [29] and C-L [195] MSPE estimators, both under the normality assumption, are known (with rigorous proofs) to have the double-goal property.

Model selection is inevitable in any model-based SAE problem. The errors from such model choice are likely to affect the properties of the MSPE estimators with one aspect of modeling influencing the MSPE estimation more than the other. To elaborate this point, let us consider a specific aspect of model choice–inclusion of small area specific random effects. Should one include area specific random effect in small area modeling? Such a component is a compromise between area specific fixed effects and no area effect and helps improving the properties of model-based estimators. For example, without such an area specific random effect, a model-based estimator may not be design-consistent, which may result in model-based estimate for an area with large sample size to deviate significantly from the corresponding design-based estimate, especially if area specific auxiliary variables fail to capture variation across the areas. A decision to exclude small area specific random effect may be based on a significance test. But such a decision is anything but perfect and depends very much on the subjective choice of the prespecified level of significance. A reasonable MSPE estimator must incorporate the impact of model selection. For example, most of the MSPE estimators, with the exception of [196], do not attempt to capture the variation due to the model choice and there is no analytical study to examine the important second-order unbiasedness property of any of these MSPE estimators, including that of [196]. Needless to say that developing a rigorous theory for the double-goal for SAE after model selection is a challenging problem, even for a simple mixed-effects SAE model.

Now let us take a different approach. Let $\phi(\cdot)$ be a simple, continuous, one-to-one function that maps $(0, \infty)$ onto $(-\infty, \infty)$. One example of such a function, which we shall primarily consider in this paper, is the logarithm, $\phi(x) = \log(x), x \in (0, \infty)$. Mathematically, estimating MSPE is equivalent to estimating $\phi(\text{MSPE})$ (hereafter, ϕMSPE) in terms of point estimator, provided that ϕ is continuous and strictly monotone. However, an advantage of $\phi(\text{MSPE})$ is that we do not need to worry about negative values—we can simply focus on the unbiasedness. This way, we have reduced the double-goal problem to a single-goal problem. Once one obtains an estimator of ϕMSPE, one can easily convert it to an estimator of the corresponding MSPE.

It should be noted that second-order unbiasedness in the ϕ scale is not equivalent, in general, to that in the original scale. Nevertheless, it turns out that measure of uncertainty in the ϕ scale still gives sensible results in the original scale. In the SAE literature, the MSPE estimates have been used, for the most part, for two different ways. First, they are routinely used in comparing different estimates, for example, an improvement of the empirical best linear unbiased predictor (EBLUP) over the direct estimator. For such a purpose, one can equivalently use the ϕMSPE for a one-to-one (nondecreasing) function ϕ, such as the logarithm, and report the improvement in the ϕ scale. Secondly, the MSPE estimates are frequently used in constructing confidence intervals, or prediction intervals, for the small-area characteristics, say, θ_i, in the form of, for example, $\hat{\theta}_i \pm 2\sqrt{\widehat{\text{MSPE}}_i}$, where $\hat{\theta}_i$ is the (point) estimate (e.g., EBLUP) of θ_i and $\widehat{\text{MSPE}}_i$ is the corresponding MSPE estimate. In this regard, the following example shows that, even if a double-goal MSPE estimator is available so that there is no direct need to consider the logarithm, targeting the log-MSPE may not be disadvantageous in terms of the prediction intervals.

Example 4.7. Consider a special case of the Fay-Herriot model (4.9) with $x_i'\beta = \mu$, an unknown mean. Let $\hat{A}$ be the P-R estimator of A, given by (4.11). It is possible that the value of $\hat{A}$ is negative; when this happens, $\hat{A}$ is truncated at 0. This way, the P-R MSPE estimator, (4.24), is guaranteed positive, provided that D_i is positive (see Section 4.2). Now suppose that one wishes to estimate the log-MSPE of the EBLUP. An obvious estimator is $G_i(\hat{A}) = \log\{g_i(\hat{A})\}$, where $g_i(A)$ is the right side of (4.24), with $\hat{A}$ replaced by A. Note that, although $g_i(\hat{A})$ is second-order unbiased for estimating the MSPE, $G_i(\hat{A})$ is only first-order unbiased for estimating the log-MSPE. A second-order unbiased estimator of the log-MSPE can be obtained via the

Table 4.1 *Summary Statistics for Coverage Probabilities (Nominal Level 95%)*

	Q1	Q2	Q3	Mean	s.d.
True MSPE	0.949	0.950	0.952	0.950	2.06×10^{-3}
P-R Est.	0.946	0.947	0.948	0.947	2.13×10^{-3}
C-Log Est.	0.948	0.949	0.950	0.949	2.12×10^{-3}

Prasad-Rao linearization method, which is given by $G_i(\hat{A}) + C_i(\hat{A})$, where

$$C_i(A) = \left\{ \frac{D_i}{A(A+D_i)} \right\}^2 \frac{A^2(m-p) + 2A\mathrm{tr}(P_{X\perp}D) + \mathrm{tr}\{(P_{X\perp}D)^2\}}{(m-p)^2}. \qquad (4.119)$$

In the following, we focus on a special case, with $D_i = D, 1 \le i \le m$. For this special case, $G_i(A)$ does not depend on i, which will be denoted by $G(A)$, and (4.119) reduces to $C(A) = (D/A)^2/(m-1)$. Consider using the estimated MSPEs to construct 95% prediction intervals for θ_i in the form of $\hat{\theta}_i \pm 1.96\sqrt{\widehat{\mathrm{MSPE}}}$, where $\hat{\theta}_i$ is the EBLUP of θ_i, and $\widehat{\mathrm{MSPE}}$ is either the P-R MSPE estimator, which is equal to $\exp\{G(\hat{A})\}$, or $\exp\{G(\hat{A}) + C(\hat{A})\}$, obtained by converting the second-order unbiased log-MSPE estimator. A simulation is run with $\mu = 0$ and $A = D = 1$. Prediction intervals are constructed for $m = 100$ small areas, and repeated 10,000 times. The boxplots of the coverage probabilities, based on the 10,000 simulations, are presented in Figure 4.1. For this special case, [195] has obtained the exact MSPE, which is given by

$$\frac{D}{A+D} \left\{ A + \frac{D}{m} + \frac{2(m-1)D}{m(m-3)} \right\}.$$

It should be noted that the latter result is valid when $\hat{A}$ is not truncated at zero when it is negative, while, in practice, a truncated version of $\hat{A}$ is often used to make sure that it is nonnegative. Prediction intervals are also constructed, based on the same simulation runs, and the same formula except with $\widehat{\mathrm{MSPE}}$ replaced by the exact MSPE. The corresponding boxplot is also presented in Figure 4.1 for comparison. Some summary statistics are given in Table 4.1, where C-Log Est. stands for converting the second-order unbiased log-MSPE estimator, and Q1, Q2, and Q3 for the first quartile, median, and third quartile, respectively. It appears that C-Log is doing better than P-R in terms of the coverage probability. Of course, the seemingly better performance in the coverage probability is not at no cost: The C-Log prediction intervals are slightly wider than the P-R ones. The mean lengths of the prediction intervals, based on the 10,000 simulations, are 2.814 for True MSPE, 2.808 for P-R Est., and 2.828 for C-Log Est.; thus, the relative increase in the mean length of C-Log over P-R is 0.00712, or about 0.7%. In summary, the example shows that C-log is competitive to the conventional method in terms of the prediction intervals.

In this example, the P-R method is simple and works well in estimating the log-MSPE. However, the approach is not feasible to handle our current problem, which is much more complicated. More specifically, we are interested in estimating the

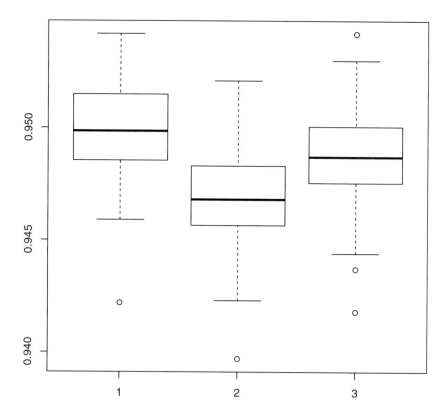

Figure 4.1 *Boxplots of coverage probabilities. Nominal Level: 95%. From left to right: Coverage probabilities of prediction intervals constructed by 1–true MSPE; 2–P-R MSPE estimator; and 3–converting the second-order unbiased Log-MSPE estimator.*

MSPE, or, equivalently, ϕMSPE, when the small area predictor is obtained after a model-selection procedure. The existing literature on inference after model selection has mainly focused on the case of independent observations (e.g., [197], Section 12 and the references therein, [198], [199]). In particular, the potential impact of model selection on MSPE has never been rigorously addressed in the SAE literature. Intuitively, there is an additional uncertainty involved in the model-selection process, that needs to be taken into account in the MSPE estimation. The P-R linearization method requires differentiability of the underlying operation. This usually holds for standard estimation and prediction procedures, but not for model selection. For example, the information criteria, such as AIC [78] and BIC [79], or the fence methods (see [116] for a review), select models from a discrete space of candidate models. Even the shrinkage methods (e.g., [117], [118]) involve continuous but non-differentiable penalty functions, such as the L^1 norm. See [200] for a review. Even if it is possible to develop a P-R type method, the derivation is tedious, and the final analytic expression is likely to be complicated. More importantly, errors often occur in the process of derivations as well as computer programming based on the lengthy expressions.

We now introduce a unified jackknife approach that is assisted Monte-Carlo simulations for the estimation of ϕMSPE. As will be seen, the approach is applicable not just to the current problem of SAE after model selection, but to a much broader class of problems to obtain nearly unbiased estimators of quantities that can be obtained via Monte-Carlo simulation, if one knows the parameters that are involved. The method is especially attractive if the quantity of interest does not carry a constraint, such as non-negativity. This will be the case for ϕMSPE. Furthermore, the Monte-Carlo jackknife method, called *McJack*, is "one-formula-for-all", which means that one needs not to re-derive the formula, as in P-R type methods, every time there is a new problem.

Recall the jackknife method described in the previous section. We call this method JLW jackknife. As far as the current method is concerned, what is important is not the full JLW theory, but rather an intermediate result. In obtaining their results, JLW showed, in particular, that (4.102) is a second-order unbiased estimator of $b(\psi)$, if the penalizers $a, a_{-j}, 1 \leq j \leq m$ in (4.105) and (4.106) satisfy certain mild conditions. In particular, those conditions are satisfied if the penalizers are zero (vectors), in which case the M-estimating equations are unbiased. Having given the proof of the result, we realize the following two facts, both are critically important to the idea of the current paper.

(I) The fact that $b(\psi)$ is an MSPE is not used anywhere in the proof (see the previous Section). In other words, as long as $b(\cdot)$ is a sufficiently smooth function, and ψ is estimated by the M-estimators, the second-order unbiased estimation of $b(\psi)$ holds. In particular, $b(\psi)$ can be ϕMSPE = log(MSPE, which is of primary interest here.

(II) More importantly, $b(\psi)$ does not have to have an analytic expression, as long as one knows how to compute it. An analytic expression would be nice, but, in the new era, the computation is typically done by a computer, perhaps, a high-powered one. In particular, suppose that, given ψ, $b(\psi)$ can be approximated by a Monte-Carlo method to an arbitrary degree of accuracy. Then, one can write a computer program, based on the Monte-Carlo, to compute $b(\cdot)$ as a function. Given this "computer-powered" function, all one needs to do is to plug $\hat{\psi}, \hat{\psi}_{-j}, 1 \leq j \leq m$ into this function to obtain the second-order unbiased estimator of $b(\psi)$.

The importance of the above observations is that they apply to virtually any kind of situation, not just the EBP. In particular, the predictor, $\hat{\xi}$, can be much more complicated than the EBP, such as an EBP obtained following a model-selection procedure. Also, the decomposition (4.100), which requies *posterior linearity* [30], and (4.101) altogether are not needed to apply these observations.

A standard Monte-Carlo approximation to $b(\psi) = \log(\text{MSPE})$ is *parametric bootstrap*. We illustrate using an example of EBLUP under a Fay-Herriot model, where the BIC [79] is used to select the fixed covariates as well as whether to include the area-specific random effects. In its general form, the model can be expressed in a way equivalent to (4.9), but more convenient for the model selection problem:

$$y_i \;\;=\;\; x_i'\beta + \sqrt{A}\,\xi_i + e_i, \tag{4.120}$$

$i = 1,\ldots,m$, where the components of x_i are to be selected from a set of candidate covariates; $\xi_i \sim N(0,1)$; if $A > 0$, the random effects are included in the model; if $A = 0$, the random effects are excluded from the model; $e_i \sim N(0,D_i)$, where $D_i, 1 \leq i \leq m$ are known; and the ξ_i's and e_i's are independent. Note that there have been further considerations regarding the choice of the random effects; see, for example, Datta et al. [183], but here we focus on a simpler situation. Let M_f denote a full model, under which x_i is the vector that includes all of the candidate covariates, and $A \geq 0$. Denote the x_i under M_f by $x_{f,i}$, and the corresponding β by β_f. Let $\psi = (\beta_f', A)'$. It is easy to see that M_f is, at least, a correct model, which means that (4.120) holds with x_i replaced by $x_{f,i}$, β replaced by β_f, and the range of A being $[0,\infty)$. Of course, some of the components of β_f may be zero, in case that the full model can be simplified, and the true A may be zero–these are the reasons for the model selection. But this does not change the fact M_f is a correct model. In particular, the true small-area mean, θ_i, can be expressed as

$$\theta_i \;\;=\;\; x_{f,i}'\beta_f + \sqrt{A}\,\xi_i. \tag{4.121}$$

On the other hand, under a candidate model, M, which corresponds to (4.120), the EBLUP of θ_i can be expressed as

$$\tilde{\theta}_i \;\;=\;\; \frac{\hat{A}}{\hat{A}+D_i}y_i + \frac{D_i}{\hat{A}+D_i}x_i'\hat{\beta}, \tag{4.122}$$

where $\hat{\beta} = \{\sum_{i=1}^m(\hat{A}+D_i)^{-1}x_ix_i'\}^{-1}\sum_{i=1}^m(\hat{A}+D_i)^{-1}x_iy_i$, and $\hat{A}$ is a consistent estimator of A obtained using a certain method (e.g., P-R, ML, REML; see Section 4.2). The BIC procedure chooses the model, M, by minimizing

$$\mathrm{BIC}(M) \;\;=\;\; -2\hat{l} + |M|\log(m), \tag{4.123}$$

where $\hat{l}$ is the maximized log-likelihood under M; $|M| = \dim(\beta) + 1$ if M includes the random effects, and $|M| = \dim(\beta)$ if M excludes the random effects. Here, for simplicity, we assume that $X = (x_i')_{1 \leq i \leq m}$ is full rank under any M. Let the minimizer of (4.123) be $\hat{M}$. We then compute the EBLUP (4.122) under $M = \hat{M}$, that is,

$$\hat{\theta}_i \;\;=\;\; \frac{\hat{A}_{\hat{M}}}{\hat{A}_{\hat{M}}+D_i}y_i + \frac{D_i}{\hat{A}_{\hat{M}}+D_i}x_{\hat{M},i}'\hat{\beta}_{\hat{M}}, \tag{4.124}$$

where $\hat{\beta}_{\hat{M}}$ and $\hat{A}_{\hat{M}}$ are the $\hat{\beta}$ and $\hat{A}$ obtained under $\hat{M}$. The MSPE of interest is

$$\mathrm{MSPE}(\hat{\theta}_i) \;\;=\;\; \mathrm{E}(\hat{\theta}_i - \theta_i)^2, \tag{4.125}$$

where θ_i is given by (4.121). It is clear that the joint distribution of $(\theta_i, y_i), 1 \leq i \leq m$ depends only on $\psi = (\beta_f', A)$. Thus, (4.125) is a function of ψ and so is its logarithm,

$$b(\psi) \;\;=\;\; \log\{\mathrm{MSPE}(\hat{\theta}_i)\}. \tag{4.126}$$

Given ψ, for the kth Monte-Carlo simulation, one first generates θ_i by (4.121) with ξ_i replaced by $\xi_i^{(k)}$, $1 \leq i \leq m$, generated independently from $N(0,1)$. Denote the generated θ_i by $\theta_i^{(k)}$. Next, let $y_i^{(k)} = \theta_i^{(k)} + e_i^{(k)}$, $1 \leq i \leq m$, where $e_i^{(k)} \sim N(0, D_i)$, $1 \leq i \leq m$, generated independently and independent with $\xi_i^{(k)}$'s. The Monte-Carlo approximation to $b(\psi)$ is

$$\tilde{b}(\psi) \;=\; \log\left[\frac{1}{K}\sum_{k=1}^{K}\left\{\hat{\theta}_i^{(k)} - \theta_i^{(k)}\right\}^2\right], \qquad (4.127)$$

where $\hat{\theta}_i^{(k)}$ is obtained the same way as the $\hat{\theta}_i$ of (4.124) except with y_i replaced by $y_i^{(k)}$, $1 \leq i \leq m$. Write the above procedure as a function, say, $\tilde{b}(\psi) = \mathbf{mcjack}(\psi)$, that computes (4.127) for every given ψ. Now suppse that $\hat{\psi}$ is an M-estimator of ψ. For example, $\hat{A}$ is the P-R estimator (See Section 4.2; truncated at zero if the expression turns out to be negative), and $\hat{\beta}_f$ is given below (11) with $x_i = x_{f,i}$, $1 \leq i \leq m$. Let $\hat{\psi}_{-j}$ be the delete-j version of $\hat{\psi}$. The McJack estimator of (4.126) is then given by

$$\widehat{b(\psi)} \;=\; \tilde{b}(\hat{\psi}) - \frac{m-1}{m}\sum_{j=1}^{m}\left\{\tilde{b}(\hat{\psi}_{-j}) - \tilde{b}(\hat{\psi})\right\}. \qquad (4.128)$$

Using the result of JLW, we can justify the second-order unbiasedness of the McJack estimator. The justification also takes into account effect of the Monte-Carlo errors. First note that, to establish a rigorous result about the unbiasedness, we need to make sure that the expectations of $\tilde{b}(\hat{\psi}_{-j})$, $0 \leq j \leq m$ exist. To avoid complicated technical conditions, we regularize these estimators as follows. Let $\tilde{s}(\psi) = \exp\{\tilde{b}(\psi)\}$, and define

$$\hat{s}(\psi) \;=\; \begin{cases} e^{-\lambda m^{\rho}}, & \text{if } \tilde{s}(\psi) < e^{-\lambda m^{\rho}}, \\ \tilde{s}(\psi), & \text{if } e^{-\lambda m^{\rho}} \leq \tilde{s}(\psi) \leq e^{\lambda m^{\rho}}, \\ e^{\lambda m^{\rho}}, & \text{if } \tilde{s}(\psi) > e^{\lambda m^{\rho}}, \end{cases}$$

and $\hat{b}(\psi) = \log\{\hat{s}(\psi)\}$, where λ, ρ are given positive numbers. Let $s(\psi)$ denote $\mathrm{MSPE}(\hat{\theta}_i)$ when ψ is the true parameter vector. We truncate $s(\cdot)$ the same way as $\tilde{s}(\cdot)$, and let $b(\psi) = \log\{s(\psi)\}$. For notation convenience, write $\hat{\psi}_{-0} = \hat{\psi}$. Also, let $F_{-0}(\psi), F_{-j}(\psi)$ denote the left sides of (4.105) and (4.106), respectively. We extend the c.u. property, defined in the previous section, as follows. The M-estimators, $\hat{\psi}_{-j}, 0 \leq j \leq m$ are said to be consistent uniformly (c.u.) at rate m^{-d} if, for any $\delta > 0$, there is a constant c_δ such that

$$P(A_{j,\delta}^c) \;\leq\; c_\delta m^{-d}, \quad 0 \leq j \leq m,$$

where $A_{j,\delta}$ is the event that $F_{-j}(\hat{\psi}_{-j}) = 0$ and $|\hat{\psi}_{-j} - \psi| \leq \delta$, with ψ being the true parameter vector. Also, write $f_i = f_i(\psi, y_i)$, $g_i = \partial f_i/\partial \psi'$, $h_{i,k} = \partial^2 f_{i,k}/\partial \psi \partial \psi'$, where $f_{i,k}$ is the kth component of f_i. Furthermore, for any function f of ψ, define

$$\|\Delta^3 f\|_w = \max_{1 \leq s,t,u \leq r}\ \sup_{|\tilde{\psi}-\psi| \leq w}\left|\frac{\partial^3 f(\tilde{\psi})}{\partial \psi_s \partial \psi_t \partial \psi_u}\right|,$$

where ψ is the true parameter vector, and $r = \dim(\psi)$. A similar definition is extended to $\|\Delta^4 f\|_w$. The spectral norm of a matrix, B, is defined as $\|B\| = \sqrt{\lambda_{\max}(B'B)}$, where $\lambda_{\max}$ denotes the largest eigenvalue. Also write $\Delta_j = a - a_{-j}$, where a, a_{-j} are the functions of ψ that appear in (4.105) and (4.106), respectively. We shall consider estimation of log-MSPE of $\hat{\theta}_i$, a predictor of θ_i after model selection, for a fixed i. Furthermore, we assume that the Monte-Carlo samples, under ψ, are generated by first generating some standard [e.g., $N(0,1)$] random variables and then plugging ψ. For example, under the full Fay-Herriot model of (4.120), y_i is generated by first generating the ξ_i's and η_i's, which are independent $N(0,1)$, and then letting $y_i = x'_{f,i}\beta_f + \sqrt{A}\xi_i + \sqrt{D_i}\eta_i$, with $\psi = (\beta'_f, A)'$. Let ξ denote the vector of the standard random variables. We first make the following general assumptions.

A1. There are $d > 2$ and $w > 0$ such that the $2d$th moments of $|f_i|$, $\|g_i\|$, $\|h_{i,k}\|$, $\|\Delta^3 f_{i,k}\|_w$, $1 \le i \le m, 1 \le k \le r$ are bounded for some $d > 2 + \rho$.

A2. For the same d and w in *A1*, a_{-j} and its up to third order partial derivatives, $0 \le j \le m$, as well as $\Delta_j, 1 \le j \le m$, all evaluated at $\tilde{\psi}$, are bounded uniformly for $|\tilde{\psi} - \psi| \le w$, where ψ is the true parameter vector, and $m^\tau(|\Delta_j| \vee \|\partial \Delta_j/\partial \psi\|), 1 \le j \le m$, evaluated at ψ, are bounded, where $\tau = (d-2)/(2d+1)$.

A3. The log-MSPE function $b(\cdot)$ of (4.126) is four-times continuously differentiable, and, for the same w in *A1*, $\|\Delta^4 b\|_w$ is bounded.

A4. $\limsup_{m\to\infty} \|\{E(\bar{g})\}^{-1}\| < \infty$ with $\bar{g} = m^{-1}\sum_{j=1}^{m} g_j$ evaluated at the true ψ.

A5. $\hat{\psi}_{-j}, 0 \le j \le m$ are c.u. at rate m^{-d} for the same d in *A1*.

A6. $\sum_{j=1}^{m} \Delta_j = O(m^{-v})$ for some $v > 0$.

Recall the way that the Monte-Carlo samples are generated specified above *A1*. Under this assumption, $\theta_i^{(k)}, \hat{\theta}_i^{(k)}, 1 \le k \le K$, generated under $\tilde{\psi}$, are functions of $\tilde{\psi}$ and ξ. The additional assumptions below are regarding the Monte-Carlo sampling.

A7. ξ is independent with the data, y.

A8. Let ψ be the true parameter vector, and w be the same as in *A1*. There are constants $0 < c_1 < c_2$ such that $c_1 \le s(\tilde{\psi}) \le c_2$ for $|\tilde{\psi} - \psi| \le w$, and random variables $G_k, 1 \le k \le K$, which do not depend on $\tilde{\psi}$, such that $|\hat{\theta}_i^{(k)} - \theta_i^{(k)}| \le G_k$ and $E(G_k^q)$ are bounded for some $q \ge 2\{2 + (\rho \vee 1)\}$.

A9. $m^2/K \to 0$, as $m \to \infty$.

Theorem 4.7. Suppose that *A1–A9* hold. Let $\widehat{b(\psi)}$ denote (4.128) with $\tilde{b}$ replaced by $\hat{b}$. Then, we have $E\{\widehat{b(\psi)} - b(\psi)\} = o(m^{-1})$, where ψ is the true ψ [hence $b(\psi)$ is the true log-MSPE], and E is with respect to both y and ξ.

The next result focuses on the special case of Fay-Herriot model (Section 4.2).

Theorem 4.8. Suppose that the true $A > 0$, and there are positive constants $0 < c_1 < c_2$ such that $c_1 \le |x_{f,i}| \le c_2, c_1 \le D_i \le c_2, 1 \le i \le m$. Furthermore, suppose that

$$\limsup_{m\to\infty} \lambda_{\min}\left(\frac{1}{m}\sum_{i=1}^{m} x_{f,i}x'_{f,i}\right) > 0, \qquad (4.129)$$

and A9 holds. Then, the conclusion of Theorem 4.7 holds.

The proofs of Theorem 4.7 and Theorem 4.8 are given in [201].

In the remaining part of this subsection, we use two simple examples to demonstrate numerical performance of McJack.

Example 4.8. As a simple demonstration, let us consider a very simple situation, which may be viewed as a special case of the Fay-Herriot model (4.9), where the components of x_i consist of an intercept, a group indicator, $x_{1,i}$, which is 0 if $1 \leq i \leq m_1 = m/2$, and 1 if $m_1 + 1 \leq i \leq m$, and potentially a third component, $x_{2,i}$, which is generated from the $N(0, 1)$ distribution, and fixed throughout the simulation. There are two candidate models: Model 1, which includes $x_{2,i}$, and Model 2: which does not include $x_{2,i}$. The model selection is carried out by BIC [79].

For this demonstration, we consider a special case that the variance of the random effects, v_i, is known to be zero, that is, $A = 0$. There have been considerations of such situations in SAE (e.g., [202]). The variance of e_i, D_i, is equal to 1 for $1 \leq i \leq m_1$, and a for $m_1 + 1 \leq i \leq m$, where the value of a is either 4 or 16. Because $A = 0$, the small area mean, θ_i, under a given model, is equal to $x_i'\beta$. The corresponding EBLUP is $\hat{\theta}_i = x_i'\hat{\beta}$, where $\hat{\beta} = (X'D^{-1}X)^{-1}X'D^{-1}y$, with $X = (x_i')_{1 \leq i \leq m}$ and $D = \mathrm{diag}(D_i, 1 \leq i \leq m)$, is the best linear unbiased estimator (BLUE) of β [see (2.47)], under the given model. Due to the unbiasedness of the BLUE, the MSPE of the EBLUP is equal to its variance, that is,

$$\mathrm{MSPE}(\hat{\theta}_i) = \mathrm{var}(\hat{\theta}_i) = x_i'(X'D^{-1}X)^{-1}x_i, \quad 1 \leq i \leq m, \quad (4.130)$$

which are known under the given model. Now suppose that the EBLUP is obtained based on the model selected by the BIC. A naive estimator of the MSPE of $\hat{\theta}_i$, which ignores model selection, would be (4.130) computed under the selected model. The naive estimator of the log-MSPE is the logarithm of the naive MSPE estimator. We compare this estimator with two competitors. The first is what we call bootstrap MSPE estimator, which corresponds to the first term in (4.128), that is, without the jackknife bias correction, where $b(\cdot)$ is the log-MSPE function. The second is the McJack estimator given by (4.128). The bootstrap and McJack estimators are computed based on $K = 1000$ Monte-Carlo samples.

A series of simulation studies were carried out with $m = 20$ and $\beta_0 = \beta_1 = 1$, where β_0 is the intercept and β_1 the slope of $x_{1,i}$, and under two different true underlying models. In the first scenario, Model 1 is the true underlying model with the slope of $x_{2,i}$, $\beta_2 = 0.5$. In the second scenario, Model 2 is the true underlying model (i.e., $\beta_2 = 0$). We present the simulated percentage relative bias (%RB), based on $N_{\mathrm{sim}} = 1000$ simulation runs, in Figures 4.2 and 4.3, where, for a given area, the %RB is defined as

$$\%\mathrm{RB} = \left[\frac{\mathrm{E}\{\log(\widehat{\mathrm{MSPE}})\} - \log(\mathrm{MSPE})}{|\log(\mathrm{MSPE})|} \right] \times 100\%, \quad (4.131)$$

MSPE is the true MSPE based on the simulations, and $\mathrm{E}\{\log(\widehat{\mathrm{MSPE}})\}$ is the mean of the estimated log-MSPE based on the simulations.

It is seen that the naive estimator significantly under estimate the log-MSPE; in fact, when Model 1 is the true model, the %RB for one of the areas is 516% in the case of $a = 4$, and there is a similar case in the case of $a = 16$. More specifically, there are some interesting trend observed. Namely, when the true model is Model 1, all of the methods seem to under-estimate the log-MSPE, but the bootstrap and McJack

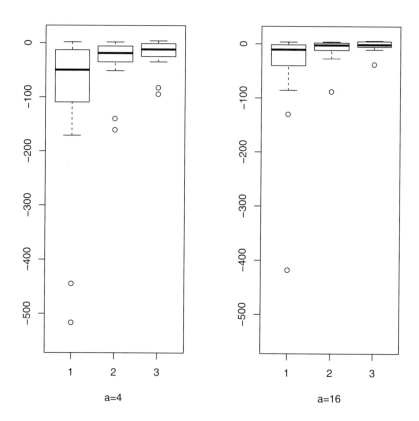

Figure 4.2 *Boxplots of %RB when Model 1 is the true model. In each plot, from left to right:*
1–naive estimator, 2–bootstrap estimator, and 3–McJack estimator, of log-MSPE.

estimators are doing much better, with McJack offering significant improvement over
the bootstrap. On the other hand, when the true model is Model 2, the naive estimator
again under-estimate the log-MSPE, but the bootstrap and McJack estimators seem
to over-estimate the log-MSPE, with McJack significantly improving the bootstrap.
The amount of underestimation by the niave estimator is less dramatic when Model
2 is the true model compared to when Model 1 is the true model. One explanation
is that the BIC is known to have the tendency to over-penalize larger models. This
would have bigger impact when Model 1 is the true model, which is the full model.
In other words, there is a higher chance of model misspecification by the BIC, which
impacts the log-MSPE estimation. To have a closer look at the numbers, we present
one set of the detailed results in Table 4.2.

Example 4.9. Datta et al. [183] proposed a method of model selection by testing
for the presence of the area-specific random effects, $v_i = \sqrt{A}\xi_i$, in the Fay–Herriot
model (4.120). This is equivalent to testing the null hypothesis $H_0 : A = 0$. The test

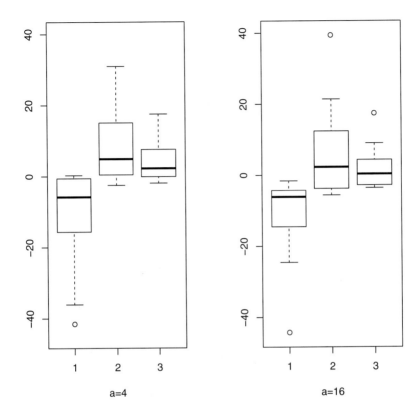

Figure 4.3 *Boxplots of %RB when Model 2 is the true model. In each plot, from left to right: 1–naive estimator, 2–bootstrap estimator, and 3–McJack estimator, of log-MSPE.*

statistic, $T = \sum_{i=1}^{m} D_i^{-1}(y_i - x_i'\hat{\beta})^2$, where $\hat{\beta}$ is the same as in Example 4.8, has a χ^2_{m-p} distribution, with $p = \text{rank}(X)$, under H_0. If H_0 is rejected, the EBLUP is used to estimate the small area mean θ_i, where in this simulation A is estimated by the P-R estimator, and the corresponding MSPE estimator is the P-R MSPE estimator; if H_0 is accepted, the estimator $\hat{\theta}_i = x_i'\hat{\beta}$ is used to estimate θ_i, and the corresponding MSPE is given by (20). Thus, if the level of significance is chosen as 0.05, the proposed MSPE estimator, denoted by DHM, is the P-R MSPE estimator if $T > \chi^2_{m-p}(0.05)$, and (4.130) if $T <= \chi^2_{m-p}(0.05)$.

We run a simulation study to compare the performance of McJack with DHM. The simulation is under the full model considered in the previous subsection (hence $p = 3$), and three different true values of A: $A = 0$, $A = 0.5$, and $A = 1$. The boxplots of %RB for these three cases are presented in Figure 4.4, with the detailed numbers for DHM and McJack given in Table 4.3. It is seen that DHM works better for the case $A = 0$, which is not surprising because, under the null hypothesis, the DHM MSPE

Table 4.2 *Log-MSPE Estimation*

Area	True log-MSPE	E(Naive Est.)	E(Bootstrap Est.)	E(McJack Est.)
1	-1.98	-2.26 (-14.0)	-1.79 (9.6)	-1.91 (3.3)
2	-1.62	-2.21 (-36.1)	-1.22 (25.0)	-1.41 (12.8)
3	-2.07	-2.27 (-9.8)	-1.95 (5.6)	-2.01 (2.9)
4	-2.20	-2.30 (-4.3)	-2.26 (-2.5)	-2.25 (-1.9)
5	-1.70	-2.22 (-30.4)	-1.33 (21.7)	-1.52 (10.7)
6	-2.05	-2.27 (-10.8)	-1.91 (6.7)	-1.97 (3.7)
7	-2.14	-2.29 (-6.9)	-2.11 (1.5)	-2.16 (-1.0)
8	-1.55	-2.20 (-41.6)	-1.11 (28.4)	-1.28 (17.5)
9	-2.19	-2.30 (-4.8)	-2.23 (-1.6)	-2.22 (-1.4)
10	-2.06	-2.27 (-10.2)	-1.94 (6.0)	-2.00 (3.2)
11	-0.91	-0.92 (-0.5)	-0.91 (0.6)	-0.91 (-0.0)
12	-0.91	-0.92 (-0.1)	-0.91 (0.2)	-0.92 (-0.1)
13	-0.76	-0.89 (-17.4)	-0.61 (19.8)	-0.69 (9.5)
14	-0.87	-0.90 (-3.7)	-0.78 (10.4)	-0.82 (5.6)
15	-0.92	-0.92 (0.3)	-0.92 (0.1)	-0.92 (-0.1)
16	-0.92	-0.92 (0.1)	-0.92 (0.1)	-0.92 (-0.1)
17	-0.74	-0.88 (-18.1)	-0.52 (30.9)	-0.62 (17.2)
18	-0.92	-0.91 (0.1)	-0.90 (2.1)	-0.91 (1.1)
19	-0.91	-0.92 (-0.6)	-0.90 (0.7)	-0.91 (-0.0)
20	-0.88	-0.91 (-3.6)	-0.84 (4.1)	-0.87 (1.5)

Note: Model 2 is True Model; $a = 4$; %RB in ( )s.

estimator is "right" 95% of times. On the other hand, McJack works significantly better in those two cases of nonzero A. Simple simulations show that, in the latter cases, the probability of rejecting the null hypothesis is about 0.26 when $A = 0.5$, and 0.44 when $A = 1$. The worst scenario seems to be the case where A is not zero but closer to zero ($A = 0.5$). There are a few "blown-up" cases under this scenario where the %RB exceeds 1000% for DHM. It is also obvious that McJack improves bootstrap in every case.

4.5.3 Bootstrapping mixed models

Linear mixed models have played a major role in SAE. Recall such a model can be expressed as (2.1) with the assumptions underneath. We change the notation a little: u for α and e for ε, and further assume that u and e are multivariate normal. Our problem of interest is the prediction of a linear mixed effect, expressed as $\zeta = c'(X\beta + Zu)$ for some known constant vector c. Such mixed effects include the small-area means, say, under a Fay–Herriot model (see Section 4.2) and genetic merits of breeding animals (e.g., [2], Section 2.6). The conditional distribution of ζ given the data y is $N(\tilde{\mu}, \tilde{\sigma}^2)$, where

$$\tilde{\mu} = c'\{X\beta + ZGZ'V^{-1}(y - X\beta)\}, \tag{4.132}$$

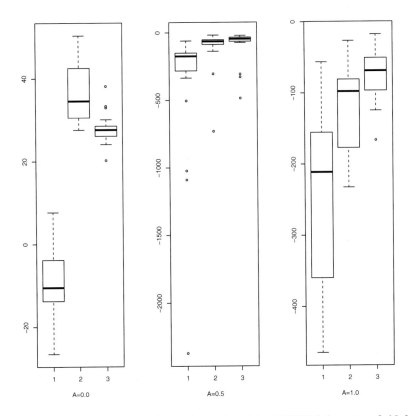

Figure 4.4 *Boxplots of %RB. In each plot, from left to right: 1–DHM, 2–bootstrap, 3–McJack. Scales are different due to the huge difference in range.*

$$\tilde{\sigma}^2 \;=\; c'Z(G - GZ'V^{-1}ZG)Z'c, \qquad\qquad (4.133)$$

and $V = \text{Var}(y) = R + ZGZ'$. Therefore, for any $\alpha \in (0, 1)$, we have

$$P\left(\left|\frac{\zeta - \tilde{\mu}}{\tilde{\sigma}}\right| \le z_{\alpha/2}\,\middle|\,y\right) \;=\; 1 - \alpha,$$

where z_α is the α-critical value of $N(0,1)$ [i.e., $\Phi(\alpha) = 1 - \alpha$, where Φ is the cdf of $N(0,1)$]. It follows that the probability is $1 - \alpha$ that the interval $I = [\tilde{\mu} - z_{\alpha/2}\tilde{\sigma}, \tilde{\mu} + z_{\alpha/2}\tilde{\sigma}]$ covers ζ. Of course, the latter is not a practical solution for the prediction interval, because there are unknown parameters involved. In addition to β, the co-variance matrices G and R typically also depend on some unknown disperson parameters, or variance components. Suppose that $G = G(\psi)$ and $R = R(\psi)$, where ψ is a q-dimensional vector of variance components, and $\theta = (\beta', \psi')'$. It is customary to replace θ by an estimator, $\hat{\theta} = (\hat{\beta}', \hat{\psi}')'$. However, with the replacement of θ by $\hat{\theta}$, the resulting (prediction) interval no longer has the coverage probability $1 - \alpha$.

Table 4.3 *DHM versus McJack in %RB*

Area	A = 0.0		A = 0.5		A = 1.0	
	DHM	McJack	DHM	McJack	DHM	McJack
1	-13.6	26.2	-216.8	-59.8	-342.0	-99.5
2	0.3	26.0	-131.0	-45.8	-343.8	-72.9
3	-3.9	27.5	-107.4	-21.1	-135.8	-30.3
4	1.1	28.2	-158.0	-37.7	-362.9	-77.6
5	-8.1	24.9	-178.9	-36.9	-191.3	-59.6
6	-6.0	30.0	-180.3	-50.5	-375.4	-166.5
7	-3.1	24.1	-210.0	-51.5	-395.5	-124.6
8	7.6	27.6	-135.3	-33.1	-464.9	-123.0
9	-10.9	27.4	-149.4	-43.0	-357.2	-108.1
10	-3.8	28.6	-163.1	-31.4	-362.8	-94.0
11	-26.5	20.2	-60.8	-21.8	-220.3	-39.4
12	-10.7	33.3	-2373.3	-486.1	-210.4	-76.5
13	-13.9	27.5	-504.8	-74.0	-94.5	-18.4
14	-10.3	32.9	-173.1	-35.3	-188.2	-64.1
15	-17.5	25.7	-1023.6	-329.5	-163.0	-58.1
16	-4.4	38.1	-154.6	-46.6	-211.9	-65.2
17	-12.1	28.4	-335.8	-48.9	-197.6	-72.8
18	-11.4	27.1	-171.2	-22.0	-148.4	-59.9
19	-14.2	27.7	-230.6	-69.6	-56.7	-17.5
20	-18.4	28.3	-1089.6	-308.1	-104.1	-43.7

To illustrate the problem more explicitly, let us consider prediction under the Fay–Herriot model. The model may be thought of, equivalently, as having two levels. In Level 1, we have, conditional on $\zeta_1, \dots, \zeta_n$, that $y_1, \dots, y_n$ are independent such that $y_i \sim N(\zeta_i, D_i)$, where $D_i > 0$ are known. In Level 2, we have $\zeta_i = x_i'\beta + u_i$, $1 \leq i \leq n$, where the x_i's are observed vectors of covariates, β is a $p \times 1$ vector of unknown regression coefficients, the u_i's are small-area-specific random effects that are independent and distributed as $N(0, A)$, and $A > 0$ is an unknown variance. A prediction interval for ζ_i based on the Level 1 model only is given by $I_{1,i} = [y_i - z_{\alpha/2}\sqrt{D_i}, y_i + z_{\alpha/2}\sqrt{D_i}]$. This interval has the right coverage probability $1 - \alpha$ but is hardly useful, because its length is too large due to the high variability of the predictor based on a single observation y_i. Alternatively, one may construct a prediction interval based on the Level 2 model only. We consider a special case.

Example 4.10. Consider the special case of the Fay–Herriot model with $D_i = 1$ and $x_i'\beta = \beta$. At Level 2, the ζ_i's are i.i.d. with the distribution $N(\beta, A)$. If β and A were known, a Level 2 prediction interval that has the coverage probability $1 - \alpha$ would be $I_2 = [\beta - z_{\alpha/2}\sqrt{A}, \beta + z_{\alpha/2}\sqrt{A}]$. Note that the interval is not area-specific (i.e., not dependent on i). Bacause β and A are unknown, we replace them by estimators. The standard estimators are $\hat{\beta} = \bar{y} = n^{-1}\sum_{i=1}^n y_i$ and $\hat{A} = \max(0, s^2 - 1)$, where $s^2 = (n-1)^{-1}\sum_{i=1}^n (y_i - \bar{y})^2$. This leads to the interval

$\tilde{I}_2 = [\hat{\beta} - z_{\alpha/2}\sqrt{\hat{A}}, \hat{\beta} + z_{\alpha/2}\sqrt{\hat{A}}]$, but, again, the coverage probability is no longer $1 - \alpha$. However, because $\hat{\beta}$ and $\hat{A}$ are consistent estimators, the coverage probability can be shown to be $1 - \alpha + o(1)$. We can do better with a simple Level 2 bootstrap: First, draw independent samples $\zeta_1^*, \ldots, \zeta_n^*$ from $N(\hat{\beta}, \hat{A})$; then draw $y_i^*, i = 1, \ldots, n$, independently such that $y_i^* \sim N(\zeta_i^*, 1)$. Compute $\hat{\beta}^*$ and $\hat{A}^*$ the same way as $\hat{\beta}$ and $\hat{A}$ except using the y_i^*'s; then determine the cutoff points t_1 and t_2 such that

$$\text{P}^* \left(\hat{\beta}^* - t_1\sqrt{\hat{A}^*} \leq \zeta_i^* \leq \hat{\beta}^* + t_2\sqrt{\hat{A}^*} \right) = 1 - \alpha,$$

where P^* denotes the bootstrap probability, evaluated using a large number of bootstrap replications. Note that, at least approximately, t_1 and t_2 do not depend on i. We then define a (Level 2) bootstrap prediction interval as $\hat{I}_2 = [\hat{\beta} - t_1\sqrt{\hat{A}}, \hat{\beta} + t_2\sqrt{\hat{A}}]$. It can be shown that $\hat{I}_2$ has a coverage probability of $1 - \alpha + o(n^{-1/2})$. It is an improvement but still not accurate enough.

Intuitively, one should do better by combining both levels of the model. For example, [203] proposed the following empirical Bayes interval for ζ_i using information of both levels:

$$I_i^C : \quad (1 - \hat{B}_i)y_i + \hat{B}_i x_i'\hat{\beta} \pm z_{\alpha/2}\sqrt{D_i(1 - \hat{B}_i)}, \tag{4.134}$$

where $\hat{B}_i$ and $\hat{\beta}$ are (consistent) estimators of $B_i = D_i/(A + D_i)$ and β, respectively. It can be seen that the center of the interval (4.134) is a weighted average of the centers of the Level 1 and Level 2 intervals. If D_i is much larger than A, then $1 - \hat{B}_i$ is expected to be close to zero. In this case, the center of I_i^C is close to the Level 2 center, but the length will be close to zero and, in particular, much smaller than that of the Level 1 interval (which is $\sqrt{D_i}$). On the other hand, if A is much larger than D_i, then the center of I_i^C is expected to be close to the Level 1 center, and the length is close to $\sqrt{D_i}$, much smaller than that of the Level 2 interval (which would be $\sqrt{A}$ if A were known). Thus, the empirical Bayes interval does seem to have the combined strength of both levels. It can be shown that, under regularity conditions, the coverage probability of I_i^C is $1 - \alpha + O(n^{-1})$. Can one do better than I_i^C? In the following, we consider a parametric bootstrap procedure, proposed and studied by [204], that leads to a prediction interval with coverage probability $1 - \alpha + O(n^{-3/2})$, provided that the number of parameters under the model is bounded.

Let $d = p + q$, the total number of fixed parameters under the model. It is not assumed that d is fixed or bounded; so it may increase with the sample size n. This is practical in many applications, where the number of fixed parameters may be comparable to the sample size. See, for example, Section 2.2. According to the discussion below (4.133), the key is to approximate the distribution of $T = (\zeta - \hat{\mu})/\hat{\sigma}$, where $\hat{\mu}$ and $\hat{\sigma}$ are estimators of $\tilde{\mu}$ and $\tilde{\sigma}$, respectively, defined by (4.132) and (4.133). For simplicity, assume that X is of full rank and $\hat{\beta} = (X'X)^{-1}X'y$ is the ordinary least squares estimator of β. Consider a bootstrap version of the linear mixed model,

$$y^* = X\hat{\beta} + Zu^* + e^*, \tag{4.135}$$

where $u^* \sim N(0, \hat{G})$, $e^* \sim N(0, \hat{R})$ with $\hat{G} = G(\hat{\psi})$ and $\hat{R} = R(\hat{\psi})$, and u^* and e^*

are independent. Here, $\hat{\psi}$ may be chosen as the REML estimator (see Section 2.2). From (4.135) we generate y^* and then obtain $\hat{\beta}^*$ and $\hat{\psi}^*$ the same way as $\hat{\beta}$ and $\hat{\psi}$, except using y^*. Next, obtain $\hat{\mu}^*$ and $\hat{\sigma}^*$ using $\hat{\beta}^*$ and $\hat{\psi}^*$, (4.132) and (4.133). Define $\zeta^* = c'(X\hat{\beta} + Zu^*)$. The distribution of $T^* = (\zeta^* - \hat{\mu}^*)/\hat{\sigma}^*$ conditional on y is the parametric bootstrap approximation to the distribution of T. Under some regularity conditions, [204] showed that

$$\sup_x |P^*(T^* \leq x) - P(T \leq x)| \;=\; O_P(d^3 n^{-3/2}), \qquad (4.136)$$

where $P^*(T^* \leq x) = P(T^* \leq x|y) \equiv F^*(x)$. As a consequence, if $d^2/n \to 0$ and for any $\alpha \in (0,1)$, let q_1 and q_2 be the real numbers such that

$$F^*(q_2) - F^*(q_1) \;=\; 1 - \alpha,$$

then, we have the following approximation of the coverage probability:

$$P(\hat{\mu} + q_1\hat{\sigma} \leq \zeta \leq \hat{\mu} + q_2\hat{\sigma}) \;=\; 1 - \alpha + O(d^3 n^{-3/2}). \qquad (4.137)$$

Going back to the special case of the Fay–Herriot model, consider the problem of constructing prediction intervals for the small-area means ζ_i. Suppose that the maximum of the diagonal elements of $P_X = X(X'X)^{-1}X'$ is $O(p/n)$ (make sense?) and the D_i's are bounded from above as well as away from zero. Then, under some regularity conditions on $\hat{A}$, we have

$$P\left\{ \zeta_i \in \left[\hat{\zeta}_i + q_{i1}\sqrt{D_i(1-\hat{B}_i)},\ \hat{\zeta}_i + q_{i2}\sqrt{D_i(1-\hat{B}_i)} \right] \right\}$$
$$= 1 - \alpha + O(p^3 n^{-3/2}), \qquad (4.138)$$

where $\hat{\zeta}_i = (1-\hat{B}_i)y_i + \hat{B}_i x_i'\hat{\beta}$, $\hat{B}_i = D_i/(\hat{A}+D_i)$, and q_{i1} and q_{i2} satisfy

$$P^*\left\{ \zeta_i^* \in \left[\hat{\zeta}_i^* + q_{i1}\sqrt{D_i(1-\hat{B}_i^*)},\ \hat{\zeta}_i^* + q_{i2}\sqrt{D_i(1-\hat{B}_i^*)} \right] \right\}$$
$$= 1 - \alpha + O_P(p^3 n^{-3/2}) \qquad (4.139)$$

with $\hat{\zeta}_i^* = (1-\hat{B}_i^*)y_i^* + \hat{B}_i^* x_i'\hat{\beta}^*$ and $\hat{B}_i^* = D_i/(\hat{A}^* + D_i)$. For example, the term $O_P(p^3 n^{-3/2})$ in (4.139) may be taken as 0 in order to determine q_{i1} and q_{i2}. Thus, in particular, if $p^3/\sqrt{n} \to 0$, the prediction interval in (14.82) is asymptotically more accurate than I_i^C of (4.138).

Regarding finite-sample performance, [204] carried out a simulation study under the special case with $m = 15$ and $x_i'\beta = 0$. The 15 small areas are divided into 5 groups with 3 areas in each group. The D_i's are the same within each group, but vary between groups according to the one of the following patterns: (a) 0.2, 0.4, 0.5, 0.6, 4.0 or (b) 0.4, 0.8, 1.0, 1.2, 8.0. Pattern (a) was considered by [180]; see Section 4.2. Pattern (b) is simply pattern (a) multiplied by 2. The true value of A is 1 for pattern (a), and $A = 2$ for pattern (b). Note that the choice of q_{i1} and q_{i2} that satisfy (4.139) is not unique. Two most commonly used choices are (1) equal-tail, in which q_{i1} and q_{i2} are

Table 4.4 *Coverage Probabilities and Lengths of Prediction Intervals*

Pattern	Cox	PR	FH	PB-1	PB-2
A1	83.1 (3.12)	92.4 (3.82)	90.4 (3.57)	96.1 (4.50)	95.7 (4.42)
A2	85.4 (2.14)	98.0 (3.19)	93.7 (2.50)	96.2 (2.83)	95.9 (2.79)
A3	85.8 (2.02)	98.0 (3.08)	93.9 (2.36)	96.0 (2.65)	95.6 (2.61)
A4	86.1 (1.89)	98.2 (2.93)	94.3 (2.19)	96.1 (2.43)	95.7 (2.39)
A5	89.7 (1.12)	97.3 (1.87)	95.2 (1.23)	95.7 (1.28)	95.3 (1.26)
B1	85.5 (4.87)	89.3 (5.35)	89.5 (5.18)	95.7 (6.55)	95.4 (6.47)
B2	83.6 (2.68)	87.3 (2.93)	86.0 (2.82)	95.2 (3.90)	94.9 (3.75)
B3	83.4 (2.49)	86.8 (2.71)	85.7 (2.60)	95.2 (3.53)	94.9 (3.49)
B4	82.9 (2.27)	86.2 (2.46)	85.0 (2.36)	95.0 (3.22)	94.5 (3.18)
B5	83.0 (1.21)	84.8 (1.29)	84.0 (1.23)	94.9 (1.72)	94.6 (1.70)

chosen such that the tail probabilities on both sides are $\alpha/2$ and (2) shortest-length, in which q_{i1} and q_{i2} are chosen to minimize the length of the prediction interval (while maintaining the coverage probability $1 - \alpha$). The parametric bootstrap (PB) prediction intervals corresponding to (1) and (2) are denoted by PB-1 and PB-2, respectively. The number of bootstrap replications used to evaluate (4.139) is 1000. In the PB procedure, the unknown variance A is estimated by the Fay–Herriot method [F-H; see Section 4.2, in particular, (4.12)], and β is estimated by the EBLUE, given by (2.47) with A replaced by its F-H estimator. The PB intervals are compared with three other competitors. The first is the Cox empirical Bayes interval (Cox) (4.134), with the P-R estimator of A (4.11). The second (PR) and third (FH) are both in the form of EBLUP $\pm 1.96\sqrt{\widehat{\text{MSPE}}}$. In the second case, $\widehat{\text{MSPE}}$ is the Prasad–Rao MSPE estimator (4.24) with the P-R estimator of A; in the third case, $\widehat{\text{MSPE}}$ is the MSPE estimator proposed by [180] [i.e., (4.26)] with the F-H estimator of A. The results, based on 10,000 simulation runs and reported by [204] in their Table 1 and Table 2, are summarized in Table 4.4, where A1–A5 correspond to the five cases of pattern (a) and B1–B5 correspond to those of pattern (b). The numbers in the main body of the table are empirical coverage probabilities, in terms of percentages, and lengths of the prediction intervals (in parentheses). The nominal coverage probability is 0.95. It is seen that, with the exception of one case, the PB methods outperform their competitors (the way to look at the table is to first consider coverage probabilities; if the latter are similar, then compare the corresponding lengths). The differences are more significant for pattern (b), where all of the competing methods have relatively poor coverage probabilities, whereas the PB intervals appear to be very accurate. These results are consistent with the theoretical findings discussed above.

Our next example of bootstrapping mixed models involves the fence method for mixed model selection, which include both linear mixed models and GLMMs. See [116] for a review. In [114], it is proposed to build the fence via the inequality

$$d_M \leq c_0 \widehat{\text{s.d.}}(d_M), \tag{4.140}$$

where $d_M = \hat{Q}(M) - \hat{Q}(M_f)$ and M_f denotes the full model, and c_0 is a tuning parameter. An adaptive fence procedure is used to choose c_0, which involves bootstrapping. The bootstrap is typically done parametrically in a similar way as in [204]. [205] proposed a simplified version of the adaptive fence. The motivation is that, in most cases, the calculation of $\hat{Q}(M)$ is fairly straightforward, but the evaluation of $\widehat{\text{s.d.}}(d_M)$ can be quite challenging. Sometimes, even if an expression can be obtained for $\widehat{\text{s.d.}}(d_M)$, its accuracy as an estimate of the standard deviation cannot be guaranteed in a finite-sample situation. In the simplified version, the estimator $\widehat{\text{s.d.}}(d_M)$ is absorbed into c_0, which is then chosen adaptively. In other words, one considers the fence inequality

$$\hat{Q}(M) - \hat{Q}(M_f) \;\; \leq \;\; c, \tag{4.141}$$

where c is chosen adaptively as follows. The idea is to let data determine the best tuning constant. First note that, ideally, one wishes to select c that maximizes the probability of choosing the optimal model. Here, to be specific, the optimal model is understood as a true model that has the minimum dimension among all of the true models. This means that one wishes to choose c that maximizes

$$P \;\; = \;\; \text{P}(M_0 = M_{\text{opt}}), \tag{4.142}$$

where M_{opt} represents the optimal model and $M_0 = M_0(c)$ is the model selected by the fence procedure with the given c. However, two things are unknown in (4.142): (i) Under what distribution should the probability P be computed? (ii) What is M_{opt}? (If we knew M_{opt}, model selection would not be necessary.)

To solve problem (i), let us assume that there is a true model among the candidate models. It follows that the full model, M_f, is a true model. Therefore, it is possible to bootstrap under M_f. For example, one may estimate the parameters under M_f and then use a model-based bootstrap to draw samples under M_f. This allows us to approximate the probability distribution P on the right side of (4.142).

To solve problem (ii), we use the idea of maximum likelihood; namely, let $p^*(M) = \text{P}^*(M_0 = M)$, where P^* denotes the empirical probability obtained by the bootstrapping. In other words, $p^*(M)$ is the sample proportion of times out of the total number of bootstrap samples that model M is selected by the fence method with the given c. Let $p^* = \max_{M \in \mathcal{M}} p^*(M)$, where $\mathcal{M}$ denotes the set of candidate models. Note that p^* depends on c. The idea is to choose c that maximizes p^*. It should be kept in mind that the maximization is not without restriction. To see this, note that if $c = 0$, then $p^* = 1$ (because when $c = 0$, the procedure always chooses M_f). Similarly, $p^* = 1$ for very large c if there is a unique model, M_*, with the minimum dimension (because when c is large enough, the procedure always chooses M_*). Therefore, what one looks for is "the peak in the middle" of the plot of p^* against c.

Here is another look at the adaptive fence. Typically, the optimal model is the one from which the data are generated; then this model should be the most likely given the data. Thus, given c, one is looking for the model (using the fence procedure) that is most supported by the data or, in other words, one that has the highest (posterior) probability. The latter is estimated by bootstrapping. Note that although the bootstrap samples are generated under M_f, they are almost the same as those generated under the optimal model. This is because, for example, the estimates corresponding to the zero parameters are expected to be close to zero, provided that the parameter estimation under M_f is consistent. One then pulls off the c that maximizes the (posterior) probability and this is the optimal choice, say, c^*.

Under suitable regularity conditions, [205] showed that the simplified adaptive fence is consistent, as in [114] for the adaptive fence. For the most part, the result states the following: (i) There is a c^* that is at least a local maximum of p^* and an approximate global maximum in the sense that the p^* at c^* goes to 1 in probability and (ii) the probability that $M_0^* = M_{\mathrm{opt}}$, where M_0^* is the model selected by the fence (4.141) with $c = c^*$ and M_{opt} is the optimal model [see (4.142)], goes to 1 as both the sample size and the number of bootstrap replications increase.

We demonstrate the consistency empirically through a simulation study. [205] used a simulation design that mimic the well-known Iowa crops data ([34]). The original data were obtained from 12 Iowa counties in the 1978 June Enumerative Survey of the U.S. Department of Agriculture as well as from land observatory satellites on crop areas involving corn and soybeans. The objective was to predict mean hectares of corn and soybeans per segment for the 12 counties using the satellite information. [34] proposed the following nested error regression model:

$$y_{ij} = x'_{ij}\beta + v_i + e_{ij}, \quad i = 1,\dots,m, j = 1,\dots,n_i, \qquad (4.143)$$

where i represents county and j represents the segment within the county; y_{ij} is the number of hectares of corn (or soybeans), v_i is a small-area-specific random effect, and e_{ij} is the sampling error. It is assumed that the random effects are independent and distributed as $N(0,\sigma_v^2)$, the sampling errors are independent and distributed as $N(0,\sigma_e^2)$, and the random effects and sampling errors are uncorrelated. For the Iowa crops data, $m = 12$ and the n_i's ranged from 1 to 6. [34] used $x'_{ij}\beta = \beta_0 + \beta_1 x_{ij1} + \beta_2 x_{ij2}$, where where x_{ij1} and x_{ij2} are the number of pixels classified as corn and soybeans, respectively, according to the satellite data. The authors did discuss, however, various model selection problems associated with the nested error regression, such as whether or not to include quadratic terms in the model. The latter had motivated a model selection problem described below.

In the simulation study, the number of clusters, m, is either 10 or 15, setting up a situation of increasing sample size. Note that the n_i's are typically small and therefore not expected to increase. Here, the n_i's are generated from a Poisson(3) distribution and fixed throughout the simulations. The random effects, v_i, and errors e_{ij}, are both generated independently from the $N(0,1)$ distribution. The components of the covariates, x_{ij}, are to be selected from x_{ijk}, $k = 0, 1,\dots,5$, where $x_{ij0} = 1$; x_{ij1} and x_{ij2} are generated independently from $N(0,1)$ and then fixed throughout; $x_{ij3} = x_{ij1}^2$, $x_{ij4} = x_{ij2}^2$, and $x_{ij5} = x_{ij1}x_{ij2}$. The simulated data are generated under two models: Model I. The model that involves the linear terms only; that is, x_{kij}, $k = 0, 1, 2$, with all of the regression coefficients equal to 1; Model II. The model that involves both the linear and the quadratic terms; that is, x_{kij}, $k = 0,\dots,5$, with all of the regression coefficients equal to 1. The measure $Q(M)$ is chosen as the negative log-likelihood function. The number of bootstrap replications is 100. Results based on 100 simulation runs are reported in Table 4.5, which shows empirical probabilities (%) of selection of the optimal model (i.e., the model from which the data is generated)

Table 4.5 *Simplified Adaptive Fence: Simulation Results*

Optimal Model	# of Clusters, m	Empirical Probability (in %)
Model I	10	82
Model I	15	99
Model II	10	98
Model II	15	100

for the simplified adaptive fence. The results show that even in these cases of fairly small sample size, the performance of the simplified adaptive fence is quite satisfactory. Note the improvement of the results when m increases, which is in line with the consistency result mentioned above.

So far, we have been talking about the parametric bootstrap. A question of interest is how to bootstrap nonparametrically in mixed model situations. It seems obvious that Efron's i.i.d. bootstrap cannot be applied directly to the data, neither do the strategies such as sieve or block bootstraps, discussed in the previous section, which require stationarity. In fact, under a mixed model, linear or generalized linear, the vector y of all of the observations is viewed as a single n-dimensional observation and there is no other replication or stationary copy of this observation (in other words, one has an i.i.d. sample of size 1). On the other hand, there are some i.i.d. random variables "inside" the model, at least for most mixed models that are practically used. For example, in most cases, the random effects are assumed to be i.i.d. However, one cannot bootstrap the random effects directly, because they are not observed—and this is a major difference from the i.i.d. situation. Hall and Maiti [31] had a clever idea on how to bootstrap the random effects nonparametrically, at least in some cases.

Consider, once again, the nested error regression model (4.143). Suppose that the random effects are i.i.d. but not normal, or at least we do not know if they are normal. Then, what can we do in order to generate the random effects? The parametric bootstrap discussed above usually requires normality, or at least a parametric distribution of the random effects, so this strategy encounters a problem. On the other hand, one cannot use Efron's (nonparametric) bootstrap because the random effects are not observed, as mentioned. An answer by Hall and Maiti: Depending on what one wants. In many cases, the quantity of interest does not involve every piece of information about the distribution of the random effects. For example, Hall and Maiti observed that the MSPE of EBLUP involves only the second and fourth moments of the random effects and errors, up to the order of $o(m^{-1})$. This means that for random effects and errors from any distributions with the same second and fourth moments, the MSPE of EBLUP, with some suitable estimators of the variance components, are different only by a term of $o(m^{-1})$. This observation leads to a seemingly simple strategy: First, estimate the second and fourth moments of the random effects and errors; then draw bootstrap samples of the random effects and errors from distributions that match the first (which is 0) and estimated second and fourth moments; given the bootstrap random effects and errors, use (4.143) to generate bootstrap data, and so on. The hope is that this leads to an MSPE estimator whose bias is $o(m^{-1})$ (i.e., second-order unbiased).

To be more precise, for any $\mu_2, \mu_4 \geq 0$ such that $\mu_2^2 \leq \mu_4$, let $D(\mu_2, \mu_4)$ denote the distribution of a random variable ξ such that $E(\xi) = 0$ and $E(\xi^j) = \mu_j, j = 2, 4$. [31] suggested using moment estimators, $\hat{\sigma}_v^2, \hat{\sigma}_e^2, \hat{\gamma}_v$, and $\hat{\gamma}_e$, of $\sigma_v^2, \sigma_e^2, \gamma_v$, and γ_e, where γ_v, γ_e are the fourth moments of v_i and e_{ij}, respectively, so that they satisfy the constraints $\hat{\sigma}_v^4 \leq \hat{\gamma}_v, \hat{\sigma}_e^4 \leq \hat{\gamma}_e$. Suppose that we are interested in prediction of the mixed effects

$$\zeta_i = X_i'\beta + u_i, \qquad (4.144)$$

where X_i is a known vector of covariates, such as the population mean of the x_{ij}'s. Denote the EBLUP of ζ_i by $\hat{\zeta}_i$. To obtain a bootstrap estimator of the MSPE of $\hat{\zeta}_i$, $\text{MSPE}_i = E(\hat{\zeta}_i - \zeta_i)^2$, we draw samples $v_i^*, i = 1, \ldots, m$, independently from $D(\hat{\sigma}_v^2, \hat{\gamma}_v)$, and $e_{ij}^*, i = 1, \ldots, m, j = 1, \ldots, n_i$, independently, from $D(\hat{\sigma}_e^2, \hat{\gamma}_e)$. Then, mimicking (4.143), define

$$y_{ij}^* = x_{ij}'\hat{\beta} + v_i^* + e_{ij}^*, \quad i = 1, \ldots, m, j = 1, \ldots, n_i,$$

where $\hat{\beta}$ is the EBLUE of β [i.e., (2.46) with σ_v^2 and σ_e^2 (in V) replaced by their estimators]. Let $\hat{\zeta}_i^*$ be the bootstrap version of $\hat{\zeta}_i$, obtained the same way except using y^* instead of y. We then define a bootstrap MSPE estimator

$$\widehat{\text{MSPE}}_i = E\{(\hat{\zeta}_i^* - \zeta_i^*)^2 | y\}, \qquad (4.145)$$

where $\zeta_i^* = X_i'\hat{\beta} + v_i^*$ [see (4.144)], and the conditional expectation is evaluated, as usual, by replications of y^*. Euation (4.145) produces an MSPE estimator whose bias is $O(m^{-1})$, not $o(m^{-1})$ (see [31], sec. 4).

To obtain an MSPE estimator whose bias is $o(m^{-1})$, we bias-correct $\widehat{\text{MSPE}}_i$ using a *double bootstrap* as follows. The first bootstrap is done above. Conditional on v^* and e^*, draw samples $v_i^{**}, i = 1, \ldots, m$ and $e_{ij}^{**}, i = 1, \ldots, m, j = 1, \ldots, n_i$ independently from $D(\hat{\sigma}_v^{*2}, \hat{\gamma}_v^*)$ and $D(\hat{\sigma}_e^{*2}, \hat{\gamma}_e^*)$, respectively, where $\hat{\sigma}_v^{*2}$, and so on are computed the same way as $\hat{\sigma}_v^2$, and so on. except using y^* instead of y. Then, similarly, define

$$y_{ij}^{**} = x_{ij}'\hat{\beta}^* + v_i^{**} + e_{ij}^{**}, \quad i = 1, \ldots, m, j = 1, \ldots, n_i,$$

and compute the double-bootstrap version of $\hat{\zeta}_i$ and $\hat{\zeta}_i^{**}$. After that, compute

$$\widehat{\text{MSPE}}_i^* = E\{(\hat{\zeta}_i^{**} - \zeta_i^{**})^2 | y^*\}, \qquad (4.146)$$

where $\zeta_i^{**} = X_i'\hat{\beta}^* + v_i^{**}$ and the conditional expectation is evaluated by replications of y^{**}. Equation (4.146) is the bootstrap analogue of (4.145) [the way to understand this is to forget that (4.145) is computed via bootstrap; then (4.146) is just a bootstrap analogue of (4.145)]. According to one of the classical usages of the bootstrap, we estimate the bias of $\widehat{\text{MSPE}}_i$ by

$$\widehat{\text{bias}}_i = E(\widehat{\text{MSPE}}_i^* | y) - \widehat{\text{MSPE}}_i,$$

where the conditional expectation is evaluated by replications of y^* [note that, after taking the conditional expectation in (4.146), $\widehat{\mathrm{MSPE}}_i^*$ is a function of y^*]. This leads to a bias-corrected MSPE estimator

$$\widehat{\mathrm{MSPE}}_i^{bc} = \widehat{\mathrm{MSPE}}_i - \widehat{\mathrm{bias}}_i = 2\widehat{\mathrm{MSPE}}_i - \mathrm{E}(\widehat{\mathrm{MSPE}}_i^*|y). \qquad (4.147)$$

Hall and Maiti [31] showed that, under regularity conditions, we have

$$\mathrm{E}(\widehat{\mathrm{MSPE}}_i^{bc}) = \mathrm{MSPE}_i + o(m^{-1}), \qquad (4.148)$$

so the bootstrap bias-corrected MSPE estimator is second-order unbiased.

Regarding the distribution $D(\mu_2,\mu_4)$, clearly, the choice is not unique. It remains a question of what is the optimal choice of such a distribution. The simplest example is, perhaps, constructed from a three-point distribution depending on a single parameter $p \in (0,1)$, defined as $\mathrm{P}(\xi = 0) = 1 - p$ and

$$\mathrm{P}\left(\xi = \frac{1}{\sqrt{p}}\right) = \mathrm{P}\left(\xi = -\frac{1}{\sqrt{p}}\right) = \frac{p}{2}.$$

It is easy to verify that $\mathrm{E}(\xi) = 0, \mathrm{E}(\xi^2) = 1$, and $\mathrm{E}(\xi^4) = 1/p$. Thus, if we let $p = \mu_2^2/\mu_4$, the distribution of $\sqrt{\mu_2}\xi$ is $D(\mu_2,\mu_4)$. Note that the inequality $\mu_2^2 \le \mu_4$ always hold with the equality if and only if ξ^2 is degenerate (i.e., a.s. a constant). Another possibility is the rescaled Student t-distribution whose degrees of freedom $v > 4$ and is not necessarily an integer. The distribution has first and third moments 0, $\mu_2 = 1$, and $\mu_4 = 3(v-2)/(v-4)$, which is always greater than 3, meaning that the tails are heavier than those of the normal distribution.

The moment-matching, double-bootstrap procedure may be computationlly intensive, and so far there have not been published comparisons of the method with other exsiting methods that also produce second-order unbiased MSPE estimators, such as the Prasad–Rao (see Section 4.1) and jackknife (e.g., Section 4.5.1). On the other hand, the idea of [31] is potentially applicable to mixed models with more complicated covariance structure, such as mixed models with crossed random effects. We conclude this section with an illustrative example.

Example 4.11. Consider a two-way random effects model

$$y_{ij} = \mu + u_i + v_j + e_{ij},$$

$i = 1,\ldots,m_1$, $j = 1,\ldots,m_2$, where μ is an unknown mean; the u_i's are i.i.d. with mean 0, variance σ_1^2, and an unknown distribution F_1; the v_j's are i.i.d. with mean 0, variance σ_2^2, and an unknown distribution F_2; the e_{ij}'s are i.i.d. with mean 0, variance σ_0^2, and an unknown distribution F_0; and u, v, and e are independent. Note that the observations are *not clustered* under this model. As a result, the jackknife method of [30] (see Section 4.5.1) may not apply. Consider a mixed effect that can be expressed as $\zeta = a_0\mu + a_1'u + a_2'v$. The BLUP of ζ can be expressed as $\tilde{\zeta} = a_0\bar{y}_{..} + a_1'\tilde{u} + a_2'\tilde{v}$, where $y_{..} = (m_1m_2)^{-1}\sum_{i=1}^{m_1}\sum_{j=1}^{m_2} y_{ij}$ and $\tilde{u}$ and $\tilde{v}$ are tbe BLUPs of u and v, respectively (e.g., Section 4.8). According to the general result of [63] (see Section 4.3), the MSPE of the EBLUP of ζ can be expressed as

$$\mathrm{MSPE} = g_1(\theta) + g_2(\theta) + g_3 + o(m_*^{-1}),$$

where $\theta = (\sigma_0^2, \sigma_1^2, \sigma_2^2)'$; $g_1(\theta) = a'(G - GZ'V^{-1}ZG)a$ with $a = (a_1', a_2')'$, $G = \mathrm{diag}(\sigma_1^2 I_{m_1}, \sigma_2^2 I_{m_2})$, $Z = (I_{m_1} \otimes 1_{m_2}\ 1_{m_1} \otimes I_{m_2})$, and $V = \sigma_0^2 I_{m_1 m_2} + ZGZ'$; $g_2(\theta) = \{a_0 - 1_{m_1 m_2}' V^{-1} ZGa\}^2 / (1_{m_1 m_2}' V^{-1} 1_{m_1 m_2})$ (one may simplify these expressions considerably in this simple case); $g_3 = \mathrm{E}(h'A^{-1}b)^2$ with $h = \partial \tilde{\zeta} / \partial \theta$, $A = \mathrm{E}(\partial^2 l / \partial \theta \partial \theta')$, l being the Gaussian restricted log-likelihood function, $b = \partial l / \partial \theta$; and $m_* = m_1 \wedge m_2$, provided that the Gaussian REML estimator of θ is used for the EBLUP (Section 2.2). Note that g_3 is not necessarily a function of θ unless the data are normal [and this is why the notation g_3, not $g_3(\theta)$, is used]. However, it is antici-pated that g_3 can be expressed as $s + o(m_*^{-1})$, where s is a term that depends only on the fourth moments of u, v, and e, in addition to θ. If this conjecture turns out to be true, then it is understandable that a similar idea of the moment-matching bootstrap should apply to this case as well.

4.6   Additional bibliographical notes and open problems

4.6.1   Bibliographical notes

Booth and Hobert [206] considered MSPE of an EBP of a mixed effect condition-ing on the data associated with the mixed effect. More specifically, the mixed effect is expressed as $\eta_i = x_i'\beta + z_i'\alpha_i$, and the EBP of η_i, $\hat{\eta}_i$, is obtained by replacing the parameters involved in the BP of η_i by their consistent estimators. Then, the condi-tional MSPE is defined as $\mathrm{E}\{(\hat{\eta}_i - \eta_i)^2 | y_i\}$, where y_i is the vector of observations associated with η_i through a GLMM (e.g., Section 3.1). The authors indicated that a second-order approximation the conditional MSPE can be obtained, although no rigorous justification is given.

Jiang [207] extended the work of [189] (see Section 4.4) to GLMMs, and ob-tained similar results about the MSPE estimation.

In a series of papers [208]–[209], Lahiri and co-authors proposed a new method of variance component estimation, called adjusted maximum likelihood, and adjusted REML. It is known that, for the Fay-Herriot model, standard methods of variance component estimation, such as maximum likelihood and REML, often produce zero estimates of the variance of the area-specific random effect, A, which is supposed to be strictly positive. Note that, when the estimator of A is equal to zero, the EBLUP of the small area mean reduces to the regression estimator [see (4.10)], which typically has an overly-shrinkage problem. In [208], the authors proposed an adjusted MLE of A that maximizes an adjusted likelihood function. The latter is a product of A and a standard likelihood function. It is clear that, when $A = 0$, the adjusted likelihood is zero. On the other hand, for any A that is positive, so is the adjusted likelihood. There-fore, in no way the maximizer of the adjusted likelihood can be zero. A key result is that, as is shown, the adjustment does not affect the mean squared error property of the estimator of A, neither does it affect the MSPE property of the corresponding EBLUP up to the second order. In [210], the authors extended the adjusted maxi-mum likelihood method to construction of a second-order efficient empirical Bayes confidence interval (c.v.). This means that the c.v. has an error of order $O(m^{-3/2})$ in the coverage probability, where m is the number of small areas. Furthermore, the proposed c.v. is carefully constructed so that it always produces an interval shorter than the corresponding direct c.v.

Lohr and Rao [211] also considered jackknife estimation of measures of uncertainty for small area predictors. The authors considered both the unconditional MSPE, as in JLW, and conditional MSPE of the EBP, $\hat{\theta}_i$, given the data, y_i, from the ith small area, that is, $\text{MSPE}(\hat{\theta}_i|y_i) = \text{E}\{(\hat{\theta}_i - \theta_i)^2|y_i\}$. The authors noted that the leading term of JLW estimator of the MSPE of the BP, (4.102), does not depend on y_i, while the "posterior variance" variance of θ_i, $\text{var}(\theta_i|y_i) = \text{E}[\{\theta_i - \text{E}(\theta_i|y_i)\}^2]$, depends on y_i. Because $\text{E}\{\text{var}(\theta_i|y_i)\}$ is equal to the MSPE of the BP, the JLW estimator has a bias of order $O(1)$ for estimating the conditional MSPE, even although its bias is $o(m^{-1})$ for estimating the unconditional MSPE. [211] proposed two modifications of JLW to come up with a second-order unbiased estimator of the conditional MSPE, and hence the unconditional MSPE (because the mean of the conditional MSPE is equal to the unconditional MSPE). Let $\tilde{\theta}_i = \text{E}(\theta_i|y_i)$ denote the BP, and $b_i(\psi) = \text{MSPE}(\tilde{\theta}_i)$. Let $g_i(\psi, y_i) = \text{var}(\theta_i|y_i, \psi)$. Rao [5] (Section 9.4) noted that, because $\text{E}\{g_i(\psi, y_i)\} = b_i(\psi)$, $g_i(\psi, y_i)$ is an unbiased estimator of $b_i(\psi)$, if ψ is known. Thus, the author suggested to modify the JLW estimator of $b_i(\psi)$,

$$b_i(\hat{\psi}) - \frac{m-1}{m} \sum_{j=1}^{m} \{g_i(\hat{\psi}_{-j}) - g_i(\hat{\psi})\},$$

by replacing the b_i function by the g_i function, that is,

$$g_i(\hat{\psi}, y_i) - \frac{m-1}{m} \sum_{j=1}^{m} \{g_i(\hat{\psi}_{-j}, y_i) - g_i(\hat{\psi}, y_i)\}. \tag{4.149}$$

The leading term of (4.149) now depends on y_i; however, as shown by [211], (4.149) is only first-order unbiased for estimating $g_i(\psi, y_i)$. The latter authors then proposed a further modification:

$$g_i(\hat{\psi}, y_i) - \sum_{j \neq i} \{g_i(\hat{\psi}_{-j}, y_i) - g_i(\hat{\psi}, y_i)\} \tag{4.150}$$

and showed that (4.150) is second-order unbiased for estimating $g_i(\psi, y_i)$; hence is second-order unbiased for estimating $b_i(\psi)$. The estimator (4.150) is then combined with the same JLW estimator of the MSAE [see (4.101)] to come up with a second-order unbiased estimator of the unconditional MSPE.

Along with their 2006 paper on nonparametric bootstrap for SAE, Hall and Maiti [212] also considered parametric bootstrap methods for estimating the MSPE as well as constructing prediction intervals for SAE. Their methods apply to two-level models, with level 1 specifying the conditional distribution of the data given the mixed effects, and level 2 specifying the distribution of the mixed effects, with both distributions being parametric. The authors argued that their double-bootstrap MSPE estimator is second-order unbiased, and their prediction interval, obtained using the double-bootstrap method, can achieve a coverage accuracy of $O(m^{-3})$.

Jiang and Nguyen [213] considered an extension of the NER model (see Example

2.6) in the sense that the variance of the errors, e_{ij}, are heteroscedastic in the following way. Suppose that the variance of y_{ij} is proportional to σ_i^2, where $\sigma_i^2, 1 \leq i \leq m$ are completely unknown. Let $u_{ij} = (y_{ij} - x_{ij}'\beta)/\sigma_i$, where $E(y_{ij}) = x_{ij}'\beta$. Then, it is assumed that u_{ij}, $i = 1,\ldots,m, j = 1,\ldots,n_i$ satisfy a NER model, $u_{ij} = \alpha_i + \varepsilon_{ij}$, where α_i's and ε_{ij}'s are independent with $\alpha_i \sim N(0, \sigma_\alpha^2)$ and $\varepsilon_{ij} \sim N(0, \sigma_\varepsilon^2)$. The authors showed that under such a heteroscedastic nested-error regression (HNER) model, the MLEs of the fixed effects and within-cluster correlation are consistent. The result implies that the EBLUP method for SAE is valid in such a case, which only depends on the within-cluster correlation under the HNER model. The authors also showed that ignoring the heteroscedasticity can lead to inconsistent estimation of the within-cluster correlation and inferior predictive performance. However, in order to estimate the MSPE of the EBLUP under the HNER model, the authors needed to add restrictions on the σ_i^2's; namely, that the latter belong to some known clusters with an unknown value for each cluster, but the number of clusters is bounded. Under such an additional assumption, the MSPE of the EBLUP can be estimated using the JLW jackknife.

4.6.2 Open problems

For the Fay-Herriot model (see Section 4.2), it is almost always assumed that the sampling variances, D_i, are known. This assumption sometimes limits the scope of applications of such a model. In fact, in some cases the sampling variances are not exactly known; however, they can be estimated using additional data, possibly much larger in size than the current data that are used to estimate the small area means. In some other cases, the D_i's are known proportionally, that is, one has $D_i = \tau^2 s_i$, where s_i is known but τ^2 is unknown. In such a case, the parameter τ^2 needs to be estimated from the current data. There are also cases where the D_i's are totally unknown. For example, [214] considered modeling the D_i's using available covariates. So far, however, not much is known about the behavior of MSPE estimation. For example, how does the estimation of the D_i's, using either the additional data or the current data, affect the order of the bias in the MSPE estimation?

There has been some growing interest in considering heteroscedastic error variances in the NER model since the paper by Jiang and Nguyen [213], which introduces the HNER model. There are, still, some restrictions on how heteroscedastic the error variances can be under the HNER model that the variances of the errors cannot vary freely. Also, in order to estimate the MSPE of the EBLUP under the HNER model, [213] added further restriction on the error variances that these variances are divided into some known groups so that, within each group, the variances have the same value, and the number of the groups is bounded. Although, from an approximation point of view, the variances could be divided into groups, some practical questions remain. For example, how would the groups be known, or how to identify such groups? and how does misspecification of the groups affect the MSPE estimation. Also, can the restriction on the number of groups being bounded be relaxed? After all, it would be more useful to develop a method of MSPE estimation that does not have such a group restriction at all.

The OBP method was discussed in Section 2.4.2. An important, and difficult problem, is to obtain second-order unbiased estimators for the area-specific MSPEs. The problem is difficult because a main motivation for OBP is that it is more robust to model misspecification than the EBLUP in terms of predictive performance. Therefore, naturally, an MSPE estimator should be derived under potential model misspecification. However, the standard asymptotic techniques in SAE, such as the Prasad-Rao linearization method (see Section 4.1) and the jackknife methods (see Section 4.5), are no longer are no longer applicable when the underlying model is misspecified. [66] used a different technique to derive a linearization MSPE estimator which is second-order unbiased. However, the latter is not guaranteed nonnegative. Furthermore, the leading term of this MSPE estimator is an $O(1)$ function of the area data, rather than all of the data. Because the area data are limited, the leading term has a relatively large variance. On the other hand, the leading term can be negative. These facts imply that there is a non-vanishing probability (as m increases) that the leading term, hence the MSPE estimator, is negative. Jiang et al. [66], [122] developed alternative bootstrap methods that produce nonnegative MSPE estimators, but these estimators are only first-order unbiased. As noted in Section 4.5.2, when considering second-order unbiased MSPE estimation, it is desirable that the estimator has the double-goal property. In fact, the task is difficult enough even for EBLUP under the assumption that the underlying model is correct. See the discussion in the first paragraph of Section 4.5.2.

A key to second-order unbiased MSPE estimation is that the estimators of the model parameters have certain nice asymptotic properties. These properties were shown to hold for ML and REML estimators under the standard linear mixed model with independent random effects (e.g., [30]). However, less is known when the random effects are correlated, such as in the case of spatial correlations among the random effects (see the discussion at the end of Section 4.3). Another situation, in which the asymptotic behavior of estimators is even less clear is GLMM, when the random effects are spatially correlated. There has been a major progress in proving consistency of the MLE for GLMM with crossed random effects (see Section 3.3). The case of spatial correlation is expected to be easier than the case of crossed random effects, under the assumption that the spatial correlation is "fast-decaying" as the distance between areas increases. The asymptotic properties of the estimators, if established, should lead to second-order unbiased MSPE estimation, with rigorous justification, for SAE with spatially correlated random effects.

Computational cost has been a main concern for resampling methods in SAE, especially for methods that involve double-bootstrapping (e.g., [31]), and McJack (to some extent; see Section 4.5.2). For example, although the double-bootstrap method of Hall and Maiti [31] is sound from an asymptotic point of view, empirical studies have shown that the finite-sample performance of the method is not quite as good as competing methods such as Prasad–Rao and JLW (e.g., [215]). It is possible that these numerical results were not accurate enough due to the very intensive computations that would have been needed to obtain more competitive results. Looking forward, computational burden will become less significant in the years to come, and eventually goes away as a result of the fast-developing computer technology.

From such a point of view, computer-intensive methods, such as double-bootstrap and McJack, are promising. However, one cannot do noting but to wait for a faster computer. Currently, there is a need to develop asymptotic equivalence to, for example, double-bootstrap, that is computationally less intensive. For example, consider the MSPE decomposition (4.2). The second term on the right side is usually of the order $O(m^{-1})$; therefore, a plug-in estimator of ψ, the vector of parameters, will result in a $o(m^{-1})$ bias. This means that, for this term, one does not need a bias-correction, hence double-bootstrap can be avoided. On the other hand, the first term on the right side of (4.2) is typically of the order $O(1)$. This term does need a bias-correction after the plug-in estimation; therefore, double-bootstrap is needed. However, luckily, the term $\mathrm{MSPE}(\tilde{\eta})$ has a simple, closed-form expression, at least under the linear mixed model. This suggests a "hybrid" between single and double bootstraps by handle these two terms differently.

Statistical modeling have played key roles in "borrowing of strength," a term that is often used in SAE (e.g., [59]). Therefore, it is not surprising that model selection in SAE has received attention in recent literature. See, for example, [216], [202], [217], and [176]. On the other hand, so far there is a lack of novel development in methods for model diagnostics for SAE. Of course, there are available methods for diagnostics of mixed effects models; see, for example, Section 2.5, but the point is to take into account the special interests of SAE in the diagnostics. For example, the goodness-of-fit test of [76] focuses on the marginal distribution of the data, but the small area means is at the conditional level (on the random effects). Should such a difference make a difference in terms of the model diagnostics? Intuitively it should, but the answer is not known.

Chapter 5

Asymptotic Analysis in Other Mixed Effects Models

In previous chapters we mainly focused on two classes of mixed effects models, namely, linear mixed models (LMM) and generalized linear mixed models (GLMM), and their applications in small area estimation. There are other types of mixed effects models, for which asymptotic analyses have also played important roles. These include nonlinear mixed effects models as well as different types of semi- and nonparametric mixed effects models. For the most part, all of these models were proposed as extensions of LMM, just like GLMM as an extension of LMM. However, the structures of these models are different, so were the initial areas of applications.

We shall begin with nonlinear mixed effects models. This class of models is relatively simpler, at least conceptually, and the coverage does not require much preliminaries. Later, after some preliminaries, we will turn our attension to semiparametric and nonparametric mixed effects models.

5.1 Nonlinear mixed effects models

Nonlinear mixed effects models occur naturally in such fields as pharmacokinetics and economics, where parameters of interest are subject-specific through a nonlinear association with the response. For example, [218] (Section 20.3) discussed a toxicokinetic model involving 15 parameters for each of six persons in a pharmacokinetic experiment. Harding and Lovenheim [219] considered a rank three demand model in behavioral economics that was used to study consumers' healthy food choices.

A nonlinear mixed effects model (NLMEM) may be expressed as

$$ y_i = f_i(X_i, \beta, \alpha_i) + \varepsilon_i, \quad j = 1, \ldots, m, \qquad (5.1) $$

where y_i is an $n_i \times 1$ vector of responses or observations for stratum i, f_i is a known function, X_i is a matrix of known covariates, β is $p \times 1$ vector of unknown population parameters, α_i is a $d \times 1$ vector of unobserved random effects, and ε_i is an $n_i \times 1$ vector of additional errors. It is typically assumed that data from different strata are independent, which holds if $(b_i, \varepsilon_i), i = 1, \ldots, m$ are independent. It is also assumed that the α_i's are independent with the ε_i's. Comparing with (2.4), it is clear in what way NLMEM extends LMM; in fact, (2.4) is a special case of (5.1) with $f_i(X_i, \beta, \alpha_i) = X_i\beta + Z_i\alpha_i$ and the function f_i is linear and characterized by Z_i.

A further specification of distributional assumptions can be parametric or semi-parametric. In the parametric setting, it is usually assumed that the conditional distribution of $y_i|\alpha_i$ has pdf $p_i(\cdot|\alpha_i, \beta)$, and α_i has pdf $g_i(\cdot|\gamma)$, where γ is a vector of dispersion parameters or variance components. For example, p_i may be multivariate normal with mean f_i and covariance matrix $\sigma^2 I_{n_i}$ or, equivalently, $\varepsilon_i \sim N(0, \sigma^2 I_{n_i})$, and g_i may be multivariate normal with mean 0 and covariance matrix G. In the semi-parametric setting, the distribution of the random effects is assumed to be non-parametric, while the parametric assumption about the conditional distribution of $y_i|\alpha_i$ is maintained.

Similar to GLMMs, likelihood-based inference about NLMEM encounters computational difficulty due to the lack of closed-form expression for the likelihood function. Existing approaches to providing computational solutions are also similar to those for the GLMMs. Namely, for the most part, there are two approaches: (i) Monte Carlo E-M algorithms; or (ii) approximate inference. In terms of asymptotic analysis, there are (much) more issues related to (ii) than to (i). Therefore, in this section, we mainly focus on approach (ii).

Demidenko [4] (Section 8.2) noted that, due to the complexity of NLMEM, "it is very difficult to establish statistical properties of estimators in the NLME model in finite samples even when the variance parameters are known, a common difficulty with all nonlinear statistical models. Therefore, asymptotic consideration (when the number of observations goes to infinity/large sample) is useful." The author then listed three asymptotic configurations in which different parts of sample sizes may or may not increase in a NLMEM:

(a) The number of strata, m, goes to infinity, but the number of observations within each stratum, n_i, remains uniformly bounded for $1 \le i \le m$.

(b) The number of strata, m, is fixed, but the number of observations within each stratum, n_i, goes to infinity for each i.

(c) The number of strata, m, goes to infinity, and so is $\min_{1 \le i \le m} n_i$.

Configuration *(a)* is the standard assumption for LMM asymptotics; see Chapter 2. While this configuration is sufficient for consistency, asymptotic normality and efficiency of the MLE in NLMEM, it is not sufficient for the approximation-based inference in (ii), as noted by [4].

To derive the MLE, we make the standard normality assumption, that is, $\alpha_i \sim N(0, G)$, where G is a $d \times d$ covariance matrix whose entries depend on a vector γ of variance components, that is, $G = G(\gamma)$. Furthermore, assume that $\varepsilon_i \sim N(0, \sigma^2 I_{n_i})$. Then, it can be shown that the log-likelihood function for estimating the unknown parameters, β, γ, σ^2, can be expressed as

$$l(\beta, \gamma, \sigma^2|y) = c - \frac{m\log(|G|) + N\log\sigma^2}{2}$$
$$+ \sum_{i=1}^{m} \log \int \exp\left[-\frac{1}{2}\left\{\frac{|y_i - f_i(X_i, \beta, u)|^2}{\sigma^2} + u'G^{-1}u\right\}\right] du, \qquad (5.2)$$

where $N = \sum_{i=1}^{m} n_i$, and c is a constant that depends on $N + md$. In spite of the integral expression in (5.2), a bottom line is that the log-likelihood is a sum of independent random variables, namely, the log-integrals in the summation which are functions of y_i (also note that the dimension of the integral is d, which is fixed). Thus, the afore-mentioned asymptotic behaviors of the MLE follow from the standard asymptotic theory for sum of independent (but not identically distributed) random variables; see, for example, Chapter 6 of [1].

As for the approximation-based inference, several approaches have been proposed, which we summarize below. Vonesh and Carter ([220]), among others, proposed first-order Taylor expansion around the mean of the random effects. We call the method FOA in the sequel. The idea is to approximate the function f_i in (5.1), in terms of its α_i argument, via the first-order Taylor expansion at the mean (vector) of the random effect, $E(\alpha_i) = 0$. This leads to

$$
\begin{aligned}
y_i &\approx f_i(X_i, \beta, 0) + \left.\frac{\partial f_i}{\partial \alpha_i}\right|_{\alpha_i=0} \alpha_i + \varepsilon_i \\
&= f_i(X_i, \beta, 0) + Z_i(\beta)\alpha_i + \varepsilon_i,
\end{aligned}
\tag{5.3}
$$

where $Z_i(\beta)$ is defined by comparing the two sides of the last equation in (5.3). If one treats the approximation in (5.3) as an exact relation, the resulting model is convenient in that it is linear in the random effects. It is easy to show that, unlike (5.2), the log-likelihood function based on the model $y_i =$ the right side of (5.3), with the normal random effects and errors, has a closed-form expression. Therefore, it is much easier, computationally, to obtain the MLE based on the latter model. However, note that this is only an approximate MLE, which we call FOA-MLE.

Lindstrom and Bates (L-B; [221]) proposed another approximation by "estimating" the random effects. It is a two-step procedure that is carried out iteratively. In the first step, one assumes that the variance components, γ, are known, hence G is known. Then, estimators of β and $\alpha_i, 1 \leq i \leq m$ are obtained by minimizing a penalized nonlinear least-squares (PNLS):

$$
\sum_{i=1}^{m} (|y_i - f_i(X_i, \beta, \alpha_i)|^2 + \alpha_i' G^{-1} \alpha_i).
\tag{5.4}
$$

Let $\tilde{\beta}, \tilde{\alpha}_i, 1 \leq i \leq m$ denote the minimizer of (5.4). Then, in the second step, one update γ based on the approximate log-likelihood function:

$$
\begin{aligned}
l_{\text{LB}}(\beta, \gamma, \sigma^2 | y) &= -\frac{1}{2} \sum_{i=1}^{m} \{\log(|\sigma^2 I_{n_i} + \hat{Z}_i G \hat{Z}_i'|) \\
&\quad + (\hat{y}_i - \hat{X}_i \beta)'(\sigma^2 I_{n_i} + \hat{Z}_i G \hat{Z}_i')^{-1}(\hat{y}_i - \hat{X}_i \beta)\},
\end{aligned}
\tag{5.5}
$$

where $\hat{X}_i = \partial f_i / \partial \beta|_{\tilde{\beta}, \tilde{\alpha}_i}, \hat{Z}_i = \partial f_i / \partial \alpha_i|_{\tilde{\beta}, \tilde{\alpha}_i}$, and

$$
\hat{y}_i = y_i - f_i(X_i, \tilde{\beta}, \tilde{\alpha}_i) + \hat{X}_i \tilde{\beta} + \hat{Z}_i \tilde{\alpha}_i.
\tag{5.6}
$$

To see how one comes up with (5.5) and (5.6), note that, by (5.1) and the first-order Taylor expansion at $\tilde{\beta}, \tilde{\alpha}_i$, one has $y_i \approx f_i(X_i, \tilde{\beta}, \tilde{\alpha}_i) + \hat{X}_i(\beta - \tilde{\beta}) + \hat{Z}_i(\alpha_i - \tilde{\alpha}_i) + \varepsilon_i$. It follows that, by defining $\hat{y}_i$ as (5.6), one has $\hat{y}_i \approx \hat{X}_i\beta + \hat{Z}_i\alpha_i + \varepsilon_i$. Thus, if the latest approximation is treated as exact relation, the log-likelihood based on $\hat{y}_i, 1 \leq i \leq m$ is given by (5.5). Wolfinger and Lin [222] showed that the L-B approximation is equivalent to a PQL approximation (see Section 3.1).

Yet another approximation is Laplace approximation. The basic idea was introduced in Section 3.1. Pinheiro and Bates ([223]) applied the method to NLMEM. Express the marginal pdf of y_i as

$$p(y_i|\beta,\gamma,\sigma^2) = \int \frac{1}{(2\pi)^{(n_i+d)/2}(\sigma^2)^{n_i/2}|G|^{1/2}} \exp\left\{-\frac{1}{2}g_i(\alpha_i,\psi,y_i)\right\} d\alpha_i,$$

where $\psi = (\beta',\gamma',\sigma^2)'$ and

$$g_i(\alpha_i,\psi,y_i) = \frac{|y_i - f_i(X_i,\beta,\alpha_i)|^2}{\sigma^2} + \alpha_i'G^{-1}\alpha_i.$$

Let $\tilde{\alpha}_i = \operatorname{argmin}_{\alpha_i} g_i(\alpha_i,\psi,y_i)$, and write $g_i'' = (\partial^2/\partial\alpha_i\partial\alpha_i')g_i(\alpha_i,\psi,y_i)$. Then, we have [e.g., [2], (3.7)]

$$\int \exp\left\{-\frac{1}{2}g_i(\alpha_i,\psi,y_i)\right\} d\alpha_i \approx c|\tilde{g}_i''|^{-1/2} \exp\left(-\frac{\tilde{g}_i}{2}\right),$$

where c is a constant depending only on d, $\tilde{g}_i = g_i(\tilde{\alpha}_i,\psi,y_i)$, and $\tilde{g}_i''$ is g_i'' with α_i replaced by $\tilde{\alpha}_i$. The Laplace approximation to the log-likelihood function is thus

$$l_{\text{La}}(\beta,\gamma,\sigma^2|y) = c - \frac{1}{2}\left(N\log\sigma^2 + m\log(|G|) + \sum_{i=1}^{m}\log(|\tilde{g}_i''|) + \sum_{i=1}^{m}\tilde{g}_i\right), \quad (5.7)$$

where c is a constant depending on the sample sizes.

A fundamental question regarding the various approximations lead to, at least, consistent estimators of the NLMEM parameters under the standard asymptotic configuration, that is, (a) above (5.2). Note that, in most practical situations of mixed effects models, there is not enough information to estimate each individual random effect consistently. In fact, this is often a main reason why these effects are considered as random effects; otherwise, they could well be treated as fixed parameters. In terms of the sample sizes, this means that the n_i's are bounded, or at least some of them are bounded. Therefore, asymptotic configurations (b) and (c) [above (5.2)] are often unrealistic in an environment where a mixed effects model is considered. Unfortunately, the answer to the fundamental question is no, that is, these approximations, in general, do not lead to consistent estimators of the model parameters. Below we demonstrate this using two simple examples.

Example 5.1. Lee [224] considered a slightly different parametrization of the covariance matrix of the random effects by assuming that $G = \sigma^2 D$, where σ^2 is the same variance of the errors, and D may depend on a vector θ of additional dispersion parameters. It can be shown that, under this parametrization, the Laplace approximation to the log-likelihood, (5.7), can be expressed as

$$l_{\text{La}}(\beta,\gamma,\sigma^2|y) = c - \frac{1}{2}\left(N\log\sigma^2 + m\log(|D|) + \sum_{i=1}^{m}\log(|\tilde{h}_i''|) + \frac{1}{\sigma^2}\sum_{i=1}^{m}\tilde{h}_i\right), \quad (5.8)$$

where c is the same constant, and

$$h_i = |y_i - f_i(X_i, \beta, \alpha_i)|^2 + \alpha_i' D^{-1} \alpha_i. \tag{5.9}$$

An advantage of this new parametrization is that σ^2 can be "solved" as a function of the data and other parameters. Namely, by differentiating (5.8) with respect to σ^2 and letting the derivative equal to zero, we get

$$\tilde{\sigma}^2 = \frac{1}{N} \sum_{i=1}^{m} \tilde{h}_i, \tag{5.10}$$

where $\tilde{h}_i$ is h_i with α_i replaced by $\tilde{\alpha}_i = \operatorname{argmin}_{\alpha_i} h_i$. Consider a very simple case that β, D are known (so that σ^2 is the only unknown parameter), and $n_i = n, 1 \le i \le m$. Then, we have $N = mn$ so that (5.10) can be expressed as $\tilde{\sigma}^2 = m^{-1} \sum_{i=1}^{m} (\tilde{h}_i/n)$. Suppose that $\tilde{\sigma}^2$ is consistent. Then, we would have

$$\frac{1}{m} \sum_{i=1}^{m} \frac{\tilde{h}_i}{n} \xrightarrow{P} \sigma^2. \tag{5.11}$$

Now consider an even special case of NLMEM, a quadratic random-effect model:

$$y_i = \alpha_i^2 + \varepsilon_i, \quad i = 1, \ldots, m, \tag{5.12}$$

where $\alpha_i, \varepsilon_i, 1 \le i \le m$ are independent $N(0, \sigma^2)$. It can be shown that, for this special case, we have

$$\tilde{h}_i = \left\{ y_i - \left(y_i - \frac{1}{2} \right)^+ \right\}^2 + \left(y_i - \frac{1}{2} \right)^+, \tag{5.13}$$

where $x^+ = \max(x, 0)$. It follows that $\tilde{h}_i, 1 \le i \le m$ are i.i.d. with finite expectation; thus, by the LNN, we have

$$\frac{1}{m} \sum_{i=1}^{m} \tilde{h}_i \xrightarrow{P} E(\tilde{h}_1). \tag{5.14}$$

Combining (5.11) and (5.14) (note that $n = 1$ in this case), one would conclude

$$E(\tilde{h}_1) = \sigma^2. \tag{5.15}$$

However, (5.15) cannot hold as a functional equation (i.e., for different values of σ^2). For example, [224] presented a plot of the difference between the two sides of (5.15) (left minus right), which is strictly increasing function of σ^2; see Fig. 3.1 of [224].

What Example 5.1 shows is something much more general, that is, ML method based on Laplace approximation of the likelihood function does not produce consistent estimators of the model parameters. What about methods other than ML that are based on the approximations? For example, based on FOA discussed above, one may use a similar IEE approach to that discussed in Section 3.2.2. Another method that has also received considerable attention is called two-stage (TS) estimatora. See, for example, [4] (Section 8.5). The method is developed based on a special form of the function f_i in (5.1), that is,

$$f_i(X_i, \beta, \alpha_i) = [f(x_{1,ij}, X_{2,i}\beta + \alpha_i)]_{1 \le j \le n_i}. \tag{5.16}$$

Then, in the first stage, one estimates each individual mixed effect,

$$\eta_i = X_{2,i}\beta + \alpha_i, \tag{5.17}$$

via nonlinear least squares, that is, by minimizing $\sum_{j=1}^{n_i}(y_{ij} - f_{ij})^2$, where f_{ij} is the jth component of the f_i given by (5.16), considered as a function of η_i. In the second stage, one estimates β by fitting (5.17), with the η_i's being the estimates from the first stage, using the I-WLS method described in Example 3.2. Note that, if $\eta_i, 1 \le i \le m$ are treated as independent observations, one has $E(\eta_i) = X_i\beta$ and $\text{var}(\eta_i) = G$. Also recall that, in the LB approximation, a PNLS method is used to estimate the parameters. The following example ([4], Section 8.9), however, shows that all three methods, the IEE method based on FOA (FOA/IEE), TS and PNLS based on LB approximation (PNLS/LB), produce inconsistent estimators of model parameters.

Example 5.2. Consider another simple case of NLMEM that can be expressed as

$$y_{ij} = e^{\beta + \alpha_i} + \varepsilon_{ij}, \tag{5.18}$$

$i = 1, \ldots, m, j = 1, \ldots, n$, where $\alpha_i, \varepsilon_{ij}$ are independent with $\alpha_i \sim N(0, \sigma^2\gamma)$ and $\varepsilon_{ij} \sim N(0, \sigma^2)$. It also assumed, for simplicity, that $\sigma^2 > 0$ and $\gamma > 0$ are known, so that β is the only unknown parameter. The FOA method leads to the approximate model, $y_{ij} = e^\beta + e^\beta \alpha_i + \varepsilon_{ij}$. If this were the true model, the covariance matrix of the data would be block diagonal with the diagonal blocks equal to $V = V(\beta) = \sigma^2(I_n + \gamma e^{2\beta}J_n)$ (see Example 2.8 for notation). Let $\hat{\beta}_0$ be an initial estimator of β. Then, an initial estimator of V is $\hat{V}_0 = V(\hat{\beta}_0)$. Using the I-WLS procedure, which is a special case of IEE discussed in Section 3.2.2, an updated estimator of β is obtained by

$$\min_\beta \sum_{i=1}^m (y_i - e^\beta 1_n)'\hat{V}_0^{-1}(y_i - e^\beta 1_n),$$

where $y_i = (y_{ij})_{1 \le j \le n}$. The solution is given by

$$\hat{\beta}_{\text{foa}} = \log\left(\frac{\sum_{i=1}^m 1_n'\hat{V}_0^{-1}y_i}{m1_n'\hat{V}_0^{-1}1_n}\right).$$

Using the expression (e.g., [21], p. 443)

$$V^{-1} = I_n - \frac{\gamma e^{2\beta}}{1 + n\gamma e^{2\beta}}J_n.$$

it is easy to show that

$$1_n'\hat{V}_0^{-1}1_n = (1 + \gamma e^{\hat{\beta}_0}n)^{-1}n,$$
$$1_n'\hat{V}_0^{-1}1_n = (1 + \gamma e^{\hat{\beta}_0}n)^{-1}y_{i\cdot},$$

where $y_{i\cdot} = \sum_{j=1}^{n} y_{ij}$. It follows that

$$\hat{\beta}_{\text{foa}} = \log(\bar{y}_{\cdot\cdot}), \tag{5.19}$$

where $\bar{y}_{\cdot\cdot} = (mn)^{-1} \sum_{i=1}^{m} y_{i\cdot}$. It turns out that the updated estimator of β does not depend on the previous estimator; therefore, the I-WLS converges in one iteration and the limiting estimator is given by (5.19). Question is: Is this a good estimator?

First of all, there is a nonzero probability that the right side of (5.19) is not well defined, if $\bar{y}_{\cdot\cdot}$ is negative. One could argue that, when the sample size is large, such a probability is vanishing, which is true. So, let us look at the asymptotic behavior of $\hat{\beta}_{\text{foa}}$. It can be shown that as $m \to \infty$, one has $\bar{y}_{\cdot\cdot} \xrightarrow{P} \exp(\beta + \gamma\sigma^2/2)$, and this holds regardless of whether $n \to \infty$ or not. It follows that $\hat{\beta}_{\text{foa}} \xrightarrow{P} \beta + \gamma\sigma^2/2$. Therefore, $\hat{\beta}_{\text{foa}}$ is an inconsistent estimator not only under the standard asymptotic configuration, (a), but also under the stronger asymptotic configuration, (c). By a similar argument, it can be shown that $\hat{\beta}_{\text{foa}}$ is also inconsistent under the asymptotic configuration (b).

Now let us consider TS for this example. The nonlinear least squares leads to $\hat{\eta}_i = \log(\bar{y}_{i\cdot}), 1 \le i \le m$. In this case, fitting (5.17) via I-WLS is the same as the least squares, leading to the estimator

$$\hat{\beta}_{\text{ts}} = \frac{1}{m} \sum_{i=1}^{m} \log(\bar{y}_{i\cdot}). \tag{5.20}$$

It should be noted that, when n is fixed, the estimator (5.20) is not well defined with probability tending to one, as m goes to infinity. To see this, note that

$$P(\bar{y}_{i\cdot} > 0, 1 \le i \le m) = P(\bar{y}_{1\cdot} > 0)^m \to 0, \tag{5.21}$$

as $m \to \infty$, because $P(\bar{y}_{1\cdot} > 0) < 1$. Thus, under the asymptotic configuration (a), $\hat{\beta}_{\text{ts}}$ is not even well defined, asymptotically, not to mention about its consistency. Now suppose that $n \to \infty$, in additional to $m \to \infty$. It is easy to show that

$$P(\bar{y}_{1\cdot} > 0) = E\left\{ \Phi\left(\frac{\sqrt{n}e^{\beta + \alpha_1}}{\sigma} \right) \right\},$$

where $\alpha_1 \sim N(0, \gamma\sigma^2)$, and $\Phi(\cdot)$ is the cdf of $N(0,1)$. Thus, from the left side of (5.20), it can be seen that if n goes to infinity sufficiently fast, the right side of (5.20) will go to 1 instead of 0. Under such a scenario, one can look at the consistency issue more seriously. By Taylor series expansion, we have

$$\log(\bar{y}_{i\cdot}) - \beta - \alpha_i = \log\left(e^{\beta + \alpha_i} + \bar{\varepsilon}_{i\cdot}\right) - \log\left(e^{\beta + \alpha_i}\right) = e^{-\beta - \alpha_i}\bar{\varepsilon}_{i\cdot} - \frac{\bar{\varepsilon}_{i\cdot}^2}{a_i^2}, \tag{5.22}$$

where a_i satisfies $|a_i - e^{\beta + \alpha_i}| \le |\bar{\varepsilon}_{i\cdot}|$. Note that

$$P\left(|\bar{\varepsilon}_{i\cdot}| < \frac{1}{2}e^{\beta + \alpha_i}, 1 \le i \le m \right) = \left\{ P\left(|\bar{\varepsilon}_{1\cdot}| < \frac{1}{2}e^{\beta + \alpha_1} \right) \right\}^m$$

$$= \left[2E\left\{ \Phi\left(\frac{\sqrt{n}}{2\sigma}e^{\beta + \sigma\xi} \right) \right\} - 1 \right]^m,$$

where $\xi \sim N(0,1)$. Thus, if we let $\mathscr{A}_{m,n,1} = \{\bar{y}_{i\cdot} > 0, 1 \leq i \leq m\}$, $\mathscr{A}_{m,n,2} = \{|\bar{\varepsilon}_{i\cdot}| < e^{\beta+\alpha_i}/2, 1 \leq i \leq m\}$, $\mathscr{A}_{m,n} = \cap_{j=1}^2 \mathscr{A}_{m,n,j}$, we have $P(\mathscr{A}_{m,n}) \to 1$ if n goes to infinity sufficiently fast, as $m \to \infty$. On the other hand, on $\mathscr{A}_{m,n}$, we have, by (5.22),

$$
\begin{aligned}
|\hat{\beta}_{ts} - \beta - \bar{\alpha}_\cdot| &= \left| \frac{1}{m} \sum_{i=1}^m \{\log(\bar{y}_{i\cdot}) - \beta - \alpha_i\} \right| \\
&= e^{-\beta} \left| \frac{1}{m} \sum_{i=1}^m e^{-\alpha_i} \bar{\varepsilon}_{i\cdot} \right| + \frac{1}{m} \sum_{i=1}^m \frac{\bar{\varepsilon}_{i\cdot}^2}{a_i^2} \\
&\leq e^{-\beta} \left| \frac{1}{m} \sum_{i=1}^m e^{-\alpha_i} \bar{\varepsilon}_{i\cdot} \right| + \frac{4e^{-2\beta}}{m} \sum_{i=1}^m e^{-2\alpha_i} \bar{\varepsilon}_{i\cdot}^2. \quad (5.23)
\end{aligned}
$$

Also note that $E(m^{-1} \sum_{i=1}^m e^{-\alpha_i} \bar{\varepsilon}_{i\cdot})^2 = \sigma^2 e^{2\sigma^2}/mn$ and $E(m^{-1} \sum_{i=1}^m e^{-2\alpha_i} \bar{\varepsilon}_{i\cdot}^2) = \sigma^2 e^{2\sigma^2}/n$. Thus, as $n \to \infty$, the right side of (5.23) converges to 0 in L^1, hence in probability (e.g., [1], p. 31). This, combined with the fact that $\bar{\alpha}_\cdot = m^{-1} \sum_{i=1}^m \alpha_i \xrightarrow{P} 0$, as $m \to \infty$, by the LNN, implies that $\hat{\beta}_{ts} \xrightarrow{P} \beta$.

In conclusion, the TS estimator is inconsistent under asymptotic configuration (a), but if n goes to infinity sufficiently fast as $m \to \infty$, the TS estimator is consistent.

Finally, we consider PNLS/LB for this example. Because σ^2 and γ are known, the procedure amounts to estimate β and $\alpha_i, 1 \leq i \leq m$ via

$$
\min_{\beta, \alpha_i, 1 \leq i \leq m} \sum_{i=1}^m \left\{ \sum_{j=1}^n (y_{ij} - e^{\beta+\alpha_i})^2 + \frac{\alpha_i^2}{\gamma} \right\}.
$$

This leads to the following estimating equations:

$$
e^{2(\beta+\alpha_i)} - \bar{y}_{i\cdot} e^{\beta+\alpha_i} + \frac{\alpha_i}{n\gamma} = 0, \quad 1 \leq i \leq m,
$$

$$
\sum_{i=1}^m \left\{ e^{2(\beta+\alpha_i)} - \bar{y}_{i\cdot} e^{\beta+\alpha_i} \right\} = 0. \quad (5.24)
$$

Let $\tilde{\alpha}_i = \phi(\beta, \bar{y}_{i\cdot}, n\gamma)$ denote the solution of the ith equation in (5.24) in terms of α_i as a function of β, where $\phi(\beta, u, v)$ denotes the solution x to the following equation:

$$
e^{2(\beta+x)} - ue^{\beta+x} + \frac{x}{v} = 0. \quad (5.25)
$$

Note that the solution to (5.25) may not be unique. Thus, [4] (p. 464) defines $\tilde{\alpha}_i$ as the limit of convergence of the Newton-Raphson procedure for solving the equation, starting from $\alpha_i^{(0)} = \log(\bar{y}_{i\cdot}) - \beta$. Then, it is easy to derive that the LB estimator of β, $\hat{\beta}_{lb}$, is the solution to the following equation:

$$
\sum_{i=1}^m \phi(\beta, \bar{y}_{i\cdot}, n\gamma) = 0. \quad (5.26)
$$

Using a similar argument as that in Example 5.1, it can be shown that $\hat{\beta}_{1b}$ is inconsistent due to the fact that

$$E\{\phi(\beta,\bar{y}_{1\cdot},n\gamma)\} \;\neq\; 0 \qquad\qquad (5.27)$$

(see Fig. 8.3 of [4]), if n is fixed and $m \to \infty$, that is, under the standard asymptotic configuration (a). However, if both $m,n \to \infty$, that is, under the asymptotic configuration (c), the LB estimator is consistent. This latter fact is due to a general result, proved in [4] (Sections 8.10 and 8.18), as follows.

Theorem 5.1 (Demidenko [4]). Under the asymptotic configuration (c) and mild regularity conditions, the MLE, TS and LB estimators are consistent and have the same asymptotic normal distribution in the sense that $\sqrt{m}(\hat{\beta} - \beta) \xrightarrow{d} N(0,\Sigma^{-1})$, where $\hat{\beta}$ is either the ML, or the TS, or the LB estimator, and

$$\Sigma \;=\; \lim_{m\to\infty} \frac{1}{m}\sum_{i=1}^{m} X_i' G^{-1} X_i,$$

assuming that the latter limit exists.

Again, recall that among the methods that have so far been examined, including ML, FOA, TS, LB/PQL, and Laplace approximation, ML is the only method that produces consistent estimators of the NLMEM parameters under the standard asymptotic configuration (a). However, there is, indeed, an alternative to ML that also produces consistent estimators under the asymptotic configuration (a). This is the IEE method, introduced in Section 3.2.2, but without using any of the approximations mentioned above in the first place. In fact, the method results in estimators that are not only consistent but also optimal in a certain sense. The set-up for IEE is similar to that described in Section 3.2. In particular, the GEE equation, assuming that $V_i = \text{Var}(y_i|X_i), 1 \le i \le m$ are known, is given by (3.19), where μ_i and $\dot{\mu}_i$ both depend on the parameter vector ψ defined above (5.7). If $V_i, 1 \le i \le m$ were known, one could estimate ψ by solving the GEE. The resulting estimator, $\tilde{\psi}$, is optimal in the sense of Section 3.2.1. On the other hand, if ψ were known, the V_i's could be estimated via MoM. This was described in Section 3.2.2, but there is an alternative approach, suggested by [225]. Recall the v_{ijk}'s defined above (3.26), and D defined below (3.25). The following assumption is made.

A1. For any $(j,k) \in D$, the number of different v_{ijk}'s is bounded, that is, there is a set of real numbers, $\mathscr{V}_{jk} = \{v((j,k,l), 1 \le l \le L_{jk}\}$, where L_{jk} is bounded, such that $v_{ijk} \in \mathscr{V}_{jk}$ for any $1 \le i \le m$ and $j,k \in J_i$.

[225] provides an equivalent condition to A1, as shown by the following lemma.

Lemma 5.1. Suppose that $|J_i|, 1 \le i \le m$ are bounded. Then, A1 holds if and only if the number of different V_i's is bounded, that is, there is a set of covariance matrices, $\mathscr{V} = \{V^{(1)},\ldots,V^{(L)}\}$, where L is bounded, such that $V_i \in \mathscr{V}, 1 \le i \le m$.

Based on the latter result, [225] defines a different MoM estimator of V_i by estimating $V^{(l)}, 1 \le l \le L$. Namely, let

$$\tilde{V}^{(l)} \;=\; \frac{1}{n_l}\sum_{i\in I_l}(y_i - \mu_i)(y_i - \mu_i)', \qquad\qquad (5.28)$$

where I_l is the set of indexes $1 \leq i \leq m$ such that $V_i = V^{(l)}$, and $n_l = |I_l|$, the cardinality of I_l. Then, define $\tilde{V}_i = \tilde{V}^{(l)}$ if $i \in I_l, 1 \leq i \leq m$. Now denote the MoM estimator of V_i defined in Section 3.2.2 by $\hat{V}_i$, $1 \leq i \leq m$. Lee [225] shows that, under certain conditions, these two MoM estimators of V_i are the same, as stated by the following lemma. Again, we continue to use the notation in Section 3.2.2.

Lemma 5.2. Suppose that $J = \{1, \ldots, b\} = \cup_{r=1}^{s} J(r)$, where $J(1), \ldots, J(s)$ are disjoint such that for each $1 \leq i \leq m$, there is $1 \leq r \leq s$ such that $J_i = J(r)$. Then, we have $\hat{V}_i = \tilde{V}_i, 1 \leq i \leq m$.

The same IEE procedure as described in Section 3.2.2 is then applied to NLMEM. Furthermore, [225] established similar results on linear convergence and asymptotic properties of the IEE for NLMEM, as follows.

Theorem 5.2 (Lee [225]). Under regularity conditions, the following hold:
(I) (Linear convergence) the results of Theorem 3.2 hold with β replaced by ψ.
(II) (Consistency) define the IEE estimator (IEEE) of ψ as the limit of the IEE convergence in (I), denoted by $\hat{\psi}$. Then, $\hat{\psi} \overset{P}{\longrightarrow} \psi$ as $m \to \infty$.
(III) (Asymptotic efficiency) let $\tilde{\psi}$ denote the solution to the GEE equation, where $V_i, 1 \leq i \leq m$ are the true covariance matrices. Then, we have $\sqrt{m}(\hat{\psi} - \tilde{\psi}) \overset{P}{\longrightarrow} 0$ as $m \to \infty$. Therefore, asymptotically, the IEEE is as efficient as the GEE estimator with the true V_i's.

5.2 Preliminaries

There are two ways to entertain this section. The first way is simply to go through it before the next section to get familiar with some of the concepts and results needed for the following sections. There is another way—and this may work more efficiently especially if the reader is relatively well trained in asymptotics, or familair with the basic elements of empirical processes. With the second approach, the reader would skip the current section at this point, and only to come back to find the relevant part(s) needed to review, when necessary. A comprehensive treatment of these materials can be found in, for example, [226].

We begin by reviewing some of the basic facts on convergence in distribution in a functional space. Consider the space of all continuous functions on $[0, 1]$, denoted by C. We can define a distance between two points, x and y in C (note that here x and y denote two continuous functions on $[0, 1]$) by

$$\rho(x, y) = \sup_{t \in [0,1]} |x(t) - y(t)|. \tag{5.29}$$

The space C, equipped with the distance ρ is a metric space, which means that ρ satisfies the following basic requirements for all $x, y, z \in C$ (to qualify as a distance, or metric):

    1. (nonnegativity) $\rho(x, y) \geq 0$;
    2. (symmetry) $\rho(x, y) = \rho(y, x)$;
    3. (triangle inequality) $\rho(x, z) \leq \rho(x, y) + \rho(y, z)$;
    4. (identity of points) $\rho(x, y) = 0$ if and only if $x = y$.
It is easy to show that the distance defined by (5.29) satisfies requirements 1–4.

We can talk about weak convergence of probability measures on the measurable space $(C, \mathscr{B})$, where $\mathscr{B}$ is the class of Borel sets in C, which is a σ-field (e.g., [1], Section A.2). A sequence of probability measures P_n converges weakly to a probability measure P, denoted by $P_n \xrightarrow{w} P$, if $P_n(B) \to P(B)$ as $n \to \infty$ for any P-continuity set B. The latter means that $P(\partial B) = 0$, where ∂B denotes the boundary of B (i.e., the set of points that are limits of sequences of points in B as well as limits of sequences of points outside B). Equivalently, $P_n \xrightarrow{w} P$ if $\int_C f \, dP_n \to \int_C f \, dP$ for all bounded, uniformly continous function f on C. The definition may be extended to situations where the function f may not be integrable, or even measuable, by using the concept of outer expectation (e.g., [227], p. 44). There is a connection between the weak convergence in C and that of all finite-dimensional distributions. Let $t_1, \ldots, t_k$ be any set of distinct points in $[0, 1]$. Let P be a probability measure on $(C, \mathscr{B})$. Then the induced probability measure

$$P\pi_{t_1, \ldots, t_k}^{-1}(A) \;=\; P\{[x(t_1), \ldots, x(t_k)] \in A\}$$

for any Borel set A in R^k is called a finite-dimensional distribution. Here, $\pi_{t_1, \ldots, t_k}$ denotes the projection that carries the point $x \in C$ to the point $[x(t_1), \ldots, x(t_k)] \in R^k$. It turns out that weak convergence in C is a little more than weak convergence of all finite-dimensional distributions; that is, $P_n \xrightarrow{w} P$ if and only if $P_n\pi_{t_1, \ldots, t_k}^{-1} \xrightarrow{w} P\pi_{t_1, \ldots, t_k}^{-1}$ for any $k \geq 1$ and any distinct points $t_1, \ldots, t_k \in [0, 1]$ plus that the sequence $P_n, n \geq 1$ is *tight*. A family of $\mathscr{P}$ of probability measures on $(C, \mathscr{B})$ is tight if for every $\varepsilon > 0$, there is a compact subset of B of C such that $P(B) > 1 - \varepsilon$ for all $P \in \mathscr{P}$. A well-known result associated with the latter concept is a probability version of the Arzelá–Ascoli theorem. The sequence $P_n, n \geq 1$, is tight in C if and only if the following two conditions hold: (i) For any $\eta > 0$, there exists $M > 0$ such that $P_n(|x(0)| > M) \leq \eta$, $n \geq 1$; and (ii) for any $\varepsilon, \eta > 0$, there exist $0 < \delta < 1$ and $N \geq 1$ such that

$$P_n \left\{ \sup_{|s-t|<\delta} |x(s) - x(t)| \geq \varepsilon \right\} \;\leq\; \eta$$

for all $n \geq N$. The concept of random variables can now be extended to C-valued random variables (i.e., random variables whose values are continuous functions on $[0, 1]$). Such a random variable is often called a stochastic process, denoted by $\xi = (\xi_t, 0 \leq t \leq 1)$, although continuity is not required for the definition. A sequence of C-valued random variables $\xi_n, n \geq 1$, converges in distribution to a C-valued random variable ξ, denoted by $\xi_n \xrightarrow{d} \xi$, if $P\xi_n^{-1} \xrightarrow{w} P\xi^{-1}$, where $P\xi_n^{-1}$ is the induced probability measures defined by

$$P\xi_n^{-1}(B) \;=\; P(\xi_n \in B)$$

for $B \in \mathscr{B}$, and $P\xi^{-1}$ is defined similarly. Note that, here, we focus on functions on $[0, 1]$, but this is without loss of generality due to the following fact.

Lemma 5.3 (The inverse transformation). Let $\xi \sim \text{Uniform}(0, 1)$ and F be a cumulative distribution function (cdf). Define

$$F^{-1}(t) \;=\; \inf\{x : F(x) \geq t\}, \quad 0 < t < 1. \tag{5.30}$$

Then $X = F^{-1}(\xi) \sim F$. In fact, $X \leq x$ if and only if $\xi \leq F(x)$.

One particular stochastic process is called *Wiener process*, or *Brownian motion*. A probability measure W on $(C, \mathcal{B})$ is called a Wiener measure if (i) for each $t \in [0,1]$, the random variable $x(t) \sim N(0,t)$ under W, that is,

$$W\{x(t) \leq \lambda\} = \frac{1}{\sqrt{2\pi t}} \int_{-\infty}^{\lambda} e^{-u^2/2t} \, du;$$

and (ii) for any $0 \leq t_0 \leq t_1 \leq \cdots \leq t_k \leq 1$, the random variables

$$x(t_1) - x(t_0), x(t_2) - x(t_1), \ldots, x(t_k) - x(t_{k-1})$$

are independent under W. Here, the random variable $x(0)$ is understood as equal to zero with probability 1 under W [i.e., $W\{x(0) = 0\} = 1$]. A C-valued random variable, denoted by $W = \{W(t), 0 \leq t \leq 1\}$, is called a Wiener process if it has the Wiener measure as its distribution [i.e., $P(W \in B) = W(B)$ for any $B \in \mathcal{B}$]. The stochastic process $U(t) = W(t) - tW(1)$ is called a *Brownian bridge*.

A stochastic process $\{x(t)\}$ is called a Gaussian process if for any $t_1 < \cdots < t_k$, the joint distribution of $x(t_1), \ldots, x(t_k)$ is (multivariate) normal. Note that Wiener process is a Gaussian process such that for any $0 \leq t_1 < \cdots < t_k \leq 1$, the random variables $W(t_2) - W(t_1), \ldots, W(t_k) - W(t_{k-1})$ are independent and distributed as $N(0, t_2 - t_1), \ldots, N(0, t_k - t_{k-1})$, respectively. It follows that a Brownian bridge is also a Gaussian process with $E\{U(t)\} = 0$ and $\text{cov}\{U(s), U(t)\} = s \wedge t - st, s, t \in [0,1]$.

Another functional space, denoted by $\mathcal{D}$, is the space of right-continuous functions on $[0,1]$ that possess left-limit at each point.

In parametric inference, such as in the previous chapters, we consider asymptotic distribution of an estimator, $\hat{\theta}$, of an unknown parameter $\theta \in R^d$ for some positive integer d. In the following sections, we shall consider estimation of θ jointly with unknown function, f. A function f is said to have bounded variation if for any $x_0 < x_1 < \cdots < x_k$ we have $\sum_{j=1}^{k} |f(x_j) - f(x_{j-1})| \leq B$ for some constant B. Let $\mathcal{BV}[0,1]$ denote the space of bounded variation functions on $[0,1]$. Let $G_{d,p}$ denote the product space of $\mathcal{BV}[0,1]$ and R^d, that is, $\{(f,\theta) : f \in \mathcal{BV}[0,1], \theta \in R^k\}$, such that

$$\|(f,\theta)\|_G \equiv \|f\|_{\text{bv}} + |\theta| \leq p, \tag{5.31}$$

where $\|f\|_{bv} = |f(0)| + \sup_{0=x_0<x_1<\cdots<x_k=1} \sum_{j=1}^{k} |f(x_j) - f(x_{j-1})|$ (and $|\theta|$ is the Euclidean norm). Here, p is supposed to be ≥ 0; in case $p = \infty$, the $\leq$ in (5.31) is replaced by $<$. Let $l^\infty(G_{d,p})$ denote the space of bounded real-valued functions on $G_{d,p}$ under the supremum norm, $\|g\| = \sup_{(f,\theta) \in G_{d,p}} |g(f,\theta)|$, $g \in l^\infty(G_{d,p})$. A map T from $l^\infty(G_{d,p})$ to $l^\infty(G_{d,p})$ is called an operator. If S, T are operators, the product operator, ST, is defined as $ST(g) = S(T(g)), g \in l^\infty(G_{d,p})$. An operator L is said to be *linear* if for any $g_1, g_2 \in l^\infty(G_{d,p})$ and $a_1, a_2 \in R$, one has

$$L(a_1 g_1 + a_2 g_2) = a_1 L(g_1) + a_2 L(g_2);$$

the operator is said to be *continuous* if $\|g - g_0\|_G \to 0$ implies $\|L(g) - L(g_0)\| \to 0$ for every $g_0 \in G_{d,p}$.

Taylor series expansion is a standard method in asymptotic analysis to obtain, for example, asymptotic distribution of an estimator (e.g., [1], Section 4.2). The following result (e.g., [228], Theorem 2) may be viewed as an extension of the Taylor series method to estimation in a functional space. Let ψ be a parameter whose value is in $\Psi \subset l^\infty(G_{d,p})$, and $\hat{\psi}$ an estimator of ψ based on i.i.d. observations $Y_1, \ldots, Y_n$. Let $S_n, n = 1, 2, \ldots$ and S be a sequence of operators: $\Psi \to l^\infty(G_{d,p})$. For any $B \subset \Psi$, let $\mathscr{L}(B)$ denote the set of all finite linear combinations of elements of B.

Theorem 5.3. Suppose that the following hold:

(a) (asymptitic distribution of score function)

$$\sqrt{n}\{S_n(\psi) - S(\psi)\} \xrightarrow{\text{d}} \mathscr{W},$$

where $\mathscr{W}$ is a tight Gaussian process on $l^\infty(G_{d,p})$;

(b) (Fréchet differentiability)

$$S(\hat{\psi}) - S(\psi) = -\dot{S}_\psi(\hat{\psi} - \psi) + o_P(n^{-1/2} \vee \|\hat{\psi} - \psi\|),$$

where $\dot{S}_\psi$ is a continuous linear operator on $\mathscr{L}\{\tilde{\psi} - \psi : \tilde{\psi} \in \Psi\}$;

(c) (invertibility) $\dot{S}_\psi$ is continuously invertible on its range;

(d) (approximation condition)

$$\|(S_n - S)(\hat{\psi}) - (S_n - S)(\psi)\| = o_P(n^{-1/2} \vee \|\hat{\psi} - \psi\|).$$

Then, we have $\sqrt{n}(\hat{\psi} - \psi) \xrightarrow{\text{d}} \dot{S}_\psi^{-1} \mathscr{W}$, where $\dot{S}_\psi^{-1}$ is the inverse operator of $\dot{S}_\psi$.

A stochastic process $P(t), t \geq 0$ is called a Poisson process if it satisfies the following: (i) $P(0) = 0$; (ii) for any $0 \leq s < t$, $P(t) - P(s)$ has a Poisson distribution with mean $\lambda(t - s)$, where λ is a positive constant; and (iii) the process has independent increments; that is, for any $n > 1$ and $0 \leq t_0 < t_1 < \cdots < t_n$, the random variables $P(t_j) - P(t_{j-1})$, $j = 1, \ldots, n$, are independent. The constant λ is called the strength, or intensity, of the Poisson process. Poisson process is a special case of counting process. The latter means a process $N(t), t \geq 0$, that represents the number of events that have occurred up to time t. Obviously, a counting process must satisfy the following: (a) the values of $N(t)$ are nonnegative integers; the function $N(t)$ is right-continuous with left-hand limit as a function of t; (c) $N(s) \leq N(t)$ if $s < t$; and (d) for $s < t$, $N(t) - N(s)$ equals the number of events that have occurred in the interval $(s,t]$. The intensity process of a counting process, $N(t)$, is defined as

$$\lambda(t) = \lim_{\Delta t \searrow 0} \frac{P\{N(t + \Delta t) - N(t) = 1 | N(s), s < t\}}{\Delta t}. \tag{5.32}$$

In survival analysis, the data typically involve survival time, denoted by T, which is assumed to be a random variable having a pdf f and cdf F. The hazard function, $h(t)$, measures the instantaneous risk in that $h(t)\Delta t$ is the probability that the individual will die in the time interval $(t, t + \Delta t]$ given that it has survived up to time t. This can be expressed as

$$1 - h(t)\Delta t = P(\text{survical up to } t + \Delta t | \text{survival up to } t)$$

$$= \quad \frac{P(\text{survical up to } t + \Delta t)}{P(\text{survival up to } t)}$$

$$= \quad \frac{P(T > t + \Delta t)}{P(T > t)}$$

$$= \quad \frac{1 - F(t + \Delta t)}{1 - F(t)}$$

$$= \quad 1 - \frac{F(t + \Delta t) - F(t)}{1 - F(t)}. \tag{5.33}$$

It follows, by canceling the 1 on both sides of (5.33), then dividing both sides by Δt and letting $\Delta \to 0$, that the hazard function can be expressed as

$$h(t) \quad = \quad \frac{f(t)}{1 - F(t)}. \tag{5.34}$$

Survival data often involve censoring, including right, left and interval censoring. For example, right censoring occurs when the survival time is not observed by the upper end of the follow-up period. If $X_1, \ldots, X_n$ are survival times and $C_1, \ldots, C_n$ are the corresponding censoring times. The lifetime of individual i is known to be X_i if and only if $X_i < C_i$. If $X_i > C_i$, the event time will be censored at C_i. Thus, what one observes is a pair of random variables, (T_i, δ_i), for individual i, where $T_i = X_i \wedge C_i$ $[a \wedge b = \min(a, b)]$ and $\delta_i = 1_{(X_i < C_i)}$.

One approach to inference with survival data is through a counting processes. Note that the sample paths of the counting process, $N(t)$, are nondecreasing, right-continuous step functions that jump whenever an event, or events, occur. The pair (T_i, δ_i) is now replaced with a pair of processes, $N_i(t), Y_i(t), i = 1, \ldots, n$, where $N_i(t) = \#$ of events observed in $[0, t]$ for individual i; $Y_i(t) = 1$, if the ith individual is at risk at time t, and 0 otherwise. The process $Y_i(t)$ is assumed to be predictable, that is, its value at time t is known given the history before t. For example, in the case of right-censored data, one has $N_i(t) = 1_{(T_i \leq t, \delta_i = 1)}$ and $Y_i(t) = 1_{(T_i \geq t)}$. For notation simplicity, we drop the subscript i when focusing on a specific individual. It can be shown that (e.g., [229])

$$E\{N(t) | \mathscr{F}_{t-}\} \quad = \quad \int_0^t h(s) Y(s) ds, \tag{5.35}$$

where $\mathscr{F}_{t-} = \sigma\{N(s), Y(s), 0 \leq s < t\}$, and $h(t)$ isthe hazard function. The process on the right side of (5.35), denoted by $\Lambda(t)$, is called the cumulative intensity process. The intensity process is defined as

$$\lambda(t) = \dot{\Lambda}(t) = h(t) Y(t). \tag{5.36}$$

A standard approach to inference with survival data is *partial likelihood*, proposed by Cox [230]. A motivation for the method is that, in some cases, the (full) likelihood function may involve distributions that one does not wish to specify. For example, such distribution may involve nuisance parameters, that is, parameters that are not of primary interest. Suppose that the data consist of pairs $(y_1, s_1), \ldots, (y_n, s_n)$. The full likelihood function can be expressed as

$$
\begin{aligned}
L = \; & f(y_1)f(s_1|y_1)f(y_2|y_1,s_1)f(s_2|y_2,y_1,s_1)\cdots \\
& f(y_n|y_{n-1},s_{n-1},\ldots,y_1,s_1)f(s_n|y_n,y_{n-1},s_{n-1},\ldots,y_1,s_1), \quad (5.37)
\end{aligned}
$$

where, with a little abuse of the notation, $f(y_1)$ denotes the pdf (or pmf) of y_1, $f(s_1|y_1)$ the conditional pdf (or pmf) of s_1 given y_1, etc., all of which depend on unknown parameters. The partial likelihood consists of the product of every other terms in (5.37), starting with $f(y_1)$, that is,

$$
L_P(\theta) = \prod_{i=1}^{n} f(y_i|y_{i-1},s_{i-1},\ldots,y_1,s_1) \quad (5.38)
$$

(the conditioning part is dropped if $i = 1$), where θ is the vector of parameters involved in L_P, which is of primary interest. Note that focusing on L_P does not mean that the rest of the terms in (5.37) do not involve θ; otherwise, inference about θ based on L_P would be equivalent to that based on L. Nevertheless, large-sample properties such as consistency and asymptotic normality of the maximum partial likelihood estimators have been established (e.g., [231]). The main advantage of L_P is that the remaining part of the likelihood function, L/L_P, may involve some unknown distributions, or nuisance parameters, which one may wish to avoid dealing with.

5.3 Frailty models

In survival analysis, frailty models have been used to model clustered data arisen from such areas as genetic and familial studies, multi-center clinical trials, group-randomized trials, as well as longitudinal studies. The frailty is usually modeled as an unobserved random variable acting multiplicatively on the baseline hazard function. So, if the hazard function for an individual with frailty 1 is $h(t)$, the hazard function for an individual with frailty z will be $zh(t)$. What is important is that different individuals can share the same frailty, and thus have correlations through the shared frailty. This is called a shared frailty model.

For the most part, there are two classes of frailty models. The first class is called gamma-frailty models, which assumes that the frailty has a gamma distribution. The consideration is mostly for mathematical convenience (e.g., [232], pp. 183–184). Under the gamma frailty model, it is assumed that one observes an aggregated counting process, $N_i(t)$, for group i, as well as a predictable process, $Y_i(t)$, for $t \in T_i \subset \{1,\ldots,T\}$, where $1 \le i \le m$ and m is the number of groups. Suppose that the intensity process for group i can be expressed as

$$
h_i(t) = z_i h(t) Y_i(t), \quad (5.39)
$$

where $h(t)$ is an unknown baseline hazard function, and z_i is a random effect that has a gamma distribution with mean equal to 1 and variance equal to θ.

The second class is called proportional hazard mixed-effects models (PHMMs).

It is assumed that the hazard function for the jth observation of the i cluster, $i = 1,\ldots,m, j = 1,\ldots,n_i$, has the expression

$$h_{ij}(t) = h_0(t)\exp(x'_{ij}\beta + z'_{ij}\alpha_i), \qquad (5.40)$$

where x_{ij}, z_{ij} are known covariate vectors, β is a vector of unknown fixed effects, and α_i is a vector of (unobserved) random effects. It is typically assumed that $\alpha_i, i = 1,\ldots,m$ are independent and distributed as $N(0,G)$, where G is a covariance matrix, which may depend on a vector θ of variance components.

Under both gamma-frailty model and PHMM, it is assumed that data from different clusters, or groups, are independent.

5.3.1 Asymptotic analysis of gamma-frailty models

Asymptotic analysis of gamma-frailty models was considered relatively earlier in the literature, partially due to the mathematical convenience of these models. In a serious of papers, Murphy [233], [228] studied asymptotic behavior of a non-parametric maximum (partial) likelihood estimators of the cumulative hazard function as well as the variance of the gamma-frailty. A statistical model can be formulated using a counting process (see the previous section). Suppose that, for each $1 \leq i \leq m$, there is a multivariate counting process $N_i(t), t \in [0, T]$ whose intensity process is given by (5.39). Suppose that $[z_i, N_i(t), Y_i(t)], i = 1,\ldots,m$ are independent. Nielsen et al. [234] showed that the partial likelihood function can be written as

$$\prod_{i=1}^{m}\left[\prod_t \{z_i h(t)Y_i(t)\}^{\Delta N_i(t)} \exp\left\{-z_i \int_0^T Y_i(t)dH(t)\right\}\right], \qquad (5.41)$$

where $\Delta N_i(t)$ is equal to 1 if there is a jump at t, and 0 otherwise, so that the product in t is a finite product of the terms $z_i h(t)Y_i(t)$ at the observed jumps, and $H(t) = \int_0^t h(u)du$ is the cumulative baseline hazard. Expression (5.41) corresponds to the conditional partial likelihood given the z_i's. It is then multiplied by the joint density of the z_i's and then integrated with respect to the z_i's, yielding the partial likelihood

$$L_P = \prod_{i=1}^{m}\frac{\prod_t[\{1 + \theta N_i(t-)\}Y_i(t)h(t)]^{\Delta N_i(t)}}{\{1 + \theta \int_0^T Y_i(t)dH(t)\}^{\theta^{-1} + N_i(T)}}, \qquad (5.42)$$

where $N_i(t-)$ denote the left-hand limit of $N_i(\cdot)$ at t. It is straightforward to show that the conditional distribution of z_i given all of the N_i, Y_i is a gamma distribution with mean

$$E(z_i|\text{all } N_i, Y_i) = \frac{1 + \theta N_i(T)}{1 + \theta \int_0^T Y_i(t)dH(t)} \qquad (5.43)$$

and variance

$$\text{var}(z_i|\text{all } N_i, Y_i) = \theta\frac{1 + \theta N_i(T)}{\{1 + \theta \int_0^T Y_i(t)dH(t)\}^2}. \qquad (5.44)$$

These expressions, of course, are not computable unless the unknown parameter, θ, and cumulative hazard, H, are known.

In order to make inference about θ and H, it is noted that there is no absolutely continuous function, H, that maximizes L_P, a fact that is well known in nonparametric maximum likelihood (e.g., [235]). One approach to "get around" this problem is to extend the parameter space for H to allow step functions. With such an extension, $\log(L_P)$ can be viewed as a function of θ and the jump sizes of H at the observed times, with the understanding that, when $\theta = 0$, the term

$$\left\{ \frac{1}{\theta} + N_i(T) \right\} \log \left\{ 1 + \theta \int_0^T Y_i(t) dH(t) \right\}$$

is replaced by its right limit at $\theta = 0$, that is, $N_i(T) + \int_0^T Y_i(t) dH(t)$. Let $\hat{\theta}$ be the θ component of the combined θ and jump sizes that maximizes $\log(L_P)$, and $\hat{H}$ be the step function whose jump sizes are the jump-size components of the maximizer. Murphy ([233]) established existence and consistency of the $(\hat{\theta}, \hat{H})$. It is assumed that $(N_i, Y_i), i = 1, \ldots, m$ are i.i.d. copies of (N, Y), such that Y is a.s. left-continuous with right-hand limit and takes on nonnegative integer values, and both N and Y are bounded.

Theorem 5.4. (Existence) if $\max_{1 \le i \le m} N_i(T) > 1$, then $(\hat{\theta}, \hat{H})$ exists and is finite.

Theorem 5.5. (Consistency) suppose that the following hold: (a) Y is a nonincreasing step function such that $P[Y(t) \ge 1]$ has at most a finite number of discontinuities on $t \in [0, T]$; or, alternatively, Y is step function with at most a bounded number of steps and an upper bound on $H(T)$ is known; (b) $\inf_{t \in (0,T)} E\{Y(t)\} > 0$; (c) $P[Y(T_1+) \ge 1] > 0$, where T_1 is the time of the first jump of N, and $Y(t+)$ denote the right-hand limit of Y at t. Then, as $m \to \infty$, we have $\hat{\theta} - \theta \to 0$ a.s. and $\sup_{t \in (0,T]} |\hat{H}(t) - H(t)| \to 0$ a.s.

The main ideas of the proof for Theorem 5.5 are similar to the ones used parametric inference. To summarize, the first step is to show that $\hat{H}$ is bounded. This is shown using the fact that $\hat{H}$ is not arbitrary—it has to maximize the partial likelihood; thus, if $\hat{H}$ is too far away from H, it cannot be the maximizer. Given that $\hat{H}$ is a bounded sequence, Helly's selection theorem (e.g., [236]) can be used to show that for any subsequence of $\hat{H}$, there is a convergent further subsequence. The final step is to show that any convergent subsequence of $(\hat{\theta}, \hat{H})$ must have the same limit, which is (θ, H). This is shown by using the fact that $L_P(\hat{\theta}, \hat{H}) - L_P(\theta, H) \ge 0$, and the limit, say, (θ^*, H^*) must satisfy $L(\theta^*, H^*) - L(\theta, H) = 0$, where L is a limiting version of L_P. One can then use the identifiability of (θ, H) to argue that the above equation must imply $(\theta^*, H^*) = (\theta, H)$.

Murphy [228] further established asymptotic normality of the estimator $(\hat{\theta}, \hat{H})$.

Theorem 5.6. (Asymptotic normality) suppose that the following are satisfied: (a) $\sup_{t \in (0,T]} |\hat{H}(t) - H(t)| \xrightarrow{P} 0$ and $|\hat{\theta} - \theta| \xrightarrow{P} 0$; (b) there exists a constant K such that $\|Y\|_{\mathrm{bv}} \le K$ and $N(T) \le K$ a.s.; (c) $\inf_{t \in (0,T]} E\{Y(t)\} > 0$; and (d) $P[Y(T_1+) \ge 1] > 0$. Then, we have $[\sqrt{n}(\hat{H} - H), \sqrt{n}(\hat{\theta} - \theta)] \xrightarrow{d} \mathscr{G}$ on $l^\infty(G_{d,p})$, where $\mathscr{G}$ is a tight Gaussian process in $l^\infty(G_{d,p})$ with mean zero.

The covariance process of the Gaussian process in Theorem 5.6 is a bit complicated, whose expression is given in [228] (pp. 186–187).

The proof of Theorem 5.6 is essentially a functional version of the standard proof to obtain asymptotic normality of an estimator in the parametric case (e.g., [1], Section 4.7). A version of the functional version can be found in [228] (Theorem 2).

Another problem related to likelihood-based inference, which is often of practical interest, is likelihood-ratio test (LRT). Recall that, here, the parameters involve a finite-dimensional one, θ, and an infinite-dimensional one, H. The null hypothesis of interest is regarding θ. For simplicity but without loss of generality, consider the case of univariate θ, and let the null hypothesis be

$$H_0 : \quad \theta \; = \; \theta_0, \tag{5.45}$$

where θ_0 is a fixed, known parameter. In the case of parametric inference, where H is also finite-dimensional, the asymptotic null distribution of the likelihood-ratio statistic, defined as

$$\log R \; = \; 2\{l(\hat{\psi}) - l_0(\hat{H}_0)\}, \tag{5.46}$$

where $\psi = (\theta, H)$, $l(\cdot)$ is the likelihood function, $\hat{\psi}$ is the MLE of ψ, $l_0(\cdot)$ is the likelihood function under the null hypothesis, and $\hat{H}_0$ is the MLE of H under the null hypothesis, is χ_1^2, where the degree of freedom is the difference between the dimension of ψ, $p = \dim(\psi)$, and the dimension of H, $q = \dim(H)$. Intuitively, the dimension of H has been subtracted from the dimension of ψ, that is,

$$p - q \; = \; 1. \tag{5.47}$$

However, there is a seeming difficulty in directly extending (5.47), because both p and q are ∞ due to the functional parameter H ($\infty - \infty$ is not well defined). To overcome this difficulty, Murphy and Van der Vaart ([237]) proposed the following idea, called semiparametric likelihood-ratio inference. The main idea is the consider a finite-dimensional path that links two infinite-dimensional points, one is $\hat{\psi}$ and the other is $\hat{\psi}_0$, the MLE of ψ under the null hypothesis. The dimension of the path is equal to the dimension of θ, which is 1 in the current case. This way, a nonparametric problem is turned into a parametric problem and the standard asymptotic technique in LRT applies. To give more detail about the latter method, let us first review the basic techniques used in deriving the asymptotic χ^2 null distribution of LRT, in the parametric setting. So, for now, imagine that H is also finite-dimensional.

The basic strategy is two second-order Taylor expansions of both l and l_0, at $\hat{\psi}$ and $\hat{H}_0$ respectively. The first derivatives in those expansions vanish, and algebraic manipulations involving the joint asymptotic (normal) distribution of $\hat{\psi} - \psi$ and $\hat{\psi}_0 - \psi_0$ yield the result, where $\psi_0 = (\theta_0, H)$ and $\hat{\psi}_0 = (\theta_0, \hat{H}_0)$. In fact, one can derive

$$\hat{\psi} - \hat{\psi}_0 = \left(\begin{array}{c} \hat{\theta} - \theta_0 \\ \hat{H} - \hat{H}_0 \end{array} \right) = \left(\begin{array}{c} 1 \\ -I_{HH}^{-1}I_{H\theta_0} + \varepsilon \end{array} \right) (\hat{\theta} - \theta_0), \tag{5.48}$$

where $I_{HH}, I_{H\theta_0}$ are the partitioned matrices in the Fisher information matrix

$$I_{\psi_0} \; = \; \left(\begin{array}{cc} I_{\theta_0\theta_0} & I_{\theta_0 H} \\ I_{H\theta_0} & I_{HH} \end{array} \right), \tag{5.49}$$

and $\varepsilon = o_P(1)$. From (5.48), one has $\hat{H} - \hat{H}_0 = (-I_{HH}^{-1}I_{H\theta_0} + \varepsilon)(\hat{\theta} - \theta_0)$. So, if one ignores ε, one has, approximately, $\hat{H}_0 = \hat{H} + I_{HH}^{-1}I_{H\theta_0}(\hat{\theta} - \theta_0)$. This motivates defining the following curve:

$$\hat{\psi}(t) = [t, \hat{H} + I_{HH}^{-1}I_{H\theta_0}(\hat{\theta} - t)]. \tag{5.50}$$

It is clear that $\hat{\psi}(\hat{\theta}) = (\hat{\theta}, \hat{H}) = \hat{\psi}$, and $\hat{\psi}(\theta_0) = (\theta_0, \hat{H}_0) = \hat{\psi}_0$, according to the above derivation. Thus, the curve (5.50) connects the two points $\hat{\psi}$ and $\hat{\psi}_0$.

The fact is, the kind of approximation leading to (5.50), and thereafter to the asymptotic χ^2 null distribution, can be extended when H is infinite-dimensional, that is, a function with certain desirable properties. In particular, the curve defined by (5.50) is called least favorable submodels in the sense that it has the smallest information about t. The readers are referred to [237] (in particular, Theorem 3.1 therein) for details. In particular, for the special case of gamma-frailty, the asymptotic distribution of the likelihood-ratio statistic (5.46) under (5.45) is χ_1^2.

5.3.2 Asymptotic analysis of proportional hazards mixed-effects models

Another type of frailty models are characterized by the expression (5.40) and the assumption that the random effects, α_i, are normally distributed. Denote the data from subject j in cluster i by $d_{ij} = (y_{ij}, \delta_{ij}, x_{ij}, z_{ij})$, where y_{ij} is a possibly right-censored failure time, and δ_{ij} is a failure-event indicator. Let $d_i = (d_{ij})_{1 \leq j \leq n_i}$ be the vector of y's for cluster i, where n_i is the number of observations for cluster i. Then, conditional on the random effect α_i, the log-likelihood for the ith cluster is

$$l_i(\beta, h_0; d_i | \alpha_i) = \sum_{j=1}^{n_i} \left[\delta_{ij} \{ \log h_0(y_{ij}) + x_{ij}'\beta + z_{ij}'\alpha_i \} \right.$$
$$\left. - H_0(y_{ij}) \exp(x_{ij}'\beta + z_{ij}'\alpha_i) \right], \tag{5.51}$$

where $H_0(t) = \int_0^t h_0(s)ds$. Thus, the log-likelihood of the observed data is

$$l(\psi) = \sum_{i=1}^{m} \log \left[\int \exp\{l_i(\beta, h_0; d_i | \alpha_i)\} \phi(\alpha_i | \Sigma) d\alpha_i \right], \tag{5.52}$$

where m is the total number of clusters, $\psi = (\beta, \Sigma, H_0)$, and $\phi(\cdot | \Sigma)$ is the multivariate normal pdf with mean 0 and covariance matrix Σ. Here, similar to the gamma-frailty case, there is no maximum of H_0 over the space of absolutely continuous functions. In order to define a nonparametric MLE, Gamst, Donohue, and Xu ([238]) extended the parameter space to include all H_0 on $[0, T]$ that are right-continuous with left limits. The modified conditional log-likelihood is

$$l_i(\beta, J_0; d_i | \alpha_i) = \sum_{j=1}^{n_i} \left[\delta_{ij} \{ \log J_0(y_{ij}) + x_{ij}'\beta + z_{ij}'\alpha_i \} \right.$$
$$\left. - H_0(y_{ij}) \exp(x_{ij}'\beta + z_{ij}'\alpha_i) \right], \tag{5.53}$$

where $J(t)$ is the size of the jump in H_0 at time t. Comparing (5.53) with (5.51), it is seen that the only difference is that the $h_0(y_{ij})$ is replaced by $J_0(y_{ij})$. The parameters under the new setting are $\varphi = (\beta, \Sigma, J_0)$, where J_0 is the vector of all of the jump sizes. The modified log-likelihood is, therefore,

$$l(\varphi) \;=\; \sum_{i=1}^{m} \log\left[\int \exp\{l_i(\beta, J_0; d_i | \alpha_i)\} \phi(\alpha_i | \Sigma) d\alpha_i \right]. \qquad (5.54)$$

In a way, the inference problem becomes parametric. Let $\hat{\varphi} = (\hat{\beta}, \hat{\Sigma}, \hat{J}_0)$ be the maximizer of $l(\varphi)$. If one defines $Y_{ij}(t) = 1_{(y_{ij} \geq t)}$, then the cumulative hazard function, H_0, can be estimated as

$$\hat{H}_0(t) \;=\; \sum_{i,j} \frac{\delta_{ij}\{1 - Y_{ij}(t)\}}{\sum_{k,l} Y_{kl}(y_{ij}) \exp(x'_{kl}\hat{\beta}) \mathrm{E}_{\hat{\varphi}}\{\exp(w'_{kl}\alpha_k) | d_k\}}. \qquad (5.55)$$

Gamst, Dohohue and Xu [238] showed that, under some regularity conditions, the estimator $\hat{\varphi}$ is consistent in the sense that $|\hat{\beta} - \beta| \to 0, \|\hat{\Sigma} - \Sigma\|_2 \to 0$ and $\sup_{t \in [0,T]} |\hat{H}_0(t) - H_0(t)| \to 0$ almost surely as $m \to \infty$, where $\|M\|_2$ denotes the Euclidean norm of matrix M, that is, $\|M\|_2 = \{\mathrm{tr}(M'M)\}^{1/2}$.

A further result of asymptotic analysis obtained by the latter authors is in terms of convergence in distribution in functional space. Their approach is similar to that of [237] when considering LRT. Namely, consider a one-dimensional interpolation between the point of true parameter and another point in the functional space. Let $d_1 = \dim(\beta)$, and Σ be $d_2 \times d_2$. Let $g = (g_1, g_2, g_3)$, where g_1 is a d_1-dimensional vector, g_2 is a $d_2(d_2 - 1)$-dimensional vector corresponding to the upper-triangular part of a symmetric matrix G, and g_3 is a function of bounded variation on $[0, T]$. Define $\beta_s = \beta + sg_1$, $\Sigma_s = \Sigma + sG$, and $H_{0,s}(t) = \int_0^t \{1 + sg_3(u)\}dH_0(u)$. By taking the derivative of $l(\psi_s)$ with respect to s, where $\psi_s = (\beta_s, \Sigma_s, H_{0,s})$, one obtains the score functional: $S_m(\psi)[g] =$

$$\frac{1}{2}\mathrm{tr}\left[\left\{ (m\Sigma)^{-1} \sum_i \mathrm{E}_{\psi}(\alpha_i \alpha'_i | d_i) - I_{d_2} \right\} \Sigma^{-1} G \right]$$

$$+ \frac{1}{m} \sum_{i,j} \left[\int_0^T \{x'_{ij}g_1 + g_3(u)\}\{dN_{ij}(u) - Y_{ij}(u)e^{x'_{ij}\beta}\mathrm{E}_{\psi}(e^{z'_{ij}\alpha_i} | d_i)dH_0(u)\} \right],$$

where $N_{ij}(t), t \in [0, T]$ is the counting process defined as above (5.41). Note that the random measure in the second term of the above expression is a martigale, and that $S_m(\hat{\psi})[g] = 0$ for all g. The martingale central limit theorem(e.g., [239]) can be employed to show that $\sqrt{m}S_m(\psi_0)$ converges in distribution to a zero-mean Gaussian process, $\mathscr{G}$ on the space of bounded real-valued functions.

Furthermore, let $S(\psi) = \mathrm{E}_{\psi_0}\{S_m(\psi)\}$. It can be shown that $S(\psi)$ is Fréchet differentiable in the sense of Theorem 5.3. Thus, using the latter theorem, it can be shown that $\sqrt{m}(\hat{\psi} - \psi_0)$ converges in distribution to the zero-mean Gaussian process $\dot{S}^{-1}(\psi_0)\mathscr{G}$, where $\dot{S}(\psi_0)$ is the linear operator involved in part (b) of Theorem 5.3. It can then be argued that the estimator $\hat{\psi}$ is semiparametric efficient in the sense of Theorem 5.1 of [240].

5.4   Joint modeling of survival and longitudinal data

In biomedical research and economical studies, there are often data collected over time from multiple individuals (e.g., patients, companies). Such data are often referred to as longitudinal data. Note that, unlike time-series data, longitudinal data are both cross-sectional and time-dependent. Analysis of longitudinal data (e.g., [7]) often focuses the mean patterns over time. However, another important feature of longitudinal data must also be explored, that is, observations collected from the same individual over time may be correlated. Such a within-subject correlation may affect, for example, standard errors of the estimators. In general, failure to incorporate correlations among the observations may lead to incorrect results, and conclusions, of statistical inference.

Meanwhile, another type of data also associated with time are often available at the same time. These are survival data, which have been the main subject of the previous section. For example, in medical studies, patients are monitored over time, and the researchers are interested in the patients' progression toward some event of interest (e.g., death) as well as the dynamics of some longitudinal biomarker processes. It is natural to consider the survival data jointly with the longitudinal data. On the other hand, there are difficulties in jointly analyzing these two types of data.

For example, survival models with time-dependent covariates often require the complete history of the longitudinal process. This is not always feasible because subjects are only measured intermittently, and the measurement times are irregular in that different subjects are measured at different time points. In the statistical literature, several approaches have been proposed to solve the inference problems associated with survival and longitudinal data. See, for example, [241] and [242]. Among various methods, the semiparametric joint likelihood approach to jointly modeling both types of data emerges as the most satisfactory. The approach was first proposed by [243], who used the nonparametric maximum likelihood estimator (NPMLE; see the previous section) for the baseline hazard function. Athough asymptotic theory has been established [244], [245], the resulting point estimator of the baseline hazard function is not consistent. For example, in [244], only the NPMLE of the cumulative baseline hazard is shown to be consistent.

An alternative approach to NPMLE was proposed by Hsieh, Ding and Wang [246] based on the method of sieves. The idea was originated by Grenander [51]. It has been used, in particular, in the study of Wald consistency of the REML estimator, discussed in Section 2.2.3. The basic idea is to consider a sequence of approximating spaces to the full parameter space, each of which is (much) simpler in certain regards. For example, the full parameter space may be unbounded, but each approximating space is bounded (Section 2.2.3, above Theorem 2.7). As another example, the full parameter space may be infinite-dimensional, but each approximating space is finite-dimensional. In fact, the latter is what we are considering in the current context. There are two main issues in implementing the method of sieves. First, the choice of the sieve space is crucial [235], [247]. Secondly, there is a sieve bias that needs to be deal with to ensure asymptotic properties. In the current case, one needs to reduce the dimension of the nonparametric part to an extent so that the estimation bias introduced by the sieve approximation is negligible in the sense that the sieve bias is of a lower order than the variance.

Let T_i denote the time to the event of interest for the ith subject, which is subject to a censoring time, C_i. It is assumed that T_i and C_i are independent in each risk set at t conditioning on the covariate history to time t. The observed data for the ith subject is (y_i, δ_i) with $y_i = T_i \wedge C_i$ $[a \wedge b = \min(a,b)]$ and $\delta_i = 1_{(T_i \leq C_i)}$. For simplicity, assume that there is only one time-dependent covariate process, $X_i(t)$, that the hazard function follows the Cox model:

$$h(t|X_{i,\leq t}) = h_0(t)\exp\{\gamma X_i(t)\}, \tag{5.56}$$

where γ is an unknown parameter, $X_{i,\leq t} = \{X_i(s), 0 \leq s \leq t\}$ denotes the covariate process of subject i up to time t, and h_0 is the unknown baseline hazard function. However, the $X_i(t)$ in (5.56) is not observed. Instead, one observes a process $x_i = (x_{ij})_{1\leq j\leq n_i}$ at intermittently scheduled time points, $t_{i1} < t_{i2} < \cdots < t_{in_i}$ so that the following relation holds:

$$x_{ij} = X_i(t_{ij}) + e_i(t_{ij}), \tag{5.57}$$

where the measurement errors $e_{ij} = e(t_{ij})$ are independent and distributed as $N(0, \sigma_e^2)$, and are independent with the covariate process $X_i(t)$. The observed x_i is only up to y_i, hence $t_{in_i} \leq y_i$, and the number of observed time points n_i is random for each i. Note that the partial likelihood approach, discussed in Section 5.3.1, is not applicable here because it requires the complete history $X_{i,\leq t}$. As a resolution, a linear mixed model is assumed for the unobserved process $X_i(t)$, that is,

$$X_i(t) = b_i'\psi(t), \tag{5.58}$$

where $\psi(t) = [\psi_j(t)]_{0\leq j\leq q-1}$ is a q-dimensional function whose components are linearly independent, and $b_i = (b_{ij})_{0\leq j\leq q-1}$ is a vector of random effects. It is assumed that $b_i, i = 1,\ldots,m$ are independent and distributed as $N(\beta,\Sigma)$, where m is the number of subjects, β is an unknown vector of parameters, and Σ is an unknown covariance matrix. We consider an example.

Example 5.3. If $q = 2$, $\psi_0(t) = 1$, $\psi_1(t) = t$, the process $X_i(t)$ is assumed to be a linear function of t whose coefficients are random and subject-specific. Furthermore, if we let $b_{ij} = \beta_j + \alpha_{ij}, j = 0, 1$, where

$$\alpha_i = \begin{pmatrix} \alpha_{i0} \\ \alpha_{i1} \end{pmatrix} \sim N\left[\begin{pmatrix} 0 \\ 0 \end{pmatrix}, \begin{pmatrix} \sigma_0^2 & \rho\sigma_0\sigma_1 \\ \rho\sigma_0\sigma_1 & \sigma_1^2 \end{pmatrix}\right],$$

where σ_0^2, σ_1^2 are unknown variances, and ρ is an unknown correlation coefficients, then model (5.58) can be expressed as

$$X_i(t) = b_{i0} + b_{i1}t = \beta_0 + \beta_1 t + \alpha_{i0} + \alpha_{i1}t.$$

Here, α_{i0} and α_{i1} are called a random intercept and random slope, respectively.

Under the assumed model, the likelihood function based on the observed data, $(y_i, \delta_i, x_i, t_i, n_i)$, where $t_i = (t_{ij})_{1 \le j \le n_i}$, can be worked out. It has the expression

$$L(\theta) = \prod_{i=1}^{m} \int \left\{ \prod_{j=1}^{n_i} f(x_{ij}|b_i, \sigma_e^2) \right\} f(y_i, \delta_i|b_i, \gamma, h_0) \phi(b_i|\beta, \Sigma) db_i, \qquad (5.59)$$

where $\phi(\cdot|\beta, \Sigma)$ denotes the q-dimensional numtivariate normal distribution with mean vector β and covariance matrix Σ,

$$f(x_{ij}|b_i, \sigma_e^2) = \frac{1}{\sqrt{2\pi\sigma_e^2}} \exp\left[-\frac{\{x_{ij} - b_i'\psi(t_{ij})\}^2}{2\sigma_e^2} \right],$$

$$f(y_i, \delta_i|b_i, \gamma, h_0) = [h_0(y_i)\exp\{\gamma b_i'\psi(y_i)\}]^{\delta_i} \exp\left\{ -\int_0^{y_i} h_0(t)e^{\gamma b_i'\psi(t)} dt \right\}.$$

As noted, the NPMLE of the baseline hazard function, h_0, is inconsistent. This is mainly due to the large number of parameters introduced by h_0 (see Section 5.3). In order to obtain a consistent estimator of h_0, Hsieh et al. [246] proposed to estimate h_0 within a piecewise constant sieve space, and let the latter grow as the sample size increases. Specifically, let the parameter space of h_0 be

$$\mathcal{H} = \{h_0(\cdot) : h_0 \in C^{1,d}, h_0(t) \ge c_0, t \in [0, \infty)\}, \qquad (5.60)$$

where c_0, d are positive constants such that $d \le 1$, and $C^{k,d}$ denotes the space of functions whose up to kth derivatives are Hölder continuous. A function, f, is Hölder continuous if there are constants $M \ge 0$ and $0 < a \le 1$ such that $|f(u) - f(v)| \le M|x - y|^a$ for all x, y. The sieve space is a space of step functions constructed as follows. Let K_1 be a constant such that $K_1 > c_0^{-1}$. Let $0 = t_{[1]} < t_{[2]} < \cdots < t_{[k_m]} = K_1 \log m$ be a chosen partition on $[0, K_1 \log m]$. Let $H_0(s) = \int_0^s h_0(t) dt$ be the cumulative hazard function. Note that

$$e^{-H_0(K_1 \log m)} \le \exp\left(-\int_0^{\log m/c_0} c_0 dt \right) = \frac{1}{m},$$

which means that the baseline survival probability beyond $K_1 \log m$ is bounded by $1/m$; therefore, one can focus on the interval $[0, K_1 \log m]$ when constructing the step function. The sieve space is defined as

$$\mathcal{H}_m = \left\{ h_m(\cdot) = \sum_{j=1}^{k_m} c_j D_j(\cdot) : c_0 \le c_j \le K_0 k_m, D_j(t) = 1_{(t_{[j-1]} < t \le t_{[j]}]}, \right.$$

$$\left. t \in [0, K_1 \log m], 1 \le j \le k_m \right\}, \qquad (5.61)$$

where K_0 is another constant. To approximate the full space (5.60) by the sieve space (5.61), we need to let k_m increase with m, and also the mesh size of the sieve, $\Delta_m = \sup_{1 \le j \le k_m} |t_{[j-1]} - t_{[j]}|$, go to zero as m goes to ∞. Denote the parameters restricted to the sieve space by $\theta_{[m]} = \{\beta, \Sigma, \sigma_e^2, \gamma, h_m(\cdot)\}$, and the corresponding parameter space by $\Theta_{[m]}$. Note that $\Theta_{[m]} = \Theta_{-h_0} \times \mathcal{H}_m$, where Θ_{-h_0} is the parameter space for

$(\beta, \Sigma, \sigma_e^2, \gamma)$ (i.e., parameters without h_0) and $\mathcal{H}_m$ is given by (5.61). The sieve MLE of θ, $\hat{\theta}_{[m]}$, is defined as the maximizer of $L(\theta)$ over $\Theta_{[m]}$, that is,

$$\hat{\theta}_{[m]} = \text{argmax}_{\theta_{[m]} \in \Theta_{[m]}} L(\theta_{[m]}). \tag{5.62}$$

To study the asymptotic behavior of $\hat{\theta}_{[m]}$, we can think of the "difference" $\hat{\theta}_{[m]} - \theta$ as the "sum" of two differences: $\hat{\theta}_{[m]} - \theta = \hat{\theta}_{[m]} - \theta_{[m]}^* + \theta_{[m]}^* - \theta$, where $\theta_{[m]}^*$ is the parameter vector that minimizes the Kullback-Leibler (KL) divergence:

$$\begin{aligned} \theta_{[m]}^* &= \text{argmin}_{\theta_{[m]} \in \Theta_{[m]}} \left[\frac{1}{n} \sum_{i=1}^{m} E_\theta \left\{ \log \frac{L_i(\theta)}{L_i(\theta_{[m]})} \right\} \right] \\ &= \text{argmin}_{\theta_{[m]} \in \Theta_{[m]}} KL(\theta, \theta_{[m]}), \end{aligned} \tag{5.63}$$

where $L_i(\theta)$ is the ith factor in the product of (5.59) corresponding to the contribution to the likelihood function by the ith subject. Here we use the terms difference and sum with quotes because their precise definitions are not given; but this does not prevent one from understanding the concepts intuitively. To see the connection between $\hat{\theta}_{[m]}$, note that, by defining an "observed" KL divergence as

$$\widetilde{KL}(\theta, \theta_{[m]}) = \frac{1}{n} \sum_{i=1}^{m} \log \frac{L_i(\theta)}{L_i(\theta_{[m]})},$$

the sieve MLE (5.62) is equivalent to

$$\hat{\theta}_{[m]} = \text{argmin}_{\theta_{[m]} \in \Theta_{[m]}} \widetilde{KL}(\theta, \theta_{[m]}). \tag{5.64}$$

Thus, $\hat{\theta}_{[m]}$ may be interpreted as the minimizer of the observed KL divergence, and $\theta_{[m]}^*$ is the "target" of $\hat{\theta}_{[m]}$. Note that $\theta_{[m]}^*$ is not the same as θ; in fact, the two are not even in the same space: $\theta_{[m]}^* \in \Theta_{[m]}$ and $\theta \in \Theta$. Therefore, the difference between $\theta_{[m]}^*$ and θ is nonzero. This is called the bias of the sieve approximation. The bias measures the distance between the target and the truth. Typically, a larger sieve space, $\Theta_{[m]}$, yields a smaller bias; on the other hand, a larger sieve space results in a larger variation between $\hat{\theta}_{[m]}$ and $\theta_{[m]}^*$. Thus, the bias and "variance" need to be balanced in order for the sieve MLE to be a consistent estimator of θ. It is shown in [246] that the balance can be achieved to ensure that $\hat{\theta}_{[m]}$ is not only consistent but $\sqrt{m}$-consistent.

First, let us look at the bias. From the definition of the step function, (5.61), it is seen that, for any given $t \in [0, K_1 \log m]$, the estimated value of $h_0(t)$ is $c_j D_j(t)$, if $t \in (t_{[j-1]}, t_{[j]}], 1 \le j \le m$. Let c_j^* denote the c_j component of $\theta_{[m]}^*$ [note that c_j is part of $h_m(\cdot)$]. Also recall the definition of Δ_m below (5.61). Hsieh et al. [246] shows that

$$\frac{c_j^* D_j(t)}{h_0(t)} - 1 = O(\Delta_m), \quad t \in (t_{[j-1]}, t_{[j]}], \ 1 \le j \le m. \tag{5.65}$$

Note that (5.65) is pointwise for each t. Similarly, let $\theta_{-(\gamma, h_0)}$ denote the subvector of

θ, $(\beta, \Sigma, \sigma_e^2)$, and $\theta_{[m]-(\gamma, h_0)}$ the corresponding part of $\theta_{[m]}$, etc. Then, we have

$$\theta^*_{[m]-(\gamma, h_0)} - \theta_{-(\gamma, h_0)} = O(\Delta_m^{1+d}),$$
$$\gamma_{[m]} - \gamma = O(\Delta_m^{1+d}), \qquad (5.66)$$

where d is the positive constant involved in (5.60).

Next we consider strong consistency of the sieve MLE. We need a metric measure to evaluate the distance between $\hat{\theta}_{[m]}$ and θ. Define

$$\rho(\theta_{[m]}, \theta) = \left\{ \int |h_{[m]}(t) - h_0(t)|^2 dt + \| \theta_{[m]-h_{[m]}} - \theta_{-h_0} \|^2 \right\}^{1/2}, \qquad (5.67)$$

where $h_{[m]}$ is the h_0 part of $\theta_{[m]}$, $\theta_{[m]-h_{[m]}}$ is the part of $\theta_{[m]}$ without $h_{[m]}$, etc., and $\| \cdot \|$ denotes the Euclidean norm. The following result was proved in [246], which also took into consideration the possibility that the sieve MLE may not be unique. Here, two sequences of positive constants, a_m and b_m, are asymptotically equivalent, denoted by $a_m \sim b_m$, if $\liminf_{m \to \infty} (a_m / b_m) > 0$ and $\limsup_{m \to \infty} (a_m / b_m) < \infty$.

Theorem 5.7. If $k_m < m$ and $k_m \sim m^{1/3 - \varepsilon}$ for some small $\varepsilon > 0$, then, we have (i) $\sup_{\hat{\theta}_{[m]} \in \mathcal{M}_m} \rho(\hat{\theta}_{[m]}, \theta) \to 0$ a.s., where $\mathcal{M}_m = \{ \hat{\theta}_{[m]} \in \Theta_{[m]} | L(\hat{\theta}_{[m]}) = \sup_{\theta_{[m]} \in \Theta_{[m]}} L(\theta_{[m]}) \}$ with $L(\cdot)$ defined by (5.59). (ii) If $k_m \sim m^{1/4 - \varepsilon}$ for some small $\varepsilon > 0$, then, we have

$$KL(\theta, \mathcal{M}_m) \equiv \sup_{\hat{\theta}_{[m]} \in \mathcal{M}_m} KL(\theta, \hat{\theta}_{[m]}) = O(m^{-1/4 + 2\varepsilon}),$$

and any sequence of $\hat{\theta}_{[m]} \in \mathcal{M}_m$ converges almost surely to θ at the rate of $O(m^{-1/8 + \varepsilon})$.

The proof of Theorem 5.7, given in [246], utilized the Borel-Cantelli Lemma; a maximum inequality played an important role, as is often the case in such proofs (e.g., [1], Section 5.5). On the other hand, the convergence rate, $O(m^{-1/8 + \varepsilon})$, may be too slow to be useful in the derivation of the asymptotic distribution of the MLE. The convergence rate can be improved, if one considers a finer sieve space, which is $B_m \times \mathcal{H}_m$, where B_m is a bounded subset of the parameter space for $(\beta, \Sigma, \sigma_e^2, \gamma)$, and $\mathcal{H}_m$ is defined by (5.61). For $\theta_{[m]} \in B_m \times \mathcal{H}_m$, let

$$\widehat{KL}(\theta, \theta_{[m]}) = \frac{1}{m} \sum_{i=1}^m \log \frac{L_i(\theta)}{L_i(\theta_{[m]})} \qquad (5.68)$$

[see (5.63)]. Then, we have the following result [246].

Theorem 5.8. Let $\theta^*_{[m]}$ be an interior point of a $B_m \times \mathcal{H}_m$, and that $k_m \sim m^p$ [see (5.61)], where $\varepsilon \leq p \leq 1 - \varepsilon$ for some small $\varepsilon > 0$. Then, we have

$$\widehat{KL}(\theta, \theta_{[m]}) - KL(\theta, \theta_{[m]}) = O_P\{ m^{-(1-p)/2} (\log m)^{3/2} \}$$

uniformly for $\theta_{[m]} \in B_m \times \mathcal{H}_m$. If, furthermore, θ is an isolated unique minimizer of the K-L divergence, that is,

$$\inf_{\theta^*: \rho(\theta^*, \theta) \geq \delta} KL(\theta, \theta^*) > 0$$

for every $\delta > 0$, then, we have $\hat{\theta}_{[m]} \xrightarrow{P} \theta$.

Note that, in a way, the consistency result of Theorem 5.8 is stronger than the corresponding result of Theorem 5.7, because $\rho(\hat{\theta}_{[m]}, \theta) \to 0$ does not necessarily imply point-wise convergence of $h_{[m]}(\cdot)$ to $h_0(\cdot)$.

As for the convergence rate, recall that, in the classical situation of independent observations, the typical rate of convergence is $O(m^{-1/2})$, where m is the number of observations. This convergence rate can be shown to hold for the sieve MLE of the finite-dimensional part of θ, θ_{-h_0}, but not for h_0. In the standard likelihood-based inference, the expected value of the score function, defined as the vector of derivatives of the log-likelihood, is zero under the true parameter vector. This is, however, not the case, when the sieve does not contain the true parameter (which includes h_0). However, as is often the case in asymptotic analysis, being exactly zero is not necessary, as long as the zero-anticipated quantity is close to zero at a suitable rate. This is the approach that [246] took, who showed the following.

Theorem 5.9. Suppose that the true baseline hazard $h_0 \in C^{1,d}$ with $d \geq 1/2$ and that θ_{-h_0} is an interior point of Θ_{-h_0}. Furthermore, let k_m in (5.61) satisfy $k_m \sim m^p$ with $1/3 < p < 1$. Then, we have $m^{1/2}\{\hat{\theta}_{[m]-h_{[m]}} - \theta_{-h_0}\} = O_P(1)$, and $m^{(1-p)/2}\{\hat{h}_{[m]}(t) - h_0(t)\} = O_P(1)$ for every fixed time point t, where $\hat{\theta}_{[m]-h_{[m]}}$ is the part of $\hat{\theta}_{[m]}$ without the h_0 part, and $\hat{h}_{[m]}$ is the h_0 part of $\hat{\theta}_{[m]}$.

5.5 Additional bibliographical notes and open problems

5.5.1 Bibliographical notes

There have been studies in functional mixed effects models. These include models for functional data, in which the random effects are Euclidean (i.e., not functional), and models in which both the data and the random effects are functional. Guo [248] noted that data in many cases arise as curves, such as growth curves, bormone profiles, and biomarkers measured over time. The author proposed a functional mixed effects model, in which the random effects are modeled as realizations of a Gaussian process. Suppose that there is a response curve associated with each of the m subjects. Let y_{ij} be the observed value of the ith curve at time t_{ij}, $i = 1, \ldots, m$, $j = 1, \ldots, n_i$ such that $y_{ij} = x'_{ij}\beta(t_{ij}) + z'_{ij}\alpha_i(t_{ij}) + \varepsilon_{ij}$, where $\beta(t) = [\beta_k(t)]_{1 \leq k \leq p}$ is a $p \times 1$ vector of fixed functions, $\alpha_i(t) = [\alpha_k(t)]_{1 \leq k \leq q}$ is a $q \times 1$ vector of random functions that are modeled as realizations of Gaussian processes, $A(t) = [\alpha_k(t)]_{1 \leq k \leq q}$, with zero means, $x_{ij} = (x_{ijk})_{1 \leq k \leq p}$ and $z_{ij} = (z_{ijk})_{1 \leq k \leq q}$ are design matrices that include covariates as well as dummy variables, and $\varepsilon_{ij} \sim N(0, \sigma_\varepsilon^2)$ and are independent with the α's.

In fitting the proposed model, Guo first approximated both the fixed and random functions by smoothing splines. He then used a connection built by Wahba [249], [250] to model $\beta(t)$ and $A(t)$ as

$$\beta_k(t) = b_{1k} + b_{2k}t + \lambda_{b,k}^{-1/2} \int_0^t W_{b,k}(s)ds, \quad k = 1, \ldots, p;$$

$$\alpha_k(t) = a_{1k} + a_{2k}t + \lambda_{a,k}^{-1/2} \int_0^t W_{a,k}(s)ds, \quad k = 1, \ldots, q,$$

where $(b_{1k}, b_{2k})' \sim N(0, \tau I_2)$ with $\tau \to \infty$ (diffuse prior), and $(a_{1k}, a_{2k})' \sim N(0, \sigma_k^2 D)$, D being an unknown covariance matrix, and $W_{b,k}(s), W_{a,k}(s)$ are Weiner processes. Here, a_{1k} and a_{2k} are considered random intercept and random slope, respectively.

In terms of statistical inference, the latest author was mainly interested in testing linearity of $\beta_k(t)$ and $\alpha_k(t)$, that is, $\lambda_{b,k}^{-1} = 0$, or $\lambda_{a,k}^{-1} = 0$. Using a result from [124] on likelihood ratio test for parameters on the boundary of the parameter space. More specifically, Guo [248] showed that, under the null hypothesis of $\lambda_{b,k}^{-1} = 0$, the asymptotic distribution of the likelihood ratio statistic (LRS) is a mixture of 0 and χ_1^2 with equal weight (1/2 for both). As for testing for a single component of $\beta(t)$, that is, $\beta_k(t) = 0$, it is equivalent to testing for $b_{1k} = b_{2k} = \lambda_{b,k}^{-1} = 0$. The asymptotic null distribution of the LRS is a mixture of χ_2^2 and χ_3^2 with equal weight.

Krafty et al. [251] considered functional mixed effects spectral analysis. In biomedical experiments, data are often collected from multiple subjects as time series. It is natural to use carry out time series analysis to study the effects of design covariates. One of the standard analysis in time series is spectral analysis (e.g., [252]). Here, a random effect is associated with a group of subjects, while multiple time series correspond to subjects within the group. The random effects are function-valued, and different groups are assumed to be independent. A novel contribution of the work is a mixed effects Cramér representation in the following form:

$$X_{jkt} = \int_{-1/2}^{1/2} A_0(\omega; U_{jk}) A_j(\omega; V_{jk}) e^{2\pi i t \omega} dZ_{jk}(\omega),$$

$j = 1, \ldots, N, k = 1, \ldots, n_j$, where N is the number of groups and n_j the number of time series within the jth group; X_{jkt} is the kth replicate time series of the jth group; U_{jk}, V_{jk} are vectors of covariates, $A_0(\cdot; U_{jk}), A_j(\cdot; V_{jk})$ are functions corresponding to the fixed and random effects, respectively. Through asymptotic analysis, the authors showed that, when the replicate-specific spectra are smooth, the log-periodograms converge to a functional mixed effects model. The authors also developed BLUP for the unit-specific random effects.

In another related work, Rady et al. [253] considered estimation in mixed-effects functional ANOVA models. The authors considered the following so-called mixed functional ANOVA model:

$$y_{ij}(t) = \mu(t) + \alpha_i(t) + \beta_j(t) + \varepsilon_{ij}(t), \quad i = 1, \ldots, a, j = 1, \ldots, b.$$

This is similar to a two-way random effects ANOVA model (e.g., [2], p. 5) except that everything involved is a function. However, in their analysis, the authors have focused on a fixed value of t. Such results have little interest in terms of understanding the functional relationship.

Semiparametric GLMM also has received attention in the literature. For example, Lombardía and Sperlich [254] considered an extension of GLMM in that the conditional mean of the response given the random effects can be expressed as $\mu_{ij} = E(y_{ij} | \alpha_i, T_{ij}, X_{ij}) = g\{\lambda(T_{ij}) + x_{ij}'\beta + z_{ij}'\alpha_i\}$, where α_i is a vector-valued random effect, x_{ij}, t_{ij} are observed vectors of regressors, and z_{ij} is a subvector of $(1, x_{ij}')'$. Furthermore, $\lambda(\cdot)$ is an unknown function. It is assumed that (i) the responses y_{ij} are conditionally independent given α, T, X; (ii) the random effects are i.i.d. with mean 0 and covariance matrix Σ; and (T, X) are independent with α. The authors combine likelihood approach for mixed effects models with kernel methods. In term of asymptotic properties, a point-wise asymptotic normality result regarding estimation of the function $\lambda(\cdot)$. As for measure of uncertainty, the authors proposed a bootstrap procedure and provided a theoretical justification, also in terms of asymptotic approximation of the distribution of (y, X, T) by the bootstrap distribution. The authors discussed application of their method to small area estimation (e.g., [176]), currently a very active area of research and applications.

5.5.2   Open problems

In a way, the types of mixed effects models covered in this chapter all have the same structure, that is, the data are clustered such that each cluster corresponds to a random effects, which could be vector or functional valued, and the different clusters are independent. It should be noted that there are other types on nonlinear mixed effects models, frailty model, or functional mixed effects models, which do not have this type of block-diagonal covariance tructure. See, for example, Example 4.11 and Section 3.1 for similar cases in LMM and GLMM. As in the case of GLMM, asymptotic analysis for the case of crossed random effects is much harder to carry out, from a rigorous standpoint, for the types of mixed effects models considered in this chapter.

Regarding the frailty models, there have been two major classes of models, namely, the gamma-frailty model and proportional hazards mixed-effects model, as discussed in Sections 5.3.1 and 5.3.2, respectively. In practice, one needs to know whether an assumed model is appropriate for the survival data. As a way of formal model checking, there is interest in developing a goodness-of-fit test for frailty models, including deriving its asymptotic null distribution. Such a test would be more usedul if the asymptotic null distribution is relatively simple, such as χ^2.

There is substantial interest in inference at subject level. Areas of applications include precision medicine, personalized nutrition, and small area estimation. Of particular interest is extension of mixed effects models to functional data. Although several extensions have been considered (see the previous subsection), a solid theoretical foundation has yet to be built.

Mixed model prediction, as discussed in Sections 2.4 and 3.4, have found broad applications. However, so far there has been little rigorous studies on prediction of functional mixed effects. Such prediction methods are particularly interesting in, for example, precision medicine. The National Research Council distinguishes *personalized medicine* and *precision medicine*. In precision medicine, the focus is on identifying which approaches will be more effective for which patients based on genetic, environmental and lifestyle factors. Personalized medicine is sometimes misinterpreted to imply that there exists a unique treatment or prevention for each individual. The precision medicine definition implicitly assumes that at least some grouping or partitioning of the population exists, and thus is preferred to personalized medicine to avoid any misinterpretations. Much focus on precision medicine research and application has ensued with the unveiling of President Obama's precision medicine initiative, Vice President Biden's cancer moonshot program, and individual cancer centers advertising the hope of a precision medicine-based (genomic) treatments with better outcomes, etc. In many medical studies, patients' responses to medical treatments over time are naturally presented as curves. Therefore, the random effects corresponding to the mean response curves are naturally functional. However, so far there is no rigorous method that addresses issues such as measure of uncertainty in prediction of such functional mixed effects in light of Chapter 4.

For example, in Section 2.4.3 we discussed a method called CMMP. How do we generalize the method to prediction of functional mixed effects? Some immediate issues are (i) how to identify potential groups among the training data in terms of the mean response curves, if such groups are unknown? (ii) how to match the group corresponding to a new (functional) observation to one of the groups in the training data; and (iii) how does the method work in case of misidentification of the groups. In Section 2.4.3, we showed that, asymptotically, the CMMP predictor of the mixed effect remains consistent, in spite of the group misidentification. Does a similar result hold in the functional case?

Finally, resampling methods have been developed in the context of LMM and GLMM. See, for example, Section 4.5. However, there is relatively little development besides these two types of mixed effects models. For example, it would be of practical interest to develop resampling methods to assess measure of uncertainty for functional mixed model prediction, as discussed above.

References

[1] J. Jiang, *Large Sample Techniques for Statistics.* Springer, New York, 2010.

[2] J. Jiang, *Linear and Generalized Linear Mixed Models and Their Applications.* Springer, New York, 2007.

[3] C. E. McCulloch, S. R. Searle, and J. M. Neuhaus, *Generalized, Linear, and Mixed Models.* Wiley, Hoboken, NJ, second ed., 2008.

[4] E. Demidenko, *Mixed Models: Theory and Application with R.* Wiley, New York, second ed., 2013.

[5] J. N. K. Rao, *Small Area Estimation.* Wiley, New York, 2003.

[6] P. McCullagh and J. A. Nelder, *Generalized Linear Models.* Chapman & Hall, London, second ed., 1989.

[7] P. J. Diggle, P. Heagerty, K. Y. Liang, and S. L. Zeger, *Analysis of Longitudinal Data.* Oxford Univ. Press, second ed., 2002.

[8] T. Nguyen and J. Jiang, "Restricted fence method for covariate selection in longitudinal data analysis," *Biostatistics*, vol. 13, no. 2, pp. 303–314, 2012.

[9] K. Y. Liang and S. L. Zeger, "Longitudinal data analysis using generalized linear models," *Biometrika*, vol. 73, no. 1, pp. 13–22, 1986.

[10] S. L. Zeger and K. Y. Liang, "Longitudinal data analysis for discrete and continuous outcomes," *Biometrics*, vol. 42, no. 1, pp. 121–130, 1986.

[11] L. E. L. and C. G., *Theory of Point Estimation.* Springer, New York, 1998.

[12] A. Sen and M. Srivastava, *Regression Analysis: Theory, Methods, and Applications.* Springer, New York, 1990.

[13] J. Neyman and E. Scott, "Consistent estimates based on partially consistent observations," *Econometrika*, vol. 16, no. 1, pp. 1–32, 1948.

[14] N. E. Breslow and D. G. Clayton, "Approximate inference in generalized linear mixed models," *J. Amer. Statist. Assoc.*, vol. 88, no. 421, pp. 9–25, 1993.

[15] N. E. Breslow and X. Lin, "Bias correction in generalized linear mixed models with a single component of disperson," *Biometrika*, vol. 82, no. 1, pp. 81–91, 1995.

[16] X. Lin and N. E. Breslow, "Bias correction in generalized linear mixed models with multiple components of dispersion," *J. Amer. Statist. Assoc.*, vol. 91, no. 435, pp. 1007–1016, 1996.

[17] J. Jiang, "Consistent estimators in generalized linear mixed models," *J. Amer.*

Statist. Assoc., vol. 93, no. 442, pp. 720–729, 1998.

[18] J. G. Booth and J. P. Hobert, "Maximum generalized linear mixed model likelihood with an automated Monte Carlo EM algorithm," *J. Roy. Statist. Soc. B*, vol. 61, no. 1, pp. 265–285, 1999.

[19] J. Jiang and W. Zhang, "Robust estimation in generalized linear mixed models," *Biometrika*, vol. 88, no. 3, pp. 753–765, 2001.

[20] J. Jiang, "The subset argument and consistency of MLE in GLMM: Answer to an open problem and beyond," *Ann. Statist.*, vol. 41, no. 1, pp. 177–195, 2013.

[21] S. R. Searle, G. Casella, and C. E. McCulloch, *Variance Components*. Wiley, New York, 1992.

[22] H. O. Hartley and J. N. K. Rao, "Maximum likelihood estimation for the mixed analysis of variance model," *Biometrika*, vol. 54, no. 1, pp. 93–108, 1967.

[23] J. Thompson, W. A., "The problem of negative estimates of variance components," *Ann. Math. Statist.*, vol. 33, no. 1, pp. 273–289, 1962.

[24] J. J. Miller, "Asymptotic properties of maximum likelihood estimates in the mixed model of analysis of variance," *Ann. Statist.*, vol. 5, no. 4, pp. 746–762, 1977.

[25] J. Jiang, "REML estimation: Asymptotic behavior and related topics," *Ann. Statist.*, vol. 24, no. 1, pp. 255–286, 1996.

[26] J. Jiang, "Wald consistency and the method of sieves in REML estimation," *Ann. Statist.*, vol. 25, no. 4, pp. 1781–1803, 1997.

[27] S. R. Lele, B. Dennis, and F. Lutscher, "Data cloning: Easy maximum likelihood estimation for complex ecological models using Bayesian Markov chain Monte Carlo methods," *Ecology Letters*, vol. 10, no. 7, pp. 551–563, 2007.

[28] S. R. Lele, K. Nadeem, and B. Schmuland, "Estimability and likelihood inference for generalized linear mixed models using data cloning," *J. Amer. Statist. Assoc.*, vol. 105, no. 492, pp. 1617–1625, 2010.

[29] N. G. N. Prasad and J. N. K. Rao, "The estimation of mean squared errors of small area estimators," *J. Amer. Statist. Assoc.*, vol. 85, no. 409, pp. 163–171, 1990.

[30] J. Jiang, P. Lahiri, and S. Wan, "A unified jackknife theory for empirical best prediction with M-estimation," *Ann. Statist.*, vol. 30, no. 6, pp. 1782–1810, 2002.

[31] P. Hall and T. Maiti, "Nonparametric estimation of mean-squared prediction error in nested-error regression models," *Ann. Statist.*, vol. 34, no. 4, pp. 1733–1750, 2006.

[32] C. R. Henderson, "Estimation of variance and covariance components," *Biometrics*, vol. 9, no. 2, pp. 226–252, 1953.

[33] D. A. Harville, "Maximum likelihood approaches to variance components estimation and related problems," *J. Amer. Statist. Assoc.*, vol. 72, no. 358, pp. 320–340, 1977.

[34] G. E. Battese, R. M. Harter, and W. A. Fuller, "An error-components model for prediction of county crop areas using survey and satellite data," *J. Amer. Statist. Assoc.*, vol. 83, no. 401, pp. 28–36, 1988.

[35] H. Scheffé, *The Analysis of Variance*. Wiley, New York, 1959.

[36] L. Weiss, "Asymototic properties of maximum likelihood estimators in some nonstandard cases," *J. Amer. Statist. Assoc.*, vol. 66, no. 334, pp. 345–350, 1971.

[37] L. Weiss, "Asymototic properties of maximum likelihood estimators in some nonstandard cases II," *J. Amer. Statist. Assoc.*, vol. 68, no. 342, pp. 428–430, 1973.

[38] H. Cramér, *Mathematical Methods of Statistics*. Princeton University Press, 1946.

[39] A. Wald, "Note on the consistency of the maximum likelihood estimate," *Ann. Math. Statist.*, vol. 20, no. 4, pp. 595–601, 1949.

[40] J. Wolfowitz, "On Wald's proof of the consistency of the maximum likelihood estimate," *Ann. Math. Statist.*, vol. 20, no. 4, pp. 601–602, 1949.

[41] K. Das, "Asymptotic optimality of restricted maximum likelihood estimates for the mixed model," *Calcutta Statist. Assoc. Bull.*, vol. 28, no. 109–112, pp. 125–142, 1979.

[42] N. Cressie and S. N. Lahiri, "The asymptotic distribution of REML estimators," *J. Multivariate Anal.*, vol. 45, no. 2, pp. 217–233, 1993.

[43] T. J. Sweeting, "Uniform asymptotic normality of the maximum likelihood estimator," *Ann. Statist.*, vol. 8, no. 6, pp. 1375–1381, 1980.

[44] A. M. Richardson and A. H. Welsh, "Asymptotic properties of restricted maximum likelihood (REML) estimates for hierarchical mixed linear models," *Austral. J. Statist.*, vol. 36, no. 1, pp. 31–43, 1994.

[45] P. Hall and C. C. Heyde, *Martingale Limit Theory and Its Application*. Academic Press, 1980.

[46] P. Guttorp and R. A. Lockhart, "On the asymptotic distribution of quadratic forms in uniform order statistics," *Ann. Statist.*, vol. 16, no. 1, pp. 433–449, 1988.

[47] R. Fox and M. S. Taqqu, "Noncentral limit theorems for quadractic forms in random variables having long-range dependence," *Ann. Probab.*, vol. 13, no. 2, pp. 428–446, 1985.

[48] C. R. Rao and J. Kleffe, *Estimation of Variance Components and Applications*. North-Holland, Amsterdam, 1988.

[49] W. H. Schmidt and R. Thrum, "Contributions to asymptotic theory in regression models with linear covariance structure," *Math. Operationsforsch. Statist. Ser. Statist.*, vol. 12, no. 2, pp. 243–269, 1981.

[50] L. Gan and J. Jiang, "A test for global maximum," *J. Amer. Statist. Assoc.*, vol. 94, no. 447, pp. 847–854, 1999.

[51] U. Grenander, *Abstract Inference*. Wiley, New York, 1981.

[52] J. Jiang, "Empirical method of moments and its applications," *J. Statist. Plann. Inference*, vol. 115, no. 1, pp. 69–84, 2003.

[53] L. P. Hansen, "Large sample properties of generalized method of moments estimators," *Econometrica*, vol. 50, no. 4, pp. 1029–1054, 1982.

[54] B. Efron and D. V. Hinkley, "Assessing the accuracy of the maximum likelihood estimator: Observed versus expected Fisher information," *Biometrika*, vol. 65, no. 3, pp. 457–487, 1978.

[55] J. Jiang, "Partially observed information and inference about non-Gaussian mixed linear models," *Ann. Statist.*, vol. 33, no. 6, pp. 2695–2731, 2005.

[56] J. N. Arvesen, "Jackknifing U-statistics," *Ann. Math. Statist.*, vol. 40, no. 6, pp. 2076–2100, 1969.

[57] J. N. Arvesen and T. H. Schmitz, "Robust procedures for variance component problems using the jackknife," *Biometrics*, vol. 26, no. 4, pp. 677–686, 1970.

[58] C. R. Henderson, Estimation of general, specific and maternal combining abilities in crosses among inbred lines of swine, Ph. D. Thesis. Iowa State University, Ames, Iowa, 1948.

[59] J. Jiang and P. Lahiri, "Mixed model prediction and small area estimation (with discussion)," *TEST*, vol. 15, no. 1, pp. 1–96, 2006.

[60] A. M. Mood, F. A. Graybill, and D. C. Boes, *Introduction to the Theory of Statistics*. McGraw-Hill, New York, 1974.

[61] R. E. Fay and R. A. Herriot, "Estimates of income for small places: An application of James–Stein procedures to census data," *J. Amer. Statist. Assoc.*, vol. 74, no. 366, pp. 269–277, 1979.

[62] J. Jiang, "Asymptotic properties of the empirical BLUP and BLUE in mixed linear models," *Statist. Sinica*, vol. 8, no. 3, pp. 861–885, 1998.

[63] K. Das, J. Jiang, and J. N. K. Rao, "Mean squared error of empirical predictor," *Ann. Statist.*, vol. 32, no. 2, pp. 818–840, 2004.

[64] R. N. Kackar and D. A. Harville, "Approximations for standard errors of estimators of fixed and random effects in mixed linear models," *J. Amer. Statist. Assoc.*, vol. 79, no. 388, pp. 853–862, 1984.

[65] C. R. Henderson, "Best linear unbiased estimation and prediction under a selection model," *Biometrics*, vol. 31, no. 2, pp. 423–447, 1975.

[66] J. Jiang, T. Nguyen, and J. S. Rao, "Best predictive small area estimation," *J. Amer. Statist. Assoc.*, vol. 106, no. 494, pp. 732–745, 2011.

[67] H. White, "Maximum likelihood estimation of misspecified models," *Econometrika*, vol. 50, no. 1, pp. 1–25, 1982.

[68] J. Jiang, J. S. Rao, J. Fan, and T. Nguyen, "Classified mixed model prediction," *J. Amer. Statist. Assoc.*, p. in press, 2016.

[69] N. Lange and L. Ryan, "Assessing normality in random effects models," *Ann.*

Statist., vol. 17, no. 2, pp. 624–642, 1989.

[70] J. A. Calvin and J. Sedransk, "Bayesian and frequentist predictive inference for the patterns of care studies," *J. Amer. Statist. Assoc.*, vol. 86, no. 413, pp. 36–48, 1991.

[71] J. A. Rice, *Mathematical Statistics and Data Analysis*. Duxbury Press, Belmont, CA, second ed., 1995.

[72] D. S. Moore, "Chi-square tests," *Studies in Statistics (R. V. Hogg, ed.)*, pp. 66–106, 1978.

[73] H. Chernoff and E. L. Lehmann, "The use of maximum-likelihood estimates in χ^2 tests for goodness of fit," *Ann. Math. Statist.*, vol. 25, no. 3, pp. 579–586, 1954.

[74] J. Jiang, P. Lahiri, and C. Wu, "A generalization of the Pearson's χ^2 goodness-of-fit test with estimated cell frequencies," *Sankhyā Ser. A*, vol. 63, no. 2, pp. 260–276, 2001.

[75] R. Bhatia, *Matrix Analysis*. Springer, New York, 1997.

[76] J. Jiang, "Goodness-of-fit tests for mixed model diagnostics," *Ann. Statist.*, vol. 29, no. 4, pp. 1137–1164, 2001.

[77] G. Claeskens and J. D. Hart, "Goodness-of-fit tests in mixed models," *TEST*, vol. 18, no. 2, pp. 213–239, 2009.

[78] H. Akaike, "Information theory as an extension of the maximum likelihood principle," *Second International Symposium on Information Theory* (B. N. Petrov and F. Csaki eds.), pp. 267–281, 1973.

[79] G. Schwarz, "Estimating the dimension of a model," *Ann. Statist.*, vol. 6, no. 2, pp. 461–464, 1978.

[80] L. A. Hindorff, P. Sethupathy, H. A. Junkins, E. M. Ramos, J. P. Mehta, F. S. Collins, and T. A. Manolio, "Potential etiologic and functional implications of genome-wide association loci for human diseases and traits," *Proc. Nat. Acad. Sci.*, vol. 106, no. 23, p. 9362, 2009.

[81] T. A. Manolio, F. S. Collins, N. J. Cox, D. B. Goldstein, L. A. Hindorff, D. J. Hunter, and et al., "Finding the missing heritability of complex diseases," *Nature*, vol. 461, no. 7265, pp. 747–753, 2009.

[82] P. M. Visscher, W. G. Hill, and N. R. Wray, "Heritability in the genomics era - concepts and misconceptions," *Nature Reviews Genetics*, vol. 9, no. 4, pp. 255–266, 2008.

[83] A. R. Wood, T. Esko, J. Yang, S. Vedantam, T. H. Pers, S. Gustafsson, and et al., "Defining the role of common variation in the genomic and biological architecture of adult human height," *Nature Genetics*, vol. 46, no. 11, pp. 1173–1186, 2014.

[84] B. Maher, "Personal genomes: The case of the missing heritability," *Nature*, vol. 456, no. 7218, pp. 18–21, 2008.

[85] T. A. Manolio, "Genomewide association studies and assessment of the risk

of disease," *New Eng. J. Med.*, vol. 363, no. 2, pp. 166–176, 2010.

[86] J. Yang, B. Benyamin, B. P. McEvoy, S. Gordon, A. K. Henders, D. R. Nyholt, and et al., "Common SNPs explain a large proportion of the heritability for human height," *Nature Genetics*, vol. 42, no. 7, pp. 565–569, 2010.

[87] S. Vattikuti, J. Guo, and C. C. Chow, "Heritability and genetic correlations explained by by common snps for metabolic syndrome traits," *PLoS Genetics*, vol. 8, no. 3, p. e1002637, 2012.

[88] S. H. Lee, T. R. DeCandia, S. Ripke, J. Yang, P. F. Sullivan, M. E. Goddard, and et al., "Estimating missing heritability for disease from genome-wide association studies," *Amer. J. Human Genetics*, vol. 88, no. 3, pp. 294–305, 2011.

[89] J. Jiang, C. Li, D. Paul, C. Yang, and H. Zhao, "On high-dimensional misspecified mixed model analysis in genome-wide association study," *Ann. Statist.*, vol. 44, no. 5, pp. 2127–2160, 2016.

[90] D. Paul and A. Aue, "Random matrix theory in statistics: A review," *J. Statist. Plann. Inference*, vol. 150, no. 1, pp. 1–29, 2014.

[91] Z. D. Bai, "Methodologies in spectral analysis of large dimensional random matrices, a review," *Statistica Sinica*, vol. 9, no. 3, pp. 611–677, 1999.

[92] R. Vershynin, *Introduction to the non-asymptotic analysis of of random matrices.* arXiv:1101.3027, 2011.

[93] M. Rudelson and R. Vershynin, "Hanson–Wright inequality and sub-Gaussian concentration," *Electron. Commun. Probab.*, vol. 18, no. 82, pp. 1–9, 2013.

[94] Z. D. Bai and J. W. Silverstein, *Spectral Analysis of Large Dimensional Random Matrices, 2nd ed.* Springer, New York, second ed., 2010.

[95] Z. D. Bai and Y. Q. Yin, "Limit of the smallest eigenvalue of a large dimensional sample covariance matrix," *Ann. Probab.*, vol. 21, no. 3, pp. 1275–1294, 1993.

[96] J. Fan, S. Guo, and N. Hao, "Variance estimation using refitted cross-validation in ultrahigh dimensional regression," *J. Roy. Statist. Soc. Ser. B*, vol. 74, no. 1, pp. 37–65, 2012.

[97] T. Sun and C.-H. Zhang, "Scaled sparse linear regression," *Biometrika*, vol. 99, no. 3, pp. 879–898, 2012.

[98] P. M. Visscher, B. M. A., M. I. McCarthy, and J. Yang, "Five years of GWAS discovery," *Amer. J. Human Genetics*, vol. 90, no. 1, pp. 7–24, 2012.

[99] J. Yang, S. H. Lee, M. E. Goddard, and P. M. Visscher, "GCTA: a tool for genome-wide complex trait analysis," *Amer. J. Human Genetics*, vol. 88, no. 1, pp. 76–82, 2011.

[100] D. Speed, G. Hemani, M. R. Johnson, and D. J. Balding, "Improved heritability estimation from genome-wide SNPs," *Nature Genetics*, vol. 91, no. 6, pp. 1011–1021, 2012.

[101] N. Zaitlen, P. Kraft, N. Patterson, B. Pasaniuc, G. Bhatia, S. Pollack, and A. L. Price, "Using extended genealogy to estimate components of heritabil-

ity for 23 quantitative and dichotomous traits," *PLoS Genetics*, vol. 9, no. 5, p. e1003520, 2013.

[102] K. G. Brown, "Asymptotic behavior of MINQUE-type estimators of variance components," *Ann. Statist.*, vol. 4, no. 4, pp. 746–754, 1976.

[103] T. W. Anderson, "Asymptotically efficient estimation of covariance matrices with linear structure," *Ann. Statist.*, vol. 1, no. 1, pp. 135–141, 1973.

[104] T. P. Speed, "Cumulants and partition lattices IV: a.s. convergence of generalized k-statistics," *J. Austral. Math. Soc.*, vol. 41, no. 1, pp. 79–94, 1986.

[105] P. H. Westfall, "Asymptotic normality of the ANOVA estimates of components of variance in the nonnormal, unbalanced hierarchal mixed model," *Ann. Statist.*, vol. 4, no. 1, pp. 1572–1582, 1986.

[106] S. J. Welham and R. Thompson, "Likelihood ratio tests for fixed model terms using residual maximum likelihood," *J. Roy. Statist. Soc. B*, vol. 59, no. 4, pp. 701–714, 1997.

[107] A. M. Richardson and A. H. Welsh, "Covariate screening in mixed linear models," *J. Multivariate Anal.*, vol. 58, no. 1, pp. 27–54, 1996.

[108] J. Jiang, "On robust versions of classical tests with dependent data," *Nonparametric Statistical Methods and Related Topics - A Festschrift in Honor of Professor P. K. Bhattacharya on the Occasion of His 80th Birthday* (J. Jiang, G. G. Roussas, F. J. Samaniego eds.), pp. 77–99, 2011.

[109] F. A. Graybill and C. M. Wang, "Confidence intervals for nonnegative linear combinations of variances," *J. Amer. Statist. Assoc.*, vol. 75, no. 372, pp. 869–873, 1980.

[110] F. E. Satterthwaite, "An approximate of distribution of estimates of variance components," *Biometrics Bull.*, vol. 2, no. 6, pp. 110–114, 1946.

[111] R. K. Burdick and F. A. Graybill, *Confidence Intervals on Variance Components.* Marcel Dekker, New York, 1992.

[112] Müller, J. L. S., Scealy, and A. H. Welsh, "Model selection in linear mixed models," *Statist. Sci.*, vol. 28, no. 1, pp. 135–167, 2013.

[113] J. Jiang and J. S. Rao, "Consistent procedures for mixed linear model selection," *Sankhyā Ser. A*, vol. 65, no. 1, pp. 23–42, 2003.

[114] J. Jiang, J. S. Rao, Z. Gu, and T. Nguyen, "Fence methods for mixed model selection," *Ann. Statist.*, vol. 36, no. 4, pp. 1669–1692, 2008.

[115] J. Jiang, T. Nguyen, and J. S. Rao, "Invisible fence method and the identification of differentially expressed gene sets," *Statist. Interface*, vol. 4, no. 3, pp. 403–415, 2011.

[116] J. Jiang, "The fence methods," *Advances in Statistics*, pp. 1–14, 2014.

[117] J. Fan and R. Li, "Variable selection via nonconcave penalized likelihood and its oracle properties," *J. Amer. Statist. Assoc.*, vol. 96, no. 456, pp. 1348–1360, 2001.

[118] R. J. Tibshirani, "Regression shrinkage and selection via the Lasso," *J. Roy.*

Statist. Soc. Ser. B, vol. 16, no. 2, pp. 385–395, 1996.

[119] H. D. Bondell, A. Krishna, and S. K. Ghosh, "Joint variable selection for fixed and random effects in linear mixed-effects models," *Biometrics*, vol. 66, no. 4, pp. 1069–1077, 2010.

[120] J. G. Ibrahim, H. Zhu, R. I. Garcia, and R. Guo, "Fixed and random effects selection in mixed effects models," *Biometrics*, vol. 67, no. 2, pp. 495–503, 2011.

[121] Y. Fan and R. Li, "Variable selection in linear mixed effects models," *Ann. Statist.*, vol. 40, no. 4, pp. 2043–2068, 2012.

[122] J. Jiang, T. Nguyen, and J. S. Rao, "Observed best prediction via nested-error regression with potentially misspecified mean and variance," *Survey Methodol.*, vol. 41, no. 1, pp. 37–55, 2015.

[123] C. A. Cole and S. Wolfram, "SMP: A symbolic manipulation program," *Proc. 4th ACM Symp. Symbolic Algebraic Comput.*, 1981.

[124] S. G. Self and K. Y. Liang, "Asymptotic properties of maximum likelihood estimators and likelihood ratio tests under nonstandard conditions," *J. Amer. Statist. Assoc.*, vol. 82, no. 398, pp. 605–610, 1987.

[125] D. O. Stram and J. W. Lee, "Variance components testing in the longitudinal mixed model," *Biometrics*, vol. 50, no. 4, pp. 1171–1177, 1994.

[126] J. Jiang and T. . Nguyen, "Comments on: Goodness-of-fit tests in mixed models by G. Claeskens and J. D. Hart," *TEST*, vol. 18, no. 2, pp. 248–255, 2009.

[127] P. J. Green, "Penalized likelihood for general semi-parametric regression models," *Int. Statist. Rew.*, vol. 55, no. 3, pp. 245–259, 1987.

[128] J. Jiang, "On maximum hierarchical likelihood estimators," *Commun. Statist.– Theory Meth.*, vol. 28, no. 8, pp. 1769–1775, 1999.

[129] V. P. Godambe, "An optimum property of regular maximum-likelihood estimation," *Ann. Math. Statist.*, vol. 31, no. 4, pp. 1208–1211, 1960.

[130] V. P. Godambe, *Estimating Functions*. Oxford Science, Oxford, 1991.

[131] A. Qu, B. G. Lindsay, and B. Li, "Improving generalised esstimating equations using quadratic inference functions," *Biometrika*, vol. 87, no. 4, pp. 823–836, 2000.

[132] J. Jiang and Y.-G. Wang, "Iterative estimating equations: Linear convergence and asymptotic properties," *Ann. Statist.*, vol. 35, no. 5, pp. 2233–2260, 2007.

[133] W. H. Press, S. A. Teukolsky, W. T. Vetterling, and B. P. Flannery, *Numerical Recipes in C—The Arts of Scientific Computing*. Cambridge University Press, Cambridge, second ed., 1997.

[134] D. Hand and M. Crowder, *Practical Longitudinal Data Analysis*. Chapman & Hall, London, 1996.

[135] D. McFadden, "A method of simulated moments for estimation of discrete response models without numerical integration," *Econometrika*, vol. 57, no. 5, pp. 995–1026, 1989.

[136] L. F. Lee, "On the efficiency of methods of simulated moments and maximum simulated likelihood estimation of discrete response models," *Econometric Theory*, vol. 8, no. 4, pp. 518–552, 1992.

[137] X. Lin, "Variance components testing in generalized linear models with random effects," *Biometrika*, vol. 84, no. 2, pp. 309–326, 1997.

[138] C. A. McGilchrist, "Estimation in generalized mixed models," *J. Roy. Statist. Soc. B*, vol. 56, no. 1, pp. 61–69, 1994.

[139] A. Y. C. Kuk, "Asymptotically unbiased estimation in generalized linear models with random effects," *J. Roy. Statist. Soc. B*, vol. 57, no. 2, pp. 395–407, 1995.

[140] T. P. Speed, "Comment on Robinson: Estimation of random effects," *Statist. Sci.*, vol. 6, no. 1, pp. 42–44, 1991.

[141] J. Jiang, "Mixed-effects models with random cluster sizes," *Statist. Probab. Letters*, vol. 53, no. 2, pp. 201–206, 2001.

[142] J. Hinde, "Compound Poisson regression models," *GLIM 82: Proc. International Conf. Generalized Lin. Models* (R. Gilchrist ed.), pp. 109–121, 1982.

[143] E. A. C. Crouch and D. Spiegelman, "The evaluation of integrals of the form $\int f(t)\exp(-t^2)dt$: Application to logistic normal models," *J. Amer. Statist. Assoc.*, vol. 85, no. 2, pp. 464–469, 1990.

[144] A. P. Dempster, N. M. Laird, and D. B. Rubin, "Maximum likelihood from incomplete data via the EM algorithm (with discussion)," *J. Roy. Statist. Soc. B*, vol. 39, no. 1, pp. 1–38, 1977.

[145] C. E. McCulloch, "Maximum likelihood variance components estimation for binary data," *J. Amer. Statist. Assoc.*, vol. 89, no. 425, pp. 330–335, 1994.

[146] J. S. Liu, *Monte Carlo Strategies in Scientific Computing*. Springer, New York, 2004.

[147] C. E. McCulloch, "Maximum likelihood algorithms for generalized linear mixed models," *J. Amer. Statist. Assoc.*, vol. 92, no. 1, pp. 162–170, 1997.

[148] W. R. Gilks, S. Richardson, and D. J. E. Spiegelhalter, *Markov Chain Monte Carlo in Practice*. Chapman & Hall, London, 1996.

[149] D. J. Spiegelhalter, A. Thomas, N. Best, and D. Lunn, *WinBUGS Version 1.4 User Manual*. MRC Biostatistics Unit, Institute of Public Health, London, 2004.

[150] M. Torabi, "Likelihood inference in generalized linear mixed models with two components of dispersion using data cloning," *Comput. Statist. Data Anal.*, vol. 56, no. 12, pp. 4259–4265, 2012.

[151] S. Kullback and R. A. Leibler, "On information and sufficiency," *Ann. Math. Statist.*, vol. 22, no. 1, pp. 79–86, 1951.

[152] J. Jiang, H. Jia, and H. Chen, "Maximum posterior estimation of random effects in generalized linear mixed models," *Statist. Sinica*, vol. 11, no. 1, pp. 97–120, 2001.

[153] C. R. Henderson, "Estimation of genetic parameters (abstract)," *Ann. Math. Statist.*, vol. 21, no. 2, pp. 309–310, 1950.

[154] J. Jiang, "A nonlinear Gauss-Seidel algorithm for inference about GLMM," *Comput. Statist.*, vol. 15, no. 2, pp. 229–241, 2000.

[155] Y. Lee and J. A. Nelder, "Hierarchical generalized linear models (with discussion)," *J. Roy. Statist. Soc. B*, vol. 58, no. 4, pp. 619–678, 1996.

[156] D. R. Cox and D. V. Hinkley, *Theoretical Statistics*. Chapman & Hall, London, 1974.

[157] J. Jiang, "Conditional inference about generalized linear mixed models," *Ann. Statist.*, vol. 27, no. 6, pp. 1974–2007, 1999.

[158] S. Portnoy, "Asymptotic behavior of M-estimators of p regression parameters when p^2/n is large," *Ann. Statist.*, vol. 12, no. 4, pp. 1298–1309, 1984.

[159] P. Wang, G.-F. Tsai, and A. Qu, "Conditional inference function for mixed-effects models with unspecified random-effects distribution," *J. Amer. Statist. Assoc.*, vol. 107, no. 498, pp. 725–736, 2012.

[160] P. X.-K. Song, Y. Fan, and J. D. Kalbfleisch, "Maximization by parts in likelihood inference (with discussion)," *J. Amer. Statist. Assoc.*, vol. 100, no. 462, pp. 1145–1158, 2005.

[161] J. Jiang, "Comment on Song, Fan and Kalbfleisch: Maximization by parts," *J. Amer. Statist. Assoc.*, vol. 100, no. 462, pp. 1158–1159, 2005.

[162] R. P. Agarwal, M. Meehan, and D. O'Regan, *Fixed Point Theory and Applications*. Cambridge University Press, 2001.

[163] D. Clayton, "Comment on Lee and Nelder: Hierarchical generalized linear models," *J. Roy. Statist. Soc. B*, vol. 58, no. 4, pp. 657–659, 1996.

[164] Z. Gu, Model diagnostics for generalized linear mixed models, Ph. D. Dissertation. University of California, Davis, 2008.

[165] M. Tang, Goodness-of-fit tests for generalized linear mixed models, Ph. D. Dissertation. University of Maryland, College Park, 2010.

[166] R. A. Fisher, "On the interpretation of chi-square from contingency tables, and the calculation of P," *J. Roy. Statist. Soc.*, vol. 85, no. 1, pp. 87–94, 1922.

[167] C. Dao and J. Jiang, "A modified Pearson's χ^2 test with application to generalized linear mixed model diagnostics," *Ann. Math. Sci. Appl.*, vol. 1, no. 1, pp. 195–215, 2016.

[168] S. R. Searle, *Linear Models*. Wiley, 1971.

[169] B. R. Bhat and B. N. Nagnur, "Locally asymptotically most stringent tests and Lagrangian multiplier tests of linear hypothesis," *Biometrika*, vol. 52, no. 3, pp. 459–468, 1965.

[170] B. C. Sutradhar, "On exact quasi-likelihood inference in generalized linear mixed models," *Sankhyā: Ind. J. Statist.*, vol. 66, no. 2, pp. 263–291, 2004.

[171] X. Huang, "Diagnosis of random-effect model misspecification in generalized

linear mixed models for binary responses," *Biometrics*, vol. 65, no. 2, pp. 361–368, 2009.

[172] S. Chen, Predictive modeling for clustered data with applications, Ph. D. Dissertation. University of California, Davis, 2012.

[173] K. Hu, J. Jiang, J. Choi, and A. Sim, "Analyzing high-speed network data," *International J. Statist. Probab.*, 2015, under review.

[174] M. L. Drum and P. McCullagh, "REML estimation with exact covariance in the logistic mixed model," *Biometrics*, vol. 49, no. 3, pp. 677–689, 1993.

[175] J. Meza, S. Chen, and P. Lahiri, Estimation of lifetime alcohol abuse for Nebraska counties. unpublished manuscript, 2003.

[176] J. N. K. Rao and I. Molina, *Small Area Estimation*. Wiley, New York, second ed., 2015.

[177] G. S. Datta and P. Lahiri, "A unified measure of uncertainty of estimated best linear unbiased predictors in small area estimation problems," *Statist. Sinica*, vol. 10, no. 2, pp. 613–627, 2000.

[178] P. Lahiri and J. N. K. Rao, "Robust estimation of mean squared error of small area estimators," *J. Amer. Statist. Assoc.*, vol. 90, no. 430, pp. 758–766, 1995.

[179] D. A. Harville, "Decomposition of prediction error," *J. Amer. Statist. Assoc.*, vol. 80, no. 389, pp. 132–138, 1985.

[180] G. S. Datta, J. N. K. Rao, and D. D. Smith, "On measuring the variability of small area estimators under a basic area level model," *Biometrika*, vol. 92, no. 1, pp. 183–196, 2005.

[181] L.-P. Rivest and E. Belmonte, "A conditional mean squared error of small area estimators," *Survey Methodology*, vol. 26, no. 1, pp. 67–78, 2000.

[182] G. Casella and R. L. Berger, *Statistical Inference*. Duxbury, Thomson Learning, Pacific Grove, CA, second ed., 2002.

[183] G. S. Datta, T. Kubokawa, J. N. K. Rao, and I. Molina, "Estimation of mean squared error of model-based small area estimators," *TEST*, vol. 20, no. 2, pp. 367–388, 2005.

[184] W. A. Fuller, "Prediction of true values for the measurement error model," in *Conference on Statistical Analysis of Measurement Error Models and Applications*, 1989.

[185] J. D. Opsomer, G. Claeskens, M. G. Ranalli, G. Kauermann, and F. J. Breidt, "Non-parametric small area estimation using penalized spline regression," *J. R. Statist. Soc. B*, vol. 70, no. 1, pp. 265–286, 2008.

[186] A. Petrucci and N. Salvati, "Small area estimation for spatial correlation in watershed erosion assessment," *J. Agric. Biol. Environ. Statist.*, vol. 11, no. 2, pp. 169–182, 2006.

[187] M. Wand, "Smoothing and mixed models," *Comput. Statist.*, vol. 18, no. 2, pp. 223–249, 2003.

[188] M. Torabi, "Spatial generalized linear mixed models with multivariate CAR

models for area data," *Spatial Statist.*, vol. 10, no. 1, pp. 12–26, 2014.

[189] J. Jiang and P. Lahiri, "Empirical best prediction for small area inference with binary data," *Ann. Inst. Statist. Math.*, vol. 53, no. 2, pp. 217–243, 2001.

[190] D. Malec, J. Sedransk, C. L. Moriarity, and F. B. LeClere, "Small area inference for binary variables in the National Health Interview Survey," *J. Amer. Statist. Assoc.*, vol. 92, no. 439, pp. 815–826, 1997.

[191] N. G. De Bruijn, *Asymptotic Methods in Analysis*. North-Holland, 1961.

[192] L. E. L., *Theory of Point Estimation*. Wiley, 1983.

[193] B. Efron and C. Morris, "Stein's estimation rule and its competitors: An empirical Bayes approach," *J. Amer. Statist. Assoc.*, vol. 68, no. 341, pp. 117–130, 1973.

[194] J. Jiang, P. Lahiri, and S. Wan, "Jackknifing the mean squared error of empirical best predictor: A theoretical synthesis," *Tech. Report*, 2002.

[195] S. Chen and P. Lahiri, "On the estimation of mean squared prediction error in small area estimation," *Calcutta Statist. Assoc. Bull.*, vol. 63, pp. 109–139, 2011.

[196] I. Molina, J. N. K. Rao, and G. S. Datta, "Small area estimation under a Fay-Herriot model with preliminary testing for the presence of random area effects," *Survey Methodology*, vol. 41, pp. 1–19, 2015.

[197] C. R. Rao and Y. Wu, "On model selection," *IMS Lecture Notes–Monograph Series*, vol. 38, pp. 1–57, 2001.

[198] H. Leeb, "Conditional predictive inference after model selection," *Ann. Statist.*, vol. 37, no. 5B, pp. 2833–2876, 2009.

[199] R. Berk, L. Brown, and L. Zhao, "Statistical inference after model selection," *J. Quant. Criminol.*, vol. 26, no. 2, pp. 271–236, 2010.

[200] S. Müller, J. L. Scealy, and A. H. Welsh, "Model selection in linear mixed models," *Statist. Sci.*, vol. 28, no. 1, pp. 135–167, 2013.

[201] J. Jiang, P. Lahiri, and T. Nguyen, "A unified Monte-Carlo jackknife for small area estimation after model selection," *submitted*, 2015.

[202] G. S. Datta, P. Hall, and A. Mandal, "Model selection by testing for the presence of small-area effects, and applications to area-level data," *J. Amer. Statist. Assoc.*, vol. 106, no. 493, pp. 361–374, 2011.

[203] D. R. Cox, "Prediction intervals and empirical Bayes confidence intervals," in *Perspectives in Probability and Statistics, Papers in Honor of M. S. Bartlett* (J. Gani, ed.), pp. 47–55, 1975.

[204] S. Chatterjee, P. Lahiri, and H. Li, "Parametric bootstrap approximation to the distribution of EBLUP, and related prediction intervals in linear mixed models," *Ann. Statist.*, vol. 36, no. 3, pp. 1221–1245, 2008.

[205] J. Jiang, T. Nguyen, and J. S. Rao, "A simplified adaptive fence procedure," *Statist. Probab. Lett.*, vol. 79, no. 5, pp. 625–629, 2009.

[206] J. G. Booth and J. P. Hobert, "Standard errors of prediction in generalized linear mixed models," *J. Amer. Statist. Assoc.*, vol. 93, no. 441, pp. 262–272, 1998.

[207] J. Jiang, "Empirical best prediction for small area inference based on generalized linear mixed models," *J. Statist. Planning Inference*, vol. 111, no. 1, pp. 117–127, 2003.

[208] H. Li and P. Lahiri, "An adjusted maximum likelihood method for solving small area estimation problems," *J. Multivariate Anal.*, vol. 101, no. 4, pp. 882–892, 2010.

[209] M. Yoshimori and P. Lahiri, "A new adjusted maximum likelihood method for the Fay–Herriot small area model," *J. Multivariate Anal.*, vol. 124, no. 1, pp. 281–294, 2014.

[210] M. Yoshimori and P. Lahiri, "A second-order efficient empirical Bayes confidence interval," *Ann. Statist.*, vol. 42, no. 4, pp. 1233–1261, 2014.

[211] S. L. Lohr and J. N. K. Rao, "Jackknife estimation of mean squared error of small area predictors in nonlinear mixed models," *Biometrika*, vol. 96, no. 2, pp. 457–468, 2009.

[212] P. Hall and T. Maiti, "On parametric bootstrap methods for small area prediction," *J. Roy. Statist. Soc. B*, vol. 68, no. 2, pp. 221–238, 2006.

[213] J. Jiang and T. Nguyen, "Small area estimation via heteroscedastic nested-error regression," *Canadian J. Statist.*, vol. 40, no. 3, pp. 588–603, 2012.

[214] T. Maiti, H. Ren, and S. Sinha, "Prediction error of small area predictors shrinking both means and variances," *Scand. J. Statist.*, vol. 41, no. 3, pp. 775–790, 2014.

[215] E.-T. Tang, On the estimation of the mean squared error in small area estimation and related topics, Ph. D. Dissertation. University of California, Davis, 2008.

[216] J. Jiang, T. Nguyen, and J. S. Rao, "Fence method for nonparametric small area estimation," *Survey Methodol.*, vol. 36, no. 1, pp. 3–11, 2010.

[217] D. Pfeffermann, "New important developments in small area estimation," *Statist. Sci.*, vol. 28, no. 1, pp. 40–68, 2013.

[218] A. Gelman, J. B. Carlin, H. S. Stern, and D. B. Rubin, *Bayesian Data Analysis.* Chapman & Hall/CRC, Boca Raton, FL, second ed., 2004.

[219] M. Harding and M. Lovenheim, "The effect of prices on nutrition: Comparing the impact of product- and nutrient-specific taxes," *Submitted*, 2015.

[220] E. F. Vonesh and R. L. Carter, "Mixed-effects nonlinear regression for unbalanced repeated measures," *Biometrics*, vol. 48, no. 1, pp. 1–17, 1992.

[221] M. Lindstrom and D. Bates, "Nonlinear mixed effects models for repeated measures data," *Biometrics*, vol. 46, no. 3, pp. 673–687, 1990.

[222] R. D. Wolfinger and X. Lin, "Two Taylor-series approximation methods for nonlinear mixed models," *Comput. Statist. Data Anal.*, vol. 25, no. 4, pp. 465–

490, 1997.

[223] J. Pinheiro and D. Bates, "Approximations to the log-likelihood function in nonlinear mixed-effects models," *J. Comput. Graphical Statist.*, vol. 4, no. 1, pp. 12–35, 1995.

[224] L. Lee, "Nonlinear mixed effects models," *International J. Statist. Management Sys.*, vol. 7, no. 1-2, pp. 136–145, 2012.

[225] L. Lee, Iterative estimating equation approach to nonlinear mixed effects models, Ph. D. Dissertation. University of California, Davis, CA., 2011.

[226] G. R. Shorack and J. A. Wellner, *Empirical Processes with Applications to Statistics.* Wiley, New York, 1986.

[227] D. Pollard, *Empirical Processes: Theory and Applications.* Institute of Mathematical Statistics, Hayward, CA, 1990.

[228] S. A. Murphy, "Asymptotic theory for the frailty model," *Ann. Statist.*, vol. 23, no. 1, pp. 182–198, 1995.

[229] T. R. Fleming and D. P. Harrington, *Counting Process and Survival Analysis.* Wiley, New York, 1991.

[230] D. R. Cox, "Partial likelihood," *Biometrika*, vol. 62, no. 2, pp. 269–276, 1975.

[231] P. K. Anderson and R. D. Gill, "Cox's regression model for counting processes: A large sample study," *Ann. Statist.*, vol. 10, no. 4, pp. 1100–1120, 1982.

[232] E. Parner, "Asymptotic theory for the correlated gamma-frailty model," *Ann. Statist.*, vol. 26, no. 1, pp. 183–214, 1998.

[233] S. A. Murphy, "Consistency in a proportional hazards model incorporating a random effect," *Ann. Statist.*, vol. 22, no. 2, pp. 712–731, 1994.

[234] G. G. Nielsen, R. D. Gill, P. K. Anderson, and T. I. A. Sørensen, "A counting process approach to maximum likelihood estimation in frailty models," *Scand. J. Statist.*, vol. 19, no. 1, pp. 25–44, 1994.

[235] S. Geman and C.-R. Hwang, "Nonparametric maximum likelihood estimation by the method of sieves," *Ann. Statist.*, vol. 10, no. 2, pp. 401–414, 1982.

[236] J. E. Porter, "Helly's selection principle for functions of bounded-variation," *Rocky Mountain J. Math.*, vol. 35, no. 2, pp. 675–679, 2005.

[237] S. A. Murphy and A. W. Van der Vaart, "Semiparametric likelihood ratio inference," *Ann. Statist.*, vol. 25, no. 4, pp. 1471–1509, 1997.

[238] A. Gamst, M. Donohue, and R. Xu, "Asymptotic properties and empirical evaluation of the NPMLE in the proportional hazards mixed-effects model," *Statistica Sinica*, vol. 19, no. 3, pp. 997–1011, 2009.

[239] D. Pollard, *Convergence of Stochastic Processes.* Springer, New York, 1984.

[240] P. J. Bickel, C. A. Klaasen, Y. Ritov, and J. A. Wellner, *Efficient and Adaptive Estimation for Semiparametric Models.* Springer, New York, 1993.

[241] A. A. Tsiatis and M. Davidian, "Joint modeling of longitudinal and time-to-

event data: An overview," *Statistica Sinica*, vol. 14, no. 3, pp. 809–834, 2004.

[242] G. Verbeke and M. Davidian, "Joint models for longitudinal data: Introduction and overview," in *Longitudinal Data Analysis: A Handbook of Modern Statistical Methods*, G. Fitzmaurice, M. Davidian, G. Verbeke, and G. Molenberghs ed., pp. 319–326, 2008.

[243] M. S. Wulfsohn and A. A. Tsiatis, "A joint model for survival and longitudinal data measured with error," *Biometrics*, vol. 53, no. 1, pp. 330–339, 1997.

[244] D. Zeng and J. Cai, "Asymptotic results for maximum likelihood estimators in joint analysis of repeated measurements and survival time," *Ann. Statist.*, vol. 33, no. 5, pp. 2132–2163, 2005.

[245] J. Dupuy, I. Grama, and M. Mesbah, "Asymptotic theory for Cox model with missing time-dependent covariates," *Ann. Statist.*, vol. 34, no. 2, pp. 903–924, 2006.

[246] F. Hsieh, J. Ding, and J.-L. Wang, "Method of sieves to jointly model survival and longitudinal data," *Statistica Sinica*, vol. 23, no. 3, pp. 1181–1213, 2013.

[247] X. Shen and W. H. Wong, "Convergence rate of sieve estimates," *Ann. Statist.*, vol. 22, no. 2, pp. 580–615, 1994.

[248] W. Guo, "Functional mixed effects models," *Biometrics*, vol. 58, no. 1, pp. 121–128, 2002.

[249] G. Wahba, "Improper priors, spline smoothing and the problem of guarding against model errors in regression," *J. Roy. Statist. Soc. B*, vol. 40, no. 3, pp. 364–372, 1978.

[250] G. Wahba, "Bayesian confidence intervals for the cross-validated smoothing spline," *J. Roy. Statist. Soc. B*, vol. 45, no. 1, pp. 133–150, 1983.

[251] R. T. Krafty, M. Hall, and W. Guo, "Functional mixed effects spectral analysis," *Biometrika*, vol. 98, no. 3, pp. 583–598, 2011.

[252] L. H. Koopmans, *The Spectral Analysis of Time Series*. Elsevier, 1995.

[253] E. A. Rady, N. M. Kilany, and S. A. Eliwa, "Estimation in mixed-effects functional ANOVA models," *J. Multivariate Anal.*, vol. 133, pp. 346–355, 2015.

[254] M. J. Lombardía and S. Sperlich, "Semiparametric inference in generalized mixed effects models," *J. Roy. Statist. Soc. B*, vol. 70, no. 5, pp. 913–930, 2008.

Index